We hope that you are happy with your order! If there are any issues, we would be love for the opportunity to better your experience. Please reach out to us and we will be glad to assist you. If you are happy with your purchase, we would be grateful to receive feedback for our store. To leave seller feedback:

- Find your order and select the options that best reflect your experience.
- Click Submit Feedback.

To view our policies, and to purchase with us in the future, follow the link or scan the QR code below:

www.amazon.com/shops/The_Wisdom_Shop

We are a small business that hopes your experience with our store was one that was positive and memorable. Thank you for your time and valuable feedback!

The Wisdom Shop

Thank you for ordering with us!

Current Topics in Membranes, Volume 67

Advances in Adrenergic Receptor Biology

Current Topics in Membranes, Volume 67

Series Editors

Dale J. Benos
Department of Physiology and Biophysics
University of Alabama
Birmingham, Alabama

Robert Balaban
National Heart, Lung and Blood Institute
National Institute of Health
Bethesda, Maryland

Sidney A. Simon
Department of Neurobiology
Duke University Medical Centre
Durham, North Carolina

Current Topics in Membranes, Volume 67

Advances in Adrenergic Receptor Biology

Edited by

Qin Wang
Department of Physiology and Biophysics
University of Alabama at Birmingham
Birmingham, AL, USA

ELSEVIER

AMSTERDAM • BOSTON • HEIDELBERG • LONDON
NEW YORK • OXFORD • PARIS • SAN DIEGO
SAN FRANCISCO • SINGAPORE • SYDNEY • TOKYO
Academic Press is an imprint of Elsevier

Academic Press is an imprint of Elsevier
225 Wyman Street, Waltham, MA 02451, USA
525 B Street, Suite 1900, San Diego, CA 92101-4495, USA

First edition 2011

ISBN: 978-0-12-384921-2
ISSN: 1063-5823

For information on all Academic Press publications
visit our website at elsevierdirect.com

Printed and bound in USA
11 12 13 10 9 8 7 6 5 4 3 2 1

Dedication

We dedicate this 67th volume of the Current Topics In Membranes (CTM) book series to our longtime friend and colleague Dr. Dale J. Benos Ph.D. (1950-2010). Dale was the series, and often a contributing, editor whose untimely death at the age of 60 has been a huge loss for his family, his friends, biomedical research community as well as the CTM editorial team.

Dale earned his Bachelor of Science degree in Biology in 1972 from Case Western Reserve University. He earned his Ph.D. in Physiology and Pharmacology from Duke University in 1976 with Dr. Daniel Tosteson. His Ph.D. work was well before his time using genetic mouse models in studying the ion transport in red blood cells. He followed this work with a post-doctoral fellowship in epithelial physiology at Duke in 1978 with Dr. Lazaro Mandel beginning his long standing interest in sodium transport and amiloride- sensitive ion channels. It was at Duke University where both of us met, socialized and worked with Dale on a variety of projects. Upon completing his postdoctoral work, he moved to Boston as an Andrew W. Mellon Scholar in the Laboratory of Human Reproduction and Reproductive Biology at Harvard Medical School, where he rose to the position of Associate Professor in the Department of Physiology and Biophysics.

In 1985, Dale joined the University of Alabama faculty an Associate Professor in the Department of Physiology and Biophysics. Two years later was promoted to Professor and in 1996 became department chair, a position he retained until his untimely death.

The scope of Dale's scientific interest and contributions were remarkably broad and insightful. He made major contributions to the understanding of the function of the amiloride-sensitive Na channel in epithelia as well as studies on the metabolism and function of the reproductive system from blastocysts to sperm as well as his most recent work on the role of Na transport in the functioning of the brain in health and disease. His studies were always complete and meticulously conducted challenging competitors as well as collaborators to move to the next level of understanding of any problem he engaged.

In addition to his own research, Dale also played a major role in the dissemination of scientific information as well as teaching. In addition to his work at CTM in which not only was the series editor, but also edited several volumes, Dale, also had major editorial chores on the Journal of Biological Chemistry, American Journal of Physiology, Biochem Biophys Acta (Biomembranes) and many other periodicals. He was also the President of the American Physiological Society and served on the boards of numerous scientific societies dedicated to the dissemination of biomedical science as well as the teaching both science and ethics of the next generation of scientists. Based on personal accounts and the number of teaching awards he received at the University of Alabama, Dale was a spectacular teacher getting students engaged and excited about the most fundamental aspects of physiology and membrane biophysics. Clearly this enthusiasm was also imparted on the contributing editors to CTM generating the remarkable series of scholarly contributions to the field of membrane biology over the last ~15 years.

We cannot replace Dale's friendship, enthusiasm, intellect and remarkable positive energy. Many of these traits are revealed in this picture of Dale from the 1970's at Duke where most people will recognize the enthusiasm and positive attitude but also the remarkable fact that he looked almost the same on his 60th birthday. Clearly, it was much too soon for him to leave us with so much more to contribute to unraveling nature and teaching the next generation the lessons learned by this generation. His scientific legacy is remarkable; however his real long term impact is the training of future generations. Moreover, his role in generating numerous volumes of CTM will be one of the many monuments he has left all of us. We pledge to carry on the tradition of scientific excellence in the future volumes of CTM series.

Sidney A. Simon and Robert S. Balaban

Co-series editors

Contents

Contributors xiii
Preface xv
Previous Volumes in Series xvii

CHAPTER 1 Historical Perspective for Understanding of Adrenergic Receptors
Lee E. Limbird

I. Overview 1
II. Introduction: It's All About Specificity 2
III. Adrenergic Receptor Identification 3
IV. What Makes an Agonist Different from an Antagonist? 4
V. Rhodopsin Enlightens Adrenergic Receptor Structure–Function Studies 4
VI. Arrestins 5
VII. Molecular Cloning Permits Structure–Function Analysis of Adrenergic Receptors Via Multiple Experimental Strategies 8
VIII. Gene Targeting Studies Reveal Subtype Selective Roles for Adrenergic Receptors in *In vivo* Settings 10
IX. The Multiple Interacting Proteins of Adrenergic Receptors 12
X. And That Brings Us Back to The Issue of Specificity and Therapeutic Selectivity 13
References 13

CHAPTER 2 Organizational Complexity of β-adrenergic Receptor Signaling Systems
Irina Glazkova, Katrin Altosaar, and Terence E. Hébert

I. Overview 19
II. Introduction 20

III. General Considerations Regarding GPCR Organization 21
IV. Signaling Diversity in β-Adrenergic Receptors 21
V. The Impact of Recent Crystal Structures on Our Understanding of Biased Signaling 24
VI. Larger Receptor Arrays 27
VII. Ontogeny of βAR Signaling Systems 31
VIII. Asymmetric GPCR Complexes 33
IX. Beyond the Paradigm of a Cell Surface Receptor 35
X. Conclusions 39
References 39

CHAPTER 3 β-Arrestin-Biased Signaling by the β-Adrenergic Receptors
Sudha K. Shenoy

I. Overview 51
II. Introduction 52
III. β-Arrestin-Biased Signaling by the β_2AR 54
IV. β-Arrestin-Biased Signaling by the β_1AR 59
V. Factors that Define β-Arrestin-Dependent Signaling 61
VI. Inhibitors of β-Arrestin-Dependent Signaling 69
VII. Future Perspectives 72
References 72

CHAPTER 4 Anterograde Trafficking of Nascent α_{2B}-Adrenergic Receptor: Structural Basis and Roles of Small GTPases
Guangyu Wu

I Overview 79
II Introduction 80
III The Structural Basis of α_{2B}-AR Export from the ER and the Golgi 82
IV The Role of Small GTPases in the Export Trafficking of α_{2B}-AR 89
V Conclusions and Perspectives 93
References 94

CHAPTER 5 Recording Kinetics of Adrenergic Receptor Activation in Live Cells
Jean-Pierre Vilardaga

I. Overview 101
II. Introduction 101
III. Principle of the Experiment 103
IV. Receptor Activation in Live Cells 105
References 110

CHAPTER 6 Modulation of Immune Cell Function by α_1-Adrenergic Receptor Activation
Laurel A. Grisanti, Dianne M. Perez and James E. Porter

I. Overview 113
II. Introduction 114
III. α_1-Adrenergic Receptor Expression in the Immune System 115
IV. The Role of α_1-Adrenergic Receptors in the Innate Immune System 117
V. α_1-Adrenergic Receptor Influences on the Adaptive Immune System 126
VI. α_1-Adrenergic Receptors in Immune Tissues 129
VII. α_1-Adrenergic Receptors in Disease States 133
VIII. Conclusions 134
References 135

CHAPTER 7 Pre- versus Postsynaptic Signaling by α_2-Adrenoceptors
Ralf Gilsbach, Julián Albarrán-Juárez, and Lutz Hein

I. Overview 139
II. Introduction 140
III. Functions of α_2-Adrenoceptor Subtypes 141
IV. Transgenic Dissection of Pre- Versus Postsynaptic α_2-Adrenoceptor Functions 144
V. Conclusions 154
References 155

CHAPTER 8 Genetic Variations of α_2-Adrenergic Receptors Illuminate the Diversity of Receptor Functions

Christopher Cottingham, Huaping Chen, Yunjia Chen, Yin Peng, and Qin Wang

I. Overview 162
II. Introduction 162
III. Overview of α_2AR Genetic Variants 164
IV. α_2ARs in Central Nervous System Dysfunction and Disease 167
V. α_2ARs in Cardiovascular Disease 173
VI. α_2ARs in Metabolism and Type 2 Diabetes 177
VII. α_2ARs in Other Peripheral Functions 179
VIII. Conclusions 181
References 181

CHAPTER 9 β-Adrenergic Receptor Subtype Signaling in the Heart: from Bench to the Bedside

Weizhong Zhu, Anthony Yiu-Ho Woo, Yan Zhang, Chun-Mei Cao, and Rui-Ping Xiao

I. Overview 191
II. Subtype-Specific Functional Roles of β_1AR and β_2AR in Regulating Cardiomyocyte Survival and Death 192
III. Mechanisms Underlying β_2AR-Coupled G_i Signaling 194
IV. RGS2-Mediated Termination of β_2AR-Coupled G_i Signaling and its Potential Pathogenic and Therapeutic Implications 195
V. Ligand-Directed Selective Activation of β_2AR-Coupling to G_s or G_i 197
VI. Development of β_2AR Agonists into New Drugs for the Treatment of Heart Failure 198
VII. Future Perspective 200
VIII. References 200

CHAPTER 10 β-Adrenergic Receptor, Amyloid β Peptide, and Alzheimer's Disease

Dayong Wang and Yang K. Xiang

I. Overview 205
II. Introduction 206
III. Adrenergic Signaling and APP Expression 208
IV. Activation of βAR Promotes Amyloid Process 210
V. Interaction of Adrenergic Receptor with Amyloid β Peptide 213
VI. CNS Adrenergic System in AD 216
VII. Clinic Implication of Activation of Downstream Targets by Aβ/βAR Signaling Axis in CNS 218
VIII. Adrenergic System in Peripheral Tissues 219
IX. Potential Drug Targets on Aβ-β_2AR Signaling Axis for AD Therapy 220
X. Conclusion and Future Direction 221
References 221

Index 229

Contributors

Numbers in parentheses indicate the pages on which the authors' contributions begin.

Julián Albarrán-Juárez (139) Institute of Experimental and Clinical Pharmacology, University of Freiburg, Freiburg, Germany

Katrin Altosaar (19) Department of Pharmacology and Therapeutics, McGill University, Montréal, Canada

Chun-Mei Cao (191) Institute of Molecular Medicine, Peking University, Beijing, China

Huaping Chen (161) Department of Physiology and Biophysics, University of Alabama at Birmingham, AL, USA

Yunjia Chen (161) Department of Physiology and Biophysics, University of Alabama at Birmingham, AL, USA

Christopher Cottingham (161) Department of Physiology and Biophysics, University of Alabama at Birmingham, AL, USA

Ralf Gilsbach (139) Institute of Experimental and Clinical Pharmacology, University of Freiburg, Freiburg, Germany

Irina Glazkova (19) Department of Pharmacology and Therapeutics, McGill University, Montréal, Canada

Laurel A. Grisanti (113) Department of Pharmacology, Physiology, and Therapeutics, School of Medicine and Health Sciences, University of North Dakota, Grand Forks, North Dakota, USA

Terence E. Hébert (19) Department of Pharmacology and Therapeutics, McGill University, Montréal, Canada

Lutz Hein (139) Institute of Experimental and Clinical Pharmacology, University of Freiburg, Freiburg, Germany; BIOSS Centre for Biological Signaling Studies, University of Freiburg, Freiburg, Germany

Lee E. Limbird (01) School of Natural Sciences, Mathematics, and Business and Department of Life and Physical Sciences, Fisk University, Nashville, TN, USA, and Vanderbilt University

Yin Peng (161) Department of Physiology and Biophysics, University of Alabama at Birmingham, AL, USA

Dianne M. Perez (113) Department of Molecular Cardiology, Lerner Research Institute, The Cleveland Clinic Foundation, Cleveland, Ohio, USA

James E. Porter (113) Department of Pharmacology, Physiology, and Therapeutics, School of Medicine and Health Sciences, University of North Dakota, Grand Forks, North Dakota, USA

Sudha K. Shenoy (51) Departments of Medicine and Cell Biology, Duke University Medical Center, Durham, North Carolina, USA

Jean-Pierre Vilardaga (101) Laboratory for GPCR Biology, Department of Pharmacology and Chemical Biology, University of Pittsburgh, School of Medicine, Pittsburgh, PA, USA

Dayong Wang (205) Molecular and Integrative Physiology, University of Illinois at Urbana-Champaign, Urbana, IL, USA

Qin Wang (161) Department of Physiology and Biophysics, University of Alabama at Birmingham, AL, USA

Guangyu Wu (79) Department of Pharmacology and Toxicology, Georgia Health Sciences University, Augusta, GA, USA

Yang K. Xiang (205) Molecular and Integrative Physiology and Neuroscience Program, University of Illinois at Urbana-Champaign, Urbana, IL, USA

Rui-Ping Xiao (191) Institute of Molecular Medicine, Peking University, Beijing, China

Anthony Yiu-Ho Woo (191) Institute of Molecular Medicine, Peking University, Beijing, China

Yan Zhang (191) Institute of Molecular Medicine, Peking University, Beijing, China

Weizhong Zhu (191) Center for Translational Medicine, Thomas Jefferson University, Philadelphia, USA

Preface

Adrenergic receptors (ARs) are expressed in almost all organs and tissues and regulate a large number of diverse physiological processes upon activation by epinephrine and norepinephrine. There are three families of ARs, α_1, α_2, and β-ARs, with distinct pharmacological properties and functions. Since the first identification of βARs more than three decades ago, research on ARs has led to the establishment of many fundamental concepts in G protein-coupled receptor (GPCR) pharmacology. In addition, appreciation of AR functions in the physiology of various systems and in the pathophysiology of many disease states has established these receptors as viable drug targets and resulted in the identification and development of a number of effective therapeutics. This volume of CTM is not intended to cover all aspects of AR biology, but rather focuses on the most recent findings, in a historic context, pertaining to AR activation, signaling, trafficking, and in vivo functions.

It has been a great privilege and genuine pleasure to work closely with the many AR experts who are dedicated to providing a state-of-the-art review of the recent advances in this active research field. I am particularly grateful to my formal mentor, Dr. Lee Limbird, for giving me advice on effectively managing such a significant project. I am also indebted to the Elsevier editorial staff, whose hard work has made publication of this volume smooth and efficient.

I want to specifically thank the Series Editor, Dr. Dale Benos, for his invitation to me to serve as editor and his helpful guidance in the development of this volume on AR biology. Unfortunately, Dr. Benos passed away during preparation of this work. His untimely passing is definitely a huge loss to the membrane biology field. In memory of his leadership, his many contributions to the field, and his devoted service to this journal series, I would like to dedicate this volume to him.

Qin Wang

Previous Volumes in Series

Current Topics in Membranes and Transport

Volume 23 Genes and Membranes: Transport Proteins and Receptors* (1985)
Edited by Edward A. Adelberg and Carolyn W. Slayman

Volume 24 Membrane Protein Biosynthesis and Turnover (1985)
Edited by Philip A. Knauf and John S. Cook

Volume 25 Regulation of Calcium Transport across Muscle Membranes (1985)
Edited by Adil E. Shamoo

Volume 26 Na^{+}–H^{+} Exchange, Intracellular pH, and Cell Function* (1986)
Edited by Peter S. Aronson and Walter F. Boron

Volume 27 The Role of Membranes in Cell Growth and Differentiation (1986)
Edited by Lazaro J. Mandel and Dale J. Benos

Volume 28 Potassium Transport: Physiology and Pathophysiology* (1987)
Edited by Gerhard Giebisch

Volume 29 Membrane Structure and Function (1987)
Edited by Richard D. Klausner, Christoph Kempf, and Jos van Renswoude

Volume 30 Cell Volume Control: Fundamental and Comparative Aspects in Animal Cells (1987)
Edited by R. Gilles, Arnost Kleinzeller, and L. Bolis

Volume 31 Molecular Neurobiology: Endocrine Approaches (1987)
Edited by Jerome F. Strauss, III, and Donald W. Pfaff

Volume 32 Membrane Fusion in Fertilization, Cellular Transport, and Viral Infection (1988)
Edited by Nejat Düzgünes and Felix Bronner

Volume 33 Molecular Biology of Ionic Channels* (1988)
Edited by William S. Agnew, Toni Claudio, and Frederick J. Sigworth

**Part of the series from the Yale Department of Cellular and Molecular Physiology.*

Volume 34 Cellular and Molecular Biology of Sodium Transport* (1989)
Edited by Stanley G. Schultz

Volume 35 Mechanisms of Leukocyte Activation (1990)
Edited by Sergio Grinstein and Ori D. Rotstein

Volume 36 Protein–Membrane Interactions* (1990)
Edited by Toni Claudio

Volume 37 Channels and Noise in Epithelial Tissues (1990)
Edited by Sandy I. Helman and Willy Van Driessche

Current Topics in Membranes

Volume 38 Ordering the Membrane Cytoskeleton Trilayer* (1991)
Edited by Mark S. Mooseker and Jon S. Morrow

Volume 39 Developmental Biology of Membrane Transport Systems (1991)
Edited by Dale J. Benos

Volume 40 Cell Lipids (1994)
Edited by Dick Hoekstra

Volume 41 Cell Biology and Membrane Transport Processes* (1994)
Edited by Michael Caplan

Volume 42 Chloride Channels (1994)
Edited by William B. Guggino

Volume 43 Membrane Protein–Cytoskeleton Interactions (1996)
Edited by W. James Nelson

Volume 44 Lipid Polymorphism and Membrane Properties (1997)
Edited by Richard Epand

Volume 45 The Eye's Aqueous Humor: From Secretion to Glaucoma (1998)
Edited by Mortimer M. Civan

Volume 46 Potassium Ion Channels: Molecular Structure, Function, and Diseases (1999)
Edited by Yoshihisa Kurachi, Lily Yeh Jan, and Michel Lazdunski

Volume 47 Amiloride-Sensitive Sodium Channels: Physiology and Functional Diversity (1999)
Edited by Dale J. Benos

Volume 48 Membrane Permeability: 100 Years since Ernest Overton (1999)
Edited by David W. Deamer, Arnost Kleinzeller, and Douglas M. Fambrough

Volume 49 Gap Junctions: Molecular Basis of Cell Communication in Health and Disease
Edited by Camillo Peracchia

Volume 50 Gastrointestinal Transport: Molecular Physiology
Edited by Kim E. Barrett and Mark Donowitz

Volume 51 Aquaporins
Edited by Stefan Hohmann, Søren Nielsen and Peter Agre

Volume 52 Peptide–Lipid Interactions
Edited by Sidney A. Simon and Thomas J. McIntosh

Volume 53 Calcium-Activated Chloride Channels
Edited by Catherine Mary Fuller

Volume 54 Extracellular Nucleotides and Nucleosides: Release, Receptors, and Physiological and Pathophysiological Effects
Edited by Erik M. Schwiebert

Volume 55 Chemokines, Chemokine Receptors, and Disease
Edited by Lisa M. Schwiebert

Volume 56 Basement Membranes: Cell and Molecular Biology
Edited by Nicholas A. Kefalides and Jacques P. Borel

Volume 57 The Nociceptive Membrane
Edited by Uhtaek Oh

Volume 58 Mechanosensitive Ion Channels, Part A
Edited by Owen P. Hamill

Volume 59 Mechanosensitive Ion Channels, Part B
Edited by Owen P. Hamill

Volume 60 Computational Modelling of Membrane Bilayers
Edited by Scott E. Feller

Volume 61 Free Radical Effects on Membranes
Edited by Sadis Matalon

Volume 62 The Eye's Aqueous Humor
Edited by Mortimer M. Civan

Volume 63 Membrane Protein Crystallization
Edited by Larry DeLucas

Volume 64 Leukocyte Adhesion
Edited by Klaus Ley

Volume 65 Claudins
Edited by Alan S. L. Yu

Volume 66 Structure and Function of Calcium Release Channels
Edited by Irina I. Serysheva

CHAPTER 1

Historical Perspective for Understanding of Adrenergic Receptors

Lee E. Limbird

School of Natural Sciences, Mathematics, and Business, Fisk University, Nashville, TN, USA

I. Overview
II. Introduction: it's all About Specificity
III. Adrenergic Receptor Identification
IV. What Makes an Agonist Different from an Antagonist?
V. Rhodopsin Enlightens Adrenergic Receptor Structure–Function Studies
VI. Arrestins
A. Their "First-Generation" Role as Mediators of Homologous Desensitization Evoked by Adrenergic Receptors
B. Arrestins as Adapters for Endocytosis Via Clathrin-Coated Pits
C. Arrestin Scaffolding of Non-G Protein-Dependent Signaling Makes Possible Ligand-Biased Signaling Along G Protein-Dependent and -Independent Signaling Networks
VII. Molecular Cloning Permits Structure–Function Analysis of Adrenergic Receptors via Multiple Experimental Strategies
VIII. Gene Targeting Studies Reveal Subtype Selective Roles for Adrenergic Receptors in *In Vivo* Settings
IX. The Multiple Interacting Proteins of Adrenergic Receptors
X. And That Brings us Back to the Issue of Specificity and Therapeutic Selectivity
Acknowledgments
References

I. OVERVIEW

The study of the basis of physiological responses to epinephrine and norepinephrine has resulted in the discovery of fundamental concepts in receptor theory and mechanisms and, based on more recent exploitation of molecular tools, has led to paradigm shifts in our thinking concerning receptor mechanisms. Thus, ongoing research about adrenergic receptors and their modulation of signaling pathways is anticipated to reveal not only cell-specific or disease-specific

Current Topics in Membranes, Volume 67 0065-230X/10 $35.00

DOI: 10.1016/B978-0-12-384921-2.00001-X

insights, but also more pervasive insights into the mechanisms underlying signaling specificity by endogenous and pharmacological agents. This chapter is *not* intended to be a detailed review of this expansive literature, but rather a contextual introduction to the topics of focus in the chapters in this volume. By its very nature, then, the references cited are very selective, and only representative of a much larger literature.

II. INTRODUCTION: IT'S ALL ABOUT SPECIFICITY

Early investigations into the mechanism of action of endogenous catecholamines and mimicking or blocking drugs focused on *how* selectivity of action was achieved. Earlier studies focusing on nonadrenergic systems had led to two alternative explanations for observed tissue selectivity in response to agents: the selective distribution or uptake of an agent in a tissue (e.g., lead accumulation in the nervous system as the basis for lead poisoning, Heubel, 1871) versus the specific interactive properties of an agent, based on the mutual antagonism of agonists and antagonists in eliciting muscle contraction (Langley, 1909). Despite compelling evidence favoring specific interactive properties between drugs and tissues as accounting for selectivity in tissue response, others stated it was equally probable that the limiting factor in differential effectiveness of adrenaline analogs in mimicking sympathetic functions in varying tissues could be due to a chemical process, such as the ease with which those agents reached their site of action (Barger & Dale, 1910).

It was the findings of Ahlquist which provided the first incontrovertible evidence that the specificity of action of drugs depended not on relative ease of their distribution to one versus another tissue, but rather on the existence of tissue-specific mechanisms for responding to those agents. Ahlquist was examining the mechanistic bases for physiological response to catecholamines, namely norepinephrine and epinephrine (Ahlquist, 1948). He observed that smooth muscle contraction was evoked by an order of catecholamine potency (based on dose ratios) of norepinephrine $>$ epinephrine $\gg$ isoproterenol, and referred to these effects as "alpha"; in contrast, smooth muscle *relaxation* was elicited by catecholamines with an order of potency of isoproterenol $>$ epinephrine $>$ norepinephrine , which he referred to as "beta" responses. "Beta" responses also accounted for catecholamine-evoked increases in the rate and strength of cardiac muscle contraction. These findings that the same agents could have an entirely different specificity in evoking contraction versus relaxation in the same tissue, that is, smooth muscle, are entirely inconsistent with distribution to tissues accounting for the specificity of action of biological agents. Instead, these findings were entirely consistent with the existence of tissue-specific *receptive substances* that served as the basis for selectivity of tissue response to agents.

The evolution of our understanding of adrenergic receptors led to the delineation of multiple subtypes of α and β adrenergic receptors (Bylund et al., 1994). This chapter focuses primarily on the evolution of our understanding of β-, and to a lesser extent α_2-, adrenergic receptors. Chapter 6, by Porter and colleagues, which addresses the role of α_1-adrenerigic receptor activation in the immune system, includes an introductory review of the known α_1 receptor subtypes and their physiological functions.

III. ADRENERGIC RECEPTOR IDENTIFICATION

Though ACTH receptors were the first to be identified with radioligand binding (Lefkowitz, Roth, Pricer, & Pastan, 1970), adrenergic receptors were among the first to be rigorously explored in the context of direct receptor identification and receptor-evoked signaling mechanisms. The demonstration by Sutherland and colleagues that the specificity of the cAMP-dependent effects of epinephrine (Davoren & Sutherland, 1963) were *beta adrenergic* in nature facilitated the identification of the β-adrenergic receptor *via* radioligand binding strategies. The properties of adrenergic stimulation of cAMP accumulation as the criteria for radioligand identification of binding to the physiologically relevant β-adrenergic receptor included: (1) specificity of competition for radioligand binding in parallel to the specificity for activation or blockade of β-adrenergic stimulation of cAMP accumulation, (2) saturability of binding, due to the finite number of receptors that would exist on target cells, (3) kinetics consistent with the rate of receptor activation of cAMP synthesis by adrenergic agonists and rate of reversal of that activation by antagonist agents. An additional important criterion for adrenergic receptor specificity included the stereoselectivity of agonists in competing for radioligand binding, in parallel with the preferential sensitivity of β-adrenergic receptors to the *l*- or (–) stereoisomers of epinephrine and norepinephrine for receptor activation compared to the *d*- or (+) isomers of catecholamines.

Early efforts to identify β-adrenergic receptors with the endogenous ligand, norepinephrine, demonstrated the limitations to attempting to identify a receptor with a relatively low affinity (micromolar) ligand for which there are also competing interactions for this endogenous agent in the target cell preparation being evaluated, including catabolizing enzymes and nonenzymatic chemical modification of the ligand during the incubation (Lefkowitz, Sharp, & Haber, 1973). Thus, receptor identification with radiolabeled antagonists proved to provide the most informative and broadly exploited experimental approach (Lefkowitz, Mukherjee, Coverstone, & Caron, 1974; Aurbach, Fedak, Woodard, Palmer, Hauser, & Troxler, 1974; Levitzki, Atlas, & Steer, 1974). Direct identification of physiologically relevant β-adrenergic receptors with radiolabeled antagonist binding allowed for rigorous characterization of receptor properties, including changes in receptor density and affinity that accompanied pathological changes in receptor response, and regulated appearance of receptors during organism development and tissue differentiation.

IV. WHAT MAKES AN AGONIST DIFFERENT FROM AN ANTAGONIST?

Cell fusion (Orly & Schramm, 1976) and biochemical resolution of detergent-solubilized activities (Limbird & Lefkowitz, 1977) demonstrated that the receptor and adenylyl cyclase were separable macromolecules. One prominent focus of research which followed the ability to confidently identify β-adrenergic receptors *via* radioligand binding was to determine what properties of receptor–agonist interactions were unique, and could not be mimicked by antagonist occupancy of receptors. Similar to glucagon receptors in liver preparations (Rodbell, Birnbaumer, Pohl, Krans, 1971a; Rodbell, Krans, Pohl, & Birnbaumer 1971b), β-adrenergic receptors were functionally coupled to adenylyl cyclase by GTP-binding proteins (Ross & Gilman, 1977). Agonist activation of β-adrenergic receptors was paralleled by reciprocal regulation of receptor affinity for agonists, but not for antagonists, by guanine nucleotides (Maguire, Van Arsdale, & Gilman, 1976; Lefkowitz, Mullikin, & Caron, 1976) and an agonist-specific ability to either induce or interact with a pre-existing receptor–G protein complex (Limbird & Lefkowitz, 1978; Limbird, Gill, & Lefkowitz, 1980). These findings led to the development of an early model of β-adrenergic receptor activation of effector, that is, adenylyl cyclase, *via* a ternary complex of agonist, receptor, and GTP binding protein (Lean, Stadel, & Lefkowitz, 1980). Subsequent findings led to the evolution of increasingly complex models for receptor-mediated activation of cellular signaling to account for data consistent with the existence of precoupled R–G complexes, with ligands that possess negative intrinsic activity (dubbed *inverse agonists*) and *increased* receptor affinity in the presence of guanine nucleotides, and G protein-independent receptor-evoked signaling (see Kenakin, 2004, for a review of this interdependent evolution of computational models and experimental data). The iterative interplay between experimental findings and computational modeling has enriched the field, while also illuminating the complexity that underlies adrenergic receptor-evoked activation of cellular processes.

V. RHODOPSIN ENLIGHTENS ADRENERGIC RECEPTOR STRUCTURE–FUNCTION STUDIES

The natural enrichment of rhodopsin to densities of $\sim$1000/μm^2 in the retinal surface membrane, that is 500 to 1000 fold more dense than β-adrenergic receptors in most target tissues, allowed the earlier purification of this molecule and molecular inquiry of its G protein coupling mechanisms. The impact of subsequent structural insights regarding rhodopsin on adrenergic receptor studies will be discussed below. However, the similarities of rhodopsin signaling to β-adrenergic signaling were too profound to ignore: rhodopsin accelerated guanine nucleotide binding to its associated G protein, dubbed transducin, in a highly amplified fashion (Fung, Hurley, & Stryer, 1981); conformational changes in rhodopsin evoked by light-activated changes in the conformation of covalently bound retinal led to incremental phosphorylation of rhodopsin in a manner that suggested a mechanism for visual adaptation (e.g., Kühn, McDowell, Leser, & Bader, 1977); rhodopsin phosphorylation was catalyzed by a specific kinase that differentiated between the inactive versus light-activated conformation of rhodopsin (Pfister, Kühn, & Chabre, 1983); and phosphorylated rhodopsin manifest an increase in affinity for a protein, dubbed arrestin (Kühn, Hall, & Wilden, 1984; earlier called S antigen) that appeared to functionally uncouple rhodopsin from its productive interactions with transducin. Taken together, these findings suggested a considerable functional homology between light-activated photoreceptor, phosphodiesterase and hormone-activated adenylyl cyclase systems (Yamazaki et al., 1985).

VI. ARRESTINS

A. Their "First-Generation" Role as Mediators of Homologous Desensitization Evoked by Adrenergic Receptors

The demonstrated functional consequence of arrestin interaction with phosphorylated rhodopsin in deactivation of light-activated signaling served as a harbinger for understanding the molecular underpinnings of homologous desensitization of adrenergic receptors. Studies had already clarified that activation of adrenergic receptors with agonist for long periods of time or at elevated concentrations of agonist could result in diminished signal output from the receptor, referred to as desensitization. Desensitization of receptor-mediated signaling can be either heterologous or homologous in nature. *Heterologous* desensitization occurs when activation of any receptor coupled to a shared signaling pathway leads to attenuation of not only the originally activated receptor, but also of all receptors linked to the same signaling pathways. For β-adrenergic receptors, heterologous desensitization was often due to cAMP-dependent phosphorylation of receptors which uncoupled receptors from their cognate G proteins (Benovic et al., 1985; Clark, Kunkel, Friedman, Goka, & Johnson, 1988). In contrast, *homologous* desensitization is the term used to describe feedback inhibition of signaling limited to output only from the stimulated receptor. A reasonably large literature demonstrated that homologous desensitization involved receptor uncoupling from G proteins that preceded receptor internalization from the surface prior to receptor degradation (also known as "down regulation") or receptor recycling. These events could be resolved from one another kinetically (Su, Harden, & Perkins, 1979) as well as mechanistically (Lohse, Benovic, Caron, & Lefkowitz, 1990).

It was demonstrated that, similar to rhodopsin, only the agonist-occupied or conformationally active state of the receptor is recognized for phosphorylation by a G protein-coupled receptor-directed kinase (dubbed GRKs; Benovic, Strasser, Caron, & Lefkowitz, 1986; Benovic, Staniszewski, Mayor, Caron, & Lefkowitz, 1988), now known to be the activity of a family of kinases with different regulatory properties (Krupnick & Benovic, 1998; Pitcher, Freedman, & Lefkowitz, 1998). The fact that only the activated "state" or conformation of the receptor serves as a substrate for GRKs explains how GRK-catalyzed receptor phosphorylation can serve as the molecular event initiating receptor-dependent *homologous* desensitization.

As a consequence of GRK-catalyzed phosphorylation of the agonist-bound or active state of the receptor, receptor interaction with a unique, nonvisual system arrestin, dubbed β-arrestin (or arrestin-2; visual arrestin is defined as arrestin-1) is facilitated. Interaction of the β-adrenergic receptor with arrestin competes with productive receptor–G protein interactions (Attramadal et al., 1992), thus accounting for the role of arrestins in homologous desensitization mechanisms. This topic is discussed in further detail in Chapter 3 by S. Shenoy.

B. Arrestins as Adapters for Endocytosis Via Clathrin-Coated Pits

Arrestins subsequently were revealed to have multiple functions, including serving as adapter molecules for interaction of GRK-phosphorylated receptors with clathrin-coated pit-associated endocytosis machinery (Goodman et al., 1996) *via* arrestin interactions with the β subunit of the AP2 adapter protein (Laporte et al., 1999). G protein-coupled receptors whose interactions with arrestin are transient and do not continue during endocytosis, dubbed Class A receptors (of which β_2-adrenergic receptors are the paradigmatic example), are rapidly recycled to the cell surface after dephosphorylation in internalized compartments. In contrast, G protein-coupled receptors that remain associated with arrestins upon internalization (dubbed Class B receptors), are more likely to recycle slowly or be targeted for degradation in lysosomes. Class A receptors, which also include α_{1B}-adrenergic, μ-opioid, endothelin ETA, and D1A dopamine receptors, have been reported to show preferential binding to β-arrestin 2 compared with β-arrestin 1, and no interaction with visual arrestin, whereas Class B receptors (including V2 vasopressin, angiotensin AT1A, thyrotropin-releasing hormone, neurotensin 1, and neurokinin NK1 receptors) show equal affinity for β-arrestin 1 and β-arrestin 2 and can bind to visual arrestin (Oakley,

Laporte, Holt, Caron, & Barak, 2000). Additional mechanisms contributing to these differential receptor–arrestin itineraries are discussed in more detail in Chapter 3.

In some cases, β_2-adrenergic receptors, following desensitization and receptor-mediated endocytosis, appeared to interact with and activate secondary signaling pathways, first identified for coupling to MAP kinases *via* Gi-dependent pathways. This switch of β_2AR signaling from Gs to Gi-coupled signaling, and the therapeutic attractiveness of achieving this directed signaling for cardiovascular regulation, is discussed in more detail by Xiao and colleagues in Chapter 9.

C. Arrestin Scaffolding of Non-G Protein-Dependent Signaling Makes Possible Ligand-Biased Signaling Along G Protein-Dependent and -Independent Signaling Networks

A rapidly expanding literature is revealing that arrestins, independent of G protein activation, can scaffold 7-transmembrane receptors to signaling pathways, some of which are unique from those activated by the same receptors *via* G protein activation, and some of which are shared by the G protein signaling pathways (Violin & Lefkowitz, 2007; Shenoy & Lefkowitz, 2003; Lefkowitz & Shenoy, 2005). It has become clear that pharmacological agents directed toward β_2-adrenergic receptors can preferentially evoke, or select for, receptor conformations that favor G protein-mediated signaling (which can be desensitized via arrestins) versus the arrestin-dependent, G protein-independent signaling pathways (Shenoy et al., 2006; Dewire, Ahn, Lefkowitz, & Shenoy, 2007; Luttrell & Gesty-Palmer, 2010). This preferential activation of one versus another pathway has been referred to as ligand-biased signaling (Violin & Lefkowitz, 2007; Rajagopal, Rajagopal, & Lefkowitz, 2010; Whalen, Rajagopal, & Lefkowitz, 2011), biased agonism, ligand-directed trafficking, protean agonism, or collateral efficacy, by others (Galandrin, Oligny-Longpré, Bonin, Ogawa, Galés, & Bouvier, 2008; Vaidehi & Kenakin, 2010). The array of biased ligands is continually expanding (Whalen et al., 2011), and the pharmaceutical specificity that can be achieved by such tools is discussed in considerable detail in Chapter 3 by Shenoy.

A compelling example of the therapeutic impact of such biased ligands is the finding for the β-adrenergic antagonist, carvedilol (Wisler et al., 2007). Though carvedilol is a competitive antagonist for occupancy of β-adrenergic receptors by endogenous epinephrine, carvedilol also appears to induce or stabilize conformations of the receptor that interact with arrestin and are coupled to cardioprotective signaling pathways. The ability to exploit the design and characterization of ligands with dual efficacies could lead to agents that extend the impact of receptor activation toward salutary collateral pathways. Thus, delineation of the *in vivo* consequences of ligand-biased signaling will undoubtedly lead not only to identification of disease-specific therapeutic tools but also to an understanding of how the available receptor-coupled pathways are causally involved in evoking various physiological responses.

VII. MOLECULAR CLONING PERMITS STRUCTURE–FUNCTION ANALYSIS OF ADRENERGIC RECEPTORS VIA MULTIPLE EXPERIMENTAL STRATEGIES

Many of the mechanisms of action of adrenergic receptors outlined above have been elucidated because of the ease of introducing cDNAs encoding wild-type or mutant receptors into heterologous cells to study their binding, coupling to signaling pathways, or trafficking among compartments. The most remarkable and not anticipated consequence of the original cloning of the β_2-adrenergic receptor was the 7-transmembrane topology predicted based on the amino acid sequence predicted from the cDNA cloning (Dixon et al., 1986). The molecular cloning of the β_2-adrenergic receptor was rapidly followed by genomic or cDNA cloning of a large number of adrenergic receptors whose regions of sequence involved in ligand binding, coupling to G proteins, association with arrestins (above) or other interacting proteins (below) rapidly ensued using a variety of mutagenesis strategies.

Chimeric receptors constructed between β_2 and α_2-adrenergic receptors provided insights concerning regions of sequence critical for agonist versus antagonist binding and selectivity for G protein coupling (Kobilka, Kobilka, Daniel, Regan, Caron, & Lefkowitz, 1988). Defining structure–function relationships with chimeric adrenergic receptors, similar to strategies for functional analysis of yeast mating factor receptors (Marsh & Herskowitz, 1988), permitted identification of functional "domains" within the receptor based on "gain of function" properties in the chimeric molecules. Thus, because chimeric receptors lead to *changes* in receptor selectivity in ligand binding or G protein coupling, not loss of function, the findings could be more confidently interpreted than loss-of-function mutations that can either reveal regions of sequence involved in one or another functional activity or, alternatively, result secondarily from perturbed receptor folding or trafficking to the surface rather than from specific loss of the particular function being evaluated.

Another particularly informative mutagenesis strategy was the generation of constitutively active adrenergic receptors. In the studies launched by initial reports for the β_2-adrenergic receptor (Samama et al., 1993), mutations that occurred in the distal part of the 3rd intracellular loop often led to constitutive receptor activation. Functionally, these studies permitted the rigorous mechanistic characterization of inverse agonists, which reduce ligand-independent activation of the receptor in the absence of agonist in addition to their blockade of the receptor at the orthostatic agonist binding site. These findings led to a necessary expansion of the ternary complex model (Samama et al., 1993). These mutations were centered on the residue E268 in the β_2-adrenergic receptor and at homologous residues in other adrenergic receptors. Structural data (discussed below) are consistent with the interpretation of these findings at the time, that is, that these constitutively active mutations resulted from abrogation of intrareceptor interactions that constrained the receptor structure in an inactive state.

Prior cysteine-scanning mutagenesis of the β_2-adrenergic receptor resulted in the design of a mutated β_2-adrenergic receptor structure with a single reactive Cys residue, Cys265, at which Cys-reactive fluorescent probes, such as fluorescein maleimide (FM), could be strategically introduced by covalent modification near the G protein coupling region within the receptor providing an environmentally sensitive reporter of conformational changes evoked by agonists (Gether, Lin, Ghanouni, Ballesteros, Weinstein, & Kobilka, 1997). Time-resolved fluorescent analysis of these FM-labeled β_2-adrenergic receptors revealed *unique* time-resolved fluorescent profiles for agonist versus partial agonist occupancy of the FM-Cys 265-labeled β_2-adrenergic receptor (Ghanouni et al., 2001a & Ghanouni et al., 2001b). These findings of unique conformations of the agonist- versus partial agonist-occupied β_2-adrenergic receptor are in direct contradiction of early two-state models of receptor activation, for example, the allosteric model of Monod, Wyman, and Changeux (1965) or early renditions of the ternary complex (DeLean et al., 1980). These models imagined receptors in equilibrium between two states, one active and one inactive. In such a model, the predicted difference between agonists and partial agonists, for example, would simply be the *fraction* of receptors that each ligand could induce to exist in the single active state, with full agonists evoking a higher fraction of occupied receptors into the active state compared to partial agonists. However, the findings of Gether et al. (1997) and later Ghanhouni et al. (2001) are consistent with the interpretation that receptors are more like rheostats than on–off switches (see Kobilka & Deupi, 2007), with receptors able to achieve multiple, ligand-specific conformational states. These biophysical studies are completely consistent with functional findings summarized above that 7-transmembrane receptors couple to both G protein-dependent and -independent pathways, and that ligands select or induce unique conformations that manifest ligand-dependent profiles for agonism or antagonism along each of these pathways (Kenakin & Miller, 2010). In Chapter 5, Vilardaga extends this conversation to discussions of detecting conformational changes of adrenergic receptors in live cells, including findings that correlate intrinsic efficacy of ligands and the kinetics of these ligand-directed conformational changes.

Studies of intentionally mutated receptors paralleled an interest in defining mutant alleles of adrenergic receptors that may have profound functional consequences *in vivo*, in individuals expressing one or both alleles of these receptors.

The insights evolving from those studies, particularly for α_2-adrenergic receptors, are summarized in Chapter 8, by Wang and colleagues.

At Last, a Structure for The β_2-Adrenergic Receptor

Ingeniously designed recombinant receptor proteins allowed the large-scale purification and crystallization of the β_2-adrenergic receptor (Cherezov et al., 2007; Rosenbaum et al., 2007). Many of the features deduced from the crystal structure are consistent with earlier biochemical and biophysical studies of wild-type and intentionally mutated receptors. For example, the seven transmembrane helices predicted from hydropathy plots from the initial cloning of the β_2-adrenergic receptor (Dixon et al., 1986) were confirmed. Considerable parallels were revealed between rhodopsin and β_2-adrenergic receptors, including the conserved tryptophan side chain in transmembrane helix 6 that stabilizes the inactive receptor conformation. Not expected from rhodopsin structures, however, was a short helix in extracellular loop 2 of the β_2-adrenergic receptor that permits access of extracellular ligands into the binding site, in contrast to the buried sheet in extracellular loop 2 of rhodopsin that shields the hydrophobic retinal in the binding pocket from the extracellular environment. The "ionic lock," defined in the structure of rhodopsin as a result of electrostatic and hydrogen-bonded interactions between the cytoplasmic ends of transmembrane helices 3 and 6, is a key contributor to the constrained configuration "unlocked" by activation. Thus, this "ionic lock" is closed in the inactive but not active rhodopsin structures characterized. A similar network of water-mediated hydrogen bonds is observed in the β_2-adrenergic receptor, but the lock appears to be "broken," as if the receptor is "activated" in the crystallized structure. This is somewhat surprising, since the β_2-adrenergic receptor structure was obtained liganded with carazolol, an antagonist/inverse agonist at the receptor. One interpretation of these data is that carazolol-liganded structure manifests the inverse agonist state of the receptor toward G proteins; alternatively it has been postulated that this "unlocked" structure might represent the active state for the receptor when biased toward effectors other than G proteins (Lefkowitz, Sun, & Shukla, 2008). Clarification of these possible interpretations will naturally require additional structures to be determined, in the presence of G proteins or alternative effectors, which are ongoing (Rasmussen et al., 2011; Rosenbaum et al., 2011).

VIII. GENE TARGETING STUDIES REVEAL SUBTYPE SELECTIVE ROLES FOR ADRENERGIC RECEPTORS IN *IN VIVO* SETTINGS

The intronless nature of many of the earliest cloned G protein-coupled receptors, specifically the β_2- and α_2-adrenergic receptors, facilitated the design of targeting vectors to delete or mutate these receptors *in vivo* relying on homologous recombination strategies (reviewed in Hein & Kobilka, 1997; Hein, Limbird, Eglen, & Kobilka, 1999). These studies led to the discovery of the roles of varying receptor subtypes in a variety of physiological settings and in response to many frequently used pharmacological agents. Despite the value of this approach in revealing the predominant receptor subtypes for drug effects or responses to endogenous agonists at receptors, these studies also reminded investigators that "therapeutic silver bullets," so to speak, would not be achieved simply by developing high affinity, highly selective agonists at receptors. The simplest example of this realization is the finding that α_2-adrenergic receptors of the α_{2A}-subtype are involved not only in central control of blood pressure, but also for the sedative effects of antihypertensive agents, like dexmedetomidine, clonidine, and guanfacine (MacMillan, Hein, Smith, Piascik, & Limbird, 1996; Tan, Wilson, MacMillan, Kobilka, & Limbird, 2002). These findings emphasize that although a highly selective α_{2A} agonist for treatment of hypertension might minimize side effects due to α_{2B} subtypes, which mediate pressor responses (Philipp, Brede, & Hein, 2002), these agonists would *still* be limited in their therapeutic use due to their sedative side effects, which also are mediated by the same α_{2A} subtype, effects that have sidelined agents, such as clonidine. Studies with mice heterozygous for null or mutant (D79N) alleles of the α_{2A} receptor, however, have revealed that a higher receptor density, or fractional occupancy, of α_{2A} receptors is required for the often undesirable sedative properties of α_{2A}-adrenergic agonists than for the hypotensive effects sought in antihypertensive agents (Tan et al., 2002). The finding that greater fractional occupancy of α_{2A}-adrenergic receptors is required for agonist-induced sedation than for other responses evoked by these receptors suggests that intentional development of partial agonists may permit selective therapeutic intervention without same receptor-mediated unwanted side effects of these receptors, particularly for treatment of ADHD or facilitating cognitive enhancement in the elderly (Tan et al., 2002). Pharmaceutical development that includes intrinsic efficacy measures along with other high-throughput screens will likely accelerate the development of agents that achieve desired therapeutic effect without unwanted side effects mediated by the same receptor population.

Newer Generation Genetic Manipulations Provide Insights into The Pre-versus Postsynaptic Roles of α_2-adrenergic Receptors

The earliest characterization of α-adrenergic receptor subtypes posited that α_1-adrenergic receptors were postsynaptic in nature, whereas α_2-adrenergic receptors were presynaptic in nature (Starke, Endo, & Taube, 1775). As reviewed in Chapter 7 of this volume by Hein et al., elegant transgenic strategies have revealed that not only can α_2-adrenergic receptors be both pre- and postsynaptic in localization and function but also that α_{2A}-adrenergic receptors can suppress presynaptic release of either catecholamines, acting as *autoceptors*, or of other neurotransmitters, functioning as presynaptic *heteroceptors.*

Hein and colleagues exploited an ingenious strategy of expressing the α_{2A} receptor subtype driven by the dopamine-β-hyroxylase promoter in mice null for the α_{2A} receptor. Using this approach, only autoceptor function of these receptors can be restored in the transgenic animals. Data summarized in Chapter 7 emphasize the pharmacological importance of α_2-adrenoceptors in non-adrenergic cells and neurons; Hein and colleagues posit that innovative drugs targeting cognition, depression, sedation, analgesia, as well as central cardiovascular regulation might result from focusing on strategies to modulate the function of presynaptic α_{2A}*hetero*ceptors.

IX. THE MULTIPLE INTERACTING PROTEINS OF ADRENERGIC RECEPTORS

The inherent assumption in signal transduction studies is that signaling pathways are mediated by protein–protein interactions, even if these interactions are only transitory in nature. This explains, then, the desire to identify interacting proteins for receptors, the initiating step in signaling pathways, and reveal the functional consequence of these protein–protein interactions in receptor folding, trafficking, coupling to or scaffolding with signaling molecules. Several reviews have summarized the wide variety of interacting proteins for β and α_2-adrenergic receptors (Ritter & Hall, 2009; Wang & Limbird, 2007). To date, the functional consequences for receptor interaction with many of these proteins, including receptor–receptor interactions among oligomers, remain to be elucidated in native cells. In Chapter 2, Hebert and Colleagues discuss the role of receptor interacting proteins in organization and assembly of signaling complexes, which is the key to achieve signaling specificity and diversified functional outcomes. Chapter 4 by Wu and colleagues addresses the function of both endofacial motifs and membrane-embedded sequences of receptors that confer selectivity in interacting with small molecular weight GTPases in receptor transport from compartment to compartment. These represent just a few areas of ongoing discovery concerning adrenergic receptors and their interacting proteins.

Given that there are a finite number of domains within 7-transmembrane proteins for interaction with other proteins, it is probable that receptor interactions with proteins, particularly cytosolic protein interactions with endofacial domains of the receptor, may represent competing interactions that provide another level of specificity for receptor signaling. For example, the α_{2A} receptor is capable of interacting with spinophilin (Wang et al., 2004), which thereby competes for arrestin interaction with the receptor. This competition is manifest *in vivo*, as well, where loss of spinophilin (i.e., studies in mice null for spinophilin), presumably resulting in greater probability of arrestin interaction, leads to more sensitive response to α_2 agonist lowering of blood pressure (Lu et al.,

2010) and to evoking sedation (Wang et al., 2004). These functional consequences of *enhanced* sensitivity to α_{2A} adrenergic receptor signaling when arrestin's interactions with the receptor are not impeded, for example, by spinophilin, provides further evidence that arrestin's role lies beyond its importance in homologous desensitization, as alluded to above. In terms of signaling specificity, it is also important to remember that spinophilin (also known as neurabin II; Satoh et al., 1998) is not uniformly distributed in cells, but is abundant in the dendritic spines of neurons (Allen, Ouimet, & Greengard, 1997), and along the basolateral domain of some polarized epithelial cells (Satoh et al., 1998). Thus, it is reasonable to postulate that reciprocal interactions among receptor-interacting proteins might not only modulate signaling or other receptor-involved processes but also do so in a compartment-selective fashion.

X. AND THAT BRINGS US BACK TO THE ISSUE OF SPECIFICITY AND THERAPEUTIC SELECTIVITY

A common overarching goal of the studies of adrenergic receptors, whether *in vivo* or at the level of single molecules assessed in real time, has been to understand the molecular basis for receptor action, with the intent of translating these insights into selective mechanisms for disease prevention or therapeutic intervention. Since the initial clarification in 1948 by Ahlquist that adrenergic agents could be either α or β in nature, each level of discovery concerning adrenergic receptors has led to an understanding of yet another level of biological complexity in achieving signaling specificity and its time-dependent modulation. Fortunately, each level of complexity also opens up the possibility of yet another lever for therapeutic selectivity: receptor subtypes; agonists of varying efficacy; disrupters or facilitators of selected receptor-interacting protein encounters; or ligands biased toward varying G protein-dependent or -independent pathways. This volume captures many of the exciting new areas of inquiry that lay the groundwork for future innovative therapeutic design of agents more selective in their efficacy and diminished in unwanted side effects.

Acknowledgments

The author thank Dr. Robert J. Lefkowitz, Duke University, and Dr. Brian Kobilka, Stanford University, for their critical review of this manuscript, and Dr. Qin Wang for her editorial oversight.

References

Ahlquist, R. P. (1948). A study of adrenotropic receptors. *Am J Physiol, 153*, 586–600.

Allen, P. B., Ouimet, C. C., & Greengard, P. (1997). Spinophilin, a novel protein phosphatase 1 binding protein localized to dendritic spines. *Proc Natl Acad Sci USA, 94*, 9956–9961.

Attramadal, H., Arriza, J. L., Aoki, C., Dawson, T. M., Codina, J., & Kwatra, M. M., et al., (1992). β-arrestin2, a novel member of the arrestin/β-arrestin gene family. *J Biol Chem, 267*, 17882–17890.

Aurbach, G. D., Fedak, S. A., Woodard, C. J., Palmer, J. S., Hauser, D., & Troxler, F. (1974). Beta-adrenergic receptor: Stereospecific interaction of iodinated beta-blocking agent with high affinity site. *Science, 186*, 1223–1224.

Barger, G., & Dale, H. H. (1910). Chemical structure and sympathomimetic action of amines. *J Physiol, 41*, 19–59.

Benovic, J. L., Pike, L. J., Cerione, R. A., Staniszewski, C., Yoshimasa, T., & Codina, J., et al., (1985). Phosphorylation of the mammalian beta-adrenergic receptor by cyclic AMP-dependent protein kinase. Regulation of the rate of receptor phosphorylation and dephosphorylation by agonist occupancy and effects on coupling of the receptor to the stimulatory guanine nucleotide regulatory protein. *J Biol Chem, 260*, 7094–7101.

Benovic, J. L., Strasser, R. H., Caron, M. G., & Lefkowitz, R. J. (1986). Beta-adrenergic receptor kinase: identification of a novel protein kinase that phosphorylates the agonist-occupied form of the receptor. *Proc Natl Acad Sci USA, 83*, 2797–2801.

Benovic, J. L., Staniszewski, C., Mayor Jr., F., Caron, M. G., & Lefkowitz, R. J. (1988). Beta-Adrenergic receptor kinase. Activity of partial agonists for stimulation of adenylate cyclase correlates with ability to promote receptor phosphorylation. *J Biol Chem, 263*, 3893–3897.

Bylund, D. B., Eikenberg, D. C., Hieble, J. P., Langer, S. Z., Lefkowitz, R. J., & Minneman, K. P., et al., (1994). International Union of Pharmacology nomenclature of adrenoceptors. *Pharmacol Rev, 46*, 121–136.

Cherezov, V., Rosenbaum, D. M., Hanson, M. A., Rasmussen, S. G., Thian, F. S., & Kobilka, T. S., et al., (2007). High-resolution crystal structure of an engineered human beta2-adrenergic G protein-coupled receptor. *Science, 318*, 1258–1265.

Clark, R. B., Kunkel, M. W., Friedman, J., Goka, T. J., & Johnson, J. A. (1988). Activation of cAMP-dependent protein kinase is required for heterologous desensitization of adenylyl cyclase in S49 wild-type lymphoma cells. *Proc Natl Acad Sci USA, 85*, 1442–1446.

Davoren, P. R., & Sutherland, E. W. (1963). The effect of L-epinephrine and other agents on the synthesis and release of adenosine 3′,5′-phosphate by whole pigeon erythrocytes. *J Biol Chem, 238*, 3009–3015.

De Lean, A., Stadel, J. M., & Lefkowitz, R. J. (1980). A ternary complex model explains the agonist-specific binding properties of the adenylate cyclase-coupled beta-adrenergic receptor. *J Biol Chem, 255*, 7108–7117.

Dewire, S. M., Ahn, S., Lefkowitz, R. J., & Shenoy, S. K. (2007). Beta-arrestins and cell signaling. *Annu Rev Physiol, 69*, 483–510.

Dixon, R. A., Kobilka, B. K., Strader, D. J., Benovic, J. L., Dohlman, H. G., & Frielle, T., et al., (1986). Cloning of the gene and cDNA for mammalian beta-adrenergic receptor and homology with rhodopsin. *Nature, 321*, 75–79.

Fung, B. K., Hurley, J. B., & Stryer, L. (1981). Flow of information in the light-triggered cyclic nucleotide cascade of vision. *Proc Natl Acad Sci USA, 78*, 152–156.

Galandrin, S., Oligny-Longpré, G., Bonin, H., Ogawa, K., Galés, C., & Bouvier, M. (2008). Conformational rearrangements and signaling cascades involved in ligand-biased mitogen-activated protein kinase signaling through the beta1-adrenergic receptor. *Mol Pharmacol, 74*, 162–172.

Gether, U., Lin, S., Ghanouni, P., Ballesteros, J. A., Weinstein, H., & Kobilka, B. K. (1997). Agonists induce conformational changes in transmembrane domains III and VI of the β_2 adrenoceptor. *EMBO J, 16*, 6737–6747.

Ghanouni, P., Steenhuis, J. J., Farrens, D. L., & Kobilka, B. K. (2001a). Agonist-induced conformational changes in the G-protein-coupling domain of the β_2-adrenergic receptor. *Proc Natl Acad Sci USA, 98*, 5997–6002.

Ghanouni, P., Gryczynski, Z., Steenhuis, J. J., Lee, T. W., Farrens, D. L., & Lakowicz, J. R., et al., (2001b). Functionally different agonists induce distinct conformations in the G protein coupling domain of the β_2 adrenergic receptor. *J Biol Chem, 276*, 24433–24436.

Goodman Jr., O. B., Krupnick, J. G., Santini, F., Gurevich, V. V., Penn, R. B., & Gagnon, A. W., et al., (1996). Beta-arrestin acts as a clathrin adaptor in endocytosis of the beta2-adrenergic receptor. *Nature, 383*, 447–450.

Hein, L., & Kobilka, B. K. (1997). Adrenergic receptors from molecular structure to in vivo function. *Trends Cardiovasc Med, 7*, 137–145.

Hein, L., Limbird, L. E., Eglen, R. M., & Kobilka, B. K. (1999). Gene substitution/knockout to delineate the role of alpha 2- adrenoceptor subtypes in mediating central effects of catecholamines and imidazolines. *Ann NY Acad Sci, 881*, 265–271.

Heubel, E. (1871). Pathogenese und symptome der chronischen Bleivergiftung. Hirschwald, Berlin.

Kenakin, T. (2004). Principles: Receptor theory in pharmacology. *Trends Pharmacol Sci, 25*, 186–192.

Kenakin, T., & Miller, L. J. (2010). Seven transmembrane receptors as shapeshifting proteins: The impact of allosteric modulation and functional selectivity on new drug discovery. *Pharmacol Rev, 62*, 265–304.

Kobilka, B. K., Kobilka, T. S., Daniel, K., Regan, J. W., Caron, M. G., & Lefkowitz, R. J. (1988). Chimeric alpha 2-,beta 2-adrenergic receptors: delineation of domains involved in effector coupling and ligand binding specificity. *Science, 240*, 1310–1316.

Kobilka, B. K., & Deupi, X. (2007). Conformational complexity of G-protein-coupled receptors. *Trends Pharmacol Sci, 28*, 397–406.

Krupnick, J. G., & Benovic, J. L. (1998). The role of receptor kinases and arrestins in G protein-coupled receptor regulation. *Annu Rev Pharmacol Toxicol, 38*, 289–319.

Kühn, H., McDowell, J. H., Leser, K. H., & Bader, S. (1977). Phosphorylation of rhodopsin as a possible mechanism of adaptation. *Biophys Struct Mech, 29*, 175–180.

Kühn, H., Hall, S. W., & Wilden, U. (1984). Light-induced binding of 48-kDa protein to photoreceptor membranes is highly enhanced by phosphorylation of rhodopsin. *FEBS Lett, 176*, 473–478.

Langley, J. N. (1909). On the contraction of muscle, chiefly in relation to the presence of "receptive" substances. Part IV. The effect of curare and of some other substances on the nicotine response of the sartorius and gastrocnemius muscles of the frog. *J. Physiol, 39*, 235–295.

Laporte, S. A., Oakley, R. H., Zhang, J., Holt, J. A., Ferguson, S. S., & Caron, M. G., et al., (1999). The beta2-adrenergic receptor/betaarrestin complex recruits the clathrin adaptor AP-2 during endocytosis. *Proc Natl Acad Sci USA, 96*, 3712–3717.

Lefkowitz, R. J., Mukherjee, C., Coverstone, M., & Caron, M. G. (1974). Stereospecific (^{3}H) (–)-alprenolol binding sites, beta-adrenergic receptors and adenylate cyclase. *Biochem Biophys Res Commun, 60*, 703–709.

Lefkowitz, R. J., Mullikin, D., & Caron, M. G. (1976). Regulation of beta-adrenergic receptors by guanyl-5'-yl imidodiphosphate and other purine nucleotides. *J Biol Chem, 251*, 4686–4892.

Lefkowitz, R. J., Roth, J., Pricer, W., & Pastan, I. (1970). ACTH receptors in the adrenal: specific binding of ACTH-125I and its relation to adenyl cyclase. *Proc Natl Acad Sci USA, 65*, 745–752.

Lefkowitz, R. J., Sharp, G. W., & Haber, E. (1973). Specific binding of -adrenergic catecholamines to a subcellular fraction from cardiac muscle. *J Biol Chem, 248*, 342–349.

Lefkowitz, R. J., & Shenoy, S. K. (2005). Transduction of receptor signals by beta-arrestins. *Science, 308*, 512–517.

Lefkowitz, R. J., Sun, J. P., & Shukla, A. K. (2008). A crystal clear view of the beta2-adrenergic receptor. *Nat. Biotechnol., 26*, 189–191.

Levitzki, A., Atlas, D., & Steer, M. L. (1974). The binding characteristics and number of beta-adrenergic receptors on the turkey erythrocyte. *Proc Natl Acad Sci USA, 71*, 2773–2776.

Limbird, L. E., & Lefkowitz, R. J. (1977). Resolution of beta-adrenergic receptor binding and adenylate cyclase activity by gel exclusion chromatography. *J Biol Chem, 252*, 799–802.

Limbird, L. E., & Lefkowitz, R. J. (1978). Agonist-induced increase in apparent beta-adrenergic receptor size. *Proc Natl Acad Sci USA, 75*, 228–232.

Limbird, L. E., Gill, D. M., & Lefkowitz, R. J. (1980). Agonist-promoted coupling of the beta-adrenergic receptor with the guanine nucleotide regulatory protein of the adenylate cyclase system. *Proc Natl Acad Sci USA, 77*, 775–779.

Lohse, M. J., Benovic, J. L., Caron, M. G., & Lefkowitz, R. J. (1990). Multiple pathways of rapid beta 2-adrenergic receptor desensitization. Delineation with specific inhibitors. *J Biol Chem, 265*, 3202–3211.

Lu, R., Chen, Y., Cottingham, C., Peng, N., Jiao, K., & Limbird, L. E., et al., (2010). Enhanced hypotensive, bradycardic, and hypnotic responses to alpha2-adrenergic agonists in spinophilin-null mice are accompanied by increased G protein coupling to the alpha2A-adrenergic receptor. *Mol Pharmacol, 78*, 279–286.

Luttrell, L. M., & Gesty-Palmer, D. (2010). Beyond desensitization: Physiological relevance of arrestin-dependent signaling. *Pharmacol Rev, 62*, 305–330.

MacMillan, L. B., Hein, L., Smith, M. S., Piascik, M. T., & Limbird, L. E. (1996). Central hypotensive effects of the alpha2a-adrenergic receptor subtype. *Science, 273*, 801–803.

Maguire, M. E., Van Arsdale, P. M., & Gilman, A. G. (1976). An agonist-specific effect of guanine nucleotides on binding to the beta adrenergic receptor. *Mol Pharmacol, 12*, 335–339.

Marsh, L., & Herskowitz, J. (1988). STE2 protein of *Saccharomyces kluyveri* is a member of the rhodopsin/beta-adrenergic receptor family and is responsible for recognition of the peptide ligand alpha factor. *Proc Natl Acad Sci USA, 85*, 3855–3859.

Monod, J., Wyman, J., & Changeux, J. P. (1965). On the nature of allosteric transitions: A plausible model. *J Mol Biol, 12*, 88–118.

Oakley, R. H., Laporte, S. A., Holt, J. A., Caron, M. G., & Barak, L. S. (2000). Differential affinities of visual arrestin, β arrestin1, and β arrestin2 for G protein-coupled receptors delineate two major classes of receptors. *J Biol Chem, 275*, 17201–17210.

Orly, J., & Schramm, M. (1976). Coupling of catecholamine receptor from one cell with adenylate cyclase from another cell by cell fusion. *Proc Natl Acad Sci USA, 73*, 4410–4414.

Pfister, C., Kühn, H., & Chabre, M. (1983). Interaction between photoexcited rhodopsin and peripheral enzymes in frog retinal rods. Influence on the postmetarhodopsin II decay and phosphorylation rate of rhodopsin. *Eur J Biochem, 136*, 489–499.

Philipp, M., Brede, M., & Hein, L. (2002). Physiological significance of alpha(2)-adrenergic receptor subtype diversity: one receptor is not enough. *Am J Physiol Regul Integr Comp Physiol, 283*, R287–R295.

Pitcher, J. A., Freedman, N. J., & Lefkowitz, R. J. (1998). G protein-coupled receptor kinases. *Annu Rev Biochem, 67*, 653–692.

Rajagopal, S., Rajagopal, K., & Lefkowitz, R. J. (2010). Teaching old receptors new tricks: Biasing seven-transmembrane receptors. *Nat Rev Drug Discovery, 5*, 373–386.

Ritter, S. L., & Hall, R. A. (2009). Fine-tuning of GPCR activity by receptor-interacting proteins. *Nat Rev Mol Cell Biol, 10*, 819–830.

Rodbell, M., Birnbaumer, L., Pohl, S. L., & Krans, H. M. (1971a). The glucagon-sensitive adenyl cyclase system in plasma membranes of rat liver. V. An obligatory role of guanylnucleotides in glucagon action. *J Biol Chem, 246*, 1877–1882.

Rodbell, M., Krans, H. M., Pohl, S. L., & Birnbaumer, L. (1971b). The glucagon-sensitive adenyl cyclase system in plasma membranes of rat liver. IV. Effects of guanylnucleotides on binding of 125I-glucagon. *J Biol Chem, 246*, 1872–1876.

Rosenbaum, D. M., Cherezov, V., Hanson, M. A., Rasmussenm, S. G., Thian, F. S., & Kobilka, T. S., et al., (2007). GPCR engineering yields high-resolution structural insights into beta2-adrenergic receptor function. *Science, 318*, 1266–1273.

Rosenbaum, D. M., Zhang, C., Lyons, J. A., Holl, R., Aragao, D., & Arlow, D. H., et al., (2011). Structure and function of an irreversible agonist-beta(2) adrenoceptor complex. *Nature, 469*, 236–240.

Ross, E. M., & Gilman, A. G. (1977). Reconstitution of catecholamine-sensitive adenylate cyclase activity: Interactions of solubilized components with receptor-replete membranes. *Proc Natl Acad Sci USA, 74*, 3715–3719.

Samama, P., Cotecchia, S., Costa, T., & Lefkowitz, R. J. (1993). A mutation-induced activated state of the beta2-adrenergic receptor. Extending the ternary complex model. *J Biol Chem, 268*, 4625–4636.

Satoh, A., Nakanishi, H., Obaishi, H., Wada, M., Takahashi, K., & Satoh, K., et al., (1998). Neurabin-II/spinophilin. An actin filament-binding protein with one pdz domain localized at cadherin-based cell-cell adhesion sites. *J Biol Chem, 273*, 3470–3475.

Shenoy, S. K., & Lefkowitz, R. J. (2003). Multifaceted roles of beta-arrestins in the regulation of seven-membrane-spanning receptor trafficking and signalling. *Biochem J, 375*, 503–515.

Shenoy, S. K., Drake, M. T., Nelson, C. D., Houtz, D. A., Xiao, K., & Madabushi, S., et al., (2006). Beta-arrestin-dependent, G protein-independent ERK1/2 activation by the beta2 adrenergic receptor. *J Biol Chem, 281*, 1261–1273.

Starke, K., Endo, T., & Taube, H. D. (1975). Pre- and postsynaptic components in effect of drugs with alpha adrenoceptor affinity. *Nature, 254*, 440–441.

Su, Y. F., Harden, T. K., & Perkins, J. P. (1979). Isoproterenol-induced desensitization of adenylate cyclase in human astrocytoma. cells Relation of loss of hormonal responsiveness and decrement in beta-adrenergic receptors. *J Biol Chem, 254*, 38–41.

Tan, C. M., Wilson, M. H., MacMillan, L. B., Kobilka, B. K., & Limbird, L. E. (2002). Heterozygous alpha 2A-adrenergic receptor mice unveil unique therapeutic benefits of partial agonists. *Proc Natl Acad Sci USA, 99*, 12471–12476.

Vaidehi, N., & Kenakin, T. (2010). The role of conformational ensembles of seven transmembrane receptors in functional selectivity. *Curr Opin Pharmacol, 10*, 775–781.

Violin, J. D., & Lefkowitz, R. J. (2007). Beta-arrestin-biased ligands at seven-transmembrane receptors. *Trends Pharmacol Sci, 28*, 416–422.

Wang, Q., & Limbird, L. E. (2007). Regulation of alpha2AR trafficking and signaling by interacting proteins. *Biochem Pharmacol, 73*, 1135–1145.

Wang, Q., Zhao, J., Brady, A. E., Feng, J., Allen, P. B., & Lefkowitz, R. J., et al., (2004). Spinophilin blocks arrestin actions in vitro and in vivo at G protein-coupled receptors. *Science, 304*, 1940–1944.

Whalen, E. J., Rajagopal, S., & Lefkowitz, R. J. (2011). Therapeutic potential of β-arrestin- and G protein-biased agonists. *Trends Mol Med, 17*, 126–139.

Wisler, J. W., DeWire, S. M., Whalen, E. J., Violin, J. D., Drake, M. T., & Ahn, S., et al., (2007). A unique mechanism of beta-blocker action: Carvedilol stimulates beta-arrestin signaling. *Proc Natl Acad Sci USA, 42*, 16657–16662.

Yamazaki, A., Halliday, K. R., George, J. S., Nagao, S., Kuo, C. H., & Ailsworth, K. S., et al., (1985). Homology between light-activated photoreceptor phosphodiesterase and hormone-activated adenylate cyclase systems. *Adv Cyclic Nucleotide Protein Phosphorylation Res, 19*, 113–124.

CHAPTER 2

Organizational Complexity of β-adrenergic Receptor Signaling Systems

Irina Glazkova, Katrin Altosaar, and Terence E. Hébert

Department of Pharmacology and Therapeutics, McGill University, Montréal, Canada

I. Overview
II. Introduction
III. General Considerations Regarding GPCR Organization
IV. Signaling Diversity in β-Adrenergic Receptors
V. The Impact of Recent Crystal Structures on Our Understanding of Biased Signaling
VI. Larger Receptor Arrays
  A. βAR Homodimers
  B. βAR Heterodimerization
  C. βAR Interactions with Other Receptor Classes
  D. Interactions with G Proteins and Effector Molecules
VII. Ontogeny of βAR Signaling Systems
VIII. Asymmetric GPCR Complexes
IX. Beyond the Paradigm of a Cell Surface Receptor
X. Conclusions
Acknowledgments
References

I. OVERVIEW

In recent years, we have come to appreciate the complexity of GPCR signaling in general and β-adrenergic receptor signaling in particular. Starting originally from three βAR subtypes with simple, linear signaling cascades, we can now discuss models of large receptor-based networks which provide a rich and diverse set of physiological responses depending on their complement of signaling partners. Further, the subcellular localization of these signaling complexes also enriches the diversity of phenotypic outcomes. Here, we review our understanding of the signaling repertoire controlled by βAR subtypes, and how

Current Topics in Membranes, Volume 67 0065-230X/10 $35.00

DOI: 10.1016/B978-0-12-384921-2.00001-X

signaling systems associated with the three receptors are organized, with a special focus on the cardiomyocyte that expresses all three subtypes. Finally, we explore how the diversity of signaling complexes and signaling outcomes are related.

II. INTRODUCTION

G protein-coupled signal transduction systems constitute the largest class of drug targets for the therapeutic treatment of diseases. More than 50% of prescription drugs on the market today, including 20% of the top 50 selling drugs, directly or indirectly, target these systems, accounting for over $20 billion of the pharmaceutical industry's annual sales worldwide. Furthermore, the potential for additional therapeutic drugs that target these systems is considerable since currently available drugs target pathways that are controlled by only about 5% of the 800 identified human G protein-coupled receptors (GPCRs, Fredriksson, Lagerstrom, M. C., Lundin, L. G., & Schioth, 2003; Lee et al., 2003). GPCRs regulate the activity of multiple effectors by activating heterotrimeric G proteins with distinct subunit compositions. The majority of drugs that target GPCR signal transduction systems act as either orthosteric agonists, antagonists, or inverse agonists, although allosteric ligands are becoming more important. Therapeutic strategies often require regulating activity of a subset of specific effectors, but drugs aimed at GPCRs coupled to multiple effectors most certainly lack specificity in this regard, and are usually associated with undesirable side effects. However, the presence of unique components within particular pathways (e.g., G proteins with a specific subunit composition, effector molecules, or regulatory proteins) suggests that there are correspondingly unique structural determinants involved in signal transduction that might serve as allosteric targets for therapeutic small molecule, peptidic, and/or peptidomimetic drugs with greater specificity and fewer side effects. A primary objective of current research will ultimately be to identify and exploit peptide motifs involved in these protein–protein interactions with a view toward designing allosteric, and pathway-specific modulators. In this review, we discuss what is known about signaling complexes based on β-adrenergic receptors (βAR). First, we discuss the diversity of signaling pathways regulated by the different βAR subtypes. Next, we delve into some basic notions about pharmacological efficacy which have reframed our ideas about signaling specificity and the organization of GPCR signaling systems. We then examine recent data regarding the structure of the receptor, including the implications of homo- and heterodimerization of βAR subtypes for cellular signaling. Further, we discuss the importance of βAR assembly into distinct signaling complexes when we consider that these receptors do not simply signal at the cell surface. Finally, we discuss the implications that these larger signaling complexes have for drug development and for the understanding of efficacy at the molecular level. Where relevant, we cite recent reviews that focus in more detail on specific aspects of receptor signaling we touch upon in the different sections.

III. GENERAL CONSIDERATIONS REGARDING GPCR ORGANIZATION

Our understanding of the complexity of GPCR signaling suggests that a spectrum of organizational arrangements are used by different signaling pathways, ranging from a series of sequential and transient interactions between partners to more stably assembled signalosomes that are built and trafficked to the plasma membrane or other subcellular destinations (reviewed in Lambert, 2008; Dupré, Robitaille, Rebois, & Hébert, 2009). One end of the spectrum is required where signal amplification is important and specificity is of lesser concern such as in the mammalian visual system where there is a single GPCR, rhodopsin, a single G protein heterotrimer, transducin, and a limited number of effectors such as cGMP phosphodiesterase. The other end of the spectrum reflects a need for rapid and highly specific signaling events in cells such as neurons or cardiomyocytes which express large numbers of GPCRs, G proteins, and effectors. The dynamic nature of these latter complexes would perhaps be better described by the term "meta-stable" where individual interactions between partners might be quite labile (especially when studied in isolation *in vitro*) but many of them together contribute to the overall integrity and stability of a larger complex. We could begin to discuss different signaling pathways in these terms as well. We have reached a point where characterization of these complexes using proteomic and imaging techniques can yield novel strategies for therapeutic intervention. This notion encompasses modulation of specific receptor complexes by disrupting interactions which are important for their formation, trafficking, and function or augmenting interactions that favor specific signal transduction events with peptidic or peptidomimetic compounds.

IV. SIGNALING DIVERSITY IN β-ADRENERGIC RECEPTORS

The three βAR subtypes were initially believed to comprise rather simple and linear signaling cascades involving the receptor, the Gs heterotrimer, and activation of adenylyl cyclase (AC). This organization, which required nothing but a sequential, agonist-driven interaction first between the receptor and the G protein and then between the activated G protein and the effector, was essentially similar in all cell types that expressed each receptor. It was soon appreciated that all three receptor subtypes could also interact with other G proteins such as the PTX-sensitive Gi heterotrimer (reviewed in Evans, Sato, Sarwar,

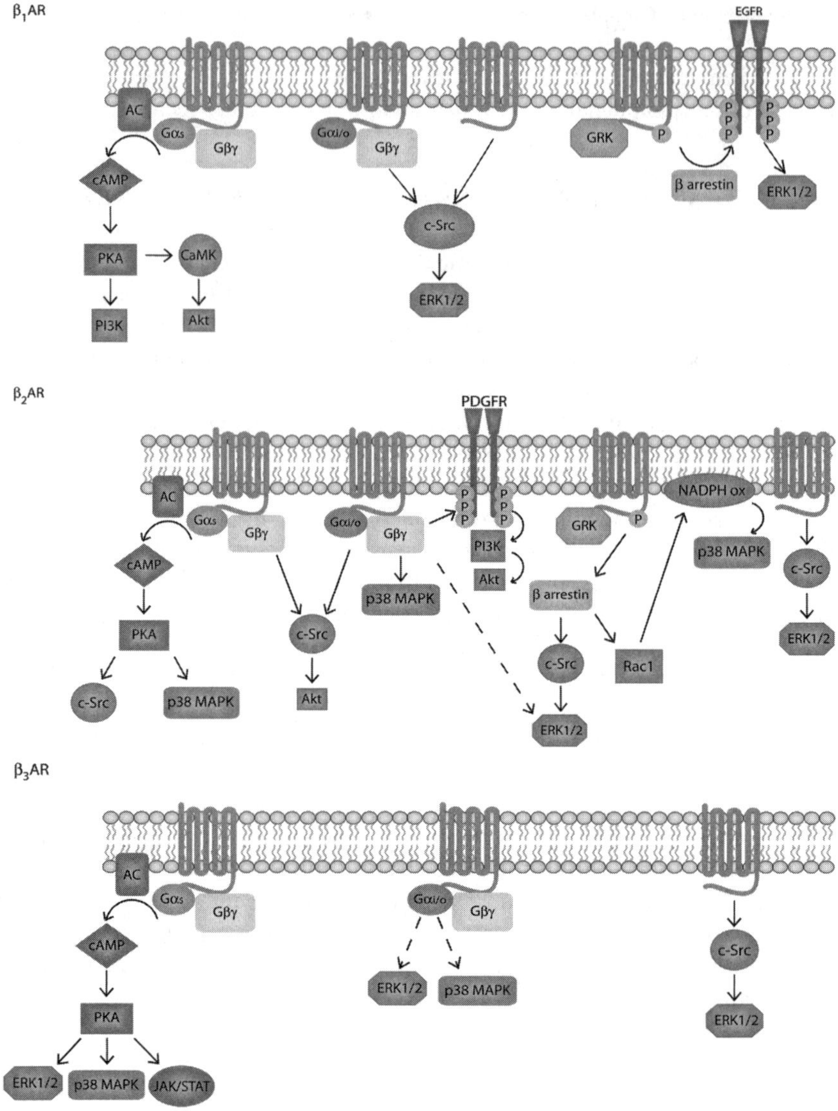

FIGURE 1 Diversity of signaling pathways modulated by different βAR subtypes. All three subtypes are coupled to both Gs and Gi under different conditions and in different cell types (see the excellent recent review from Evans et al., 2010). However, both desensitization profiles and subsequent waves of signaling depend on their relative propensities to become phosphorylated by GRKs and their affinities for β-arrestin (Suzuki et al., 1992; Shiina, Kawasaki, Nagao, & Kurose, 2000; Shiina, Nagao, & Kurose, 2001). Distinct populations of interacting partners for each receptor, again, dependent on cell and tissue type, will also set limits on potential signaling outcomes. For example, β_3AR activation of p38 occurs only in adipocytes (Cao, Medvedev, Daniel, & Collins, 2001). For simplicity, receptors are represented as monomers. However, as described in the text, receptor heterodimerization may alter this picture substantially.

Hutchinson, & Summers, 2010), presumably not simply to provide an inhibitory stimulus to the same effector enzyme, AC. Since then, it has become clear that each receptor interacts with a wide array of signaling pathways (Fig. 1), some of which depend directly on G protein-dependent signaling and others which involve agonist-dependent recruitment of G protein-coupled receptor kinases (GRKs) and β-arrestins (reviewed in Luttrell & Gesty-Palmer, 2010). Once believed to be primarily involved in the desensitization and internalization of GPCRs, it has become clear that β-arrestin-dependent signaling events enrich both the phenotypic diversity of signaling and also deliver receptor-dependent signals to distinct subcellular targets. This second wave of signaling is thought by some authors to be essentially G protein-independent (see e.g., Rajagopal, Lefkowitz, & Rockman, 2005; Patel, Noor, & Rockman, 2010).

As the rich diversity of signaling pathways for the different receptor subtypes was being appreciated, it was also appreciated that βAR could exist in an active state even in the absence of an agonist (Chidiac, Hébert, Valiquette, Dennis, & Bouvier, 1994; Samama. Pei, Costa, Cotecchia, & Lefkowitz, 1994). This constitutive activity lead to the identification of a new class of receptor ligands known as inverse agonists and an appreciation that most GPCRs existed in at least two states that could be toggled toward active by agonists and in the opposite direction by inverse agonists (see Galandrin, Oligny-Longpre, & Bouvier, 2007; Kenakin, 2007a, 2007b for review). However, even this notion turned out to be a gross oversimplification of the number of possible receptor states that might exist (see Kenakin, 2010; Vaidehi & Kenakin, 2010 for review). Recent studies have even challenged the basic definitions of pharmacological efficacy, in that ligands defined as agonists, antagonists, or inverse agonists for a given signaling pathway may not necessarily exhibit similar effects in other signaling pathways. The ability of different ligands to discriminate between signaling pathways coupled to a given GPCR has been termed "biased" agonism (Kenakin, 2010). Such biased signaling has been well demonstrated for all three βAR subtypes. For example, work from Michel Bouvier's group has shown that different βAR agonists have differing abilities to activate two signaling pathways downstream of both the β_1AR and the β_2AR, that is AC activation and ERK1/2 MAP kinase activation. The surprises came when they noted that certain neutral antagonists and even inverse agonists for the AC pathway turned out to be agonists for the ERK1/2 pathway (Azzi et al., 2003; Galandrin & Bouvier, 2006, reviewed in Galandrin et al., 2007; Evans et al., 2010; Patel et al., 2010). It has been recently shown that certain β-blockers, such as carvedilol, act as agonists for a prosurvival pathway in the heart involving the β_1AR, β-arrestin, and transactivation of the EGFR leading to MAPK activation (Noma et al., 2007; Kim et al., 2008; Tilley, Kim, Patel, Violin, & Rockman, 2009). These findings are likely to have significant clinical consequences for the development of more appropriate β-blockers for use in treating heart failure. Similar patterns of biased agonism have emerged for the β_3AR (Sato, Horinouchi, Hutchinson, Evans, & Summers, 2007; Sato, Hutchinson, Evans, & Summers, 2008). This work suggested that efficacy, in this sense can now be thought of in Cartesian terms with "*n*" dimensions for an equivalent number of signaling pathways, with individual ligands occupying a unique space (Galandrin et al., 2007). Ligands, once classified according to results obtained with a single signaling readout will now have to be reassessed according to their ability to act as biased ligands in a pathway-specific manner.

These inherent features of GPCRs most certainly reflect their large conformational flexibility, which is required for the diversity of their interactions with signaling partners at different stages of their life cycles. Unfortunately, this conformational flexibility has made structural studies considerably more difficult. Below, we describe a number of recent successes associated with efforts to stabilize GPCRs (or at least restrict the number of possible conformational states) in obtaining a diverse array of receptor structures.

V. THE IMPACT OF RECENT CRYSTAL STRUCTURES ON OUR UNDERSTANDING OF BIASED SIGNALING

A number of studies have been published in the last few years providing structural insight into how GPCRs in general and βAR in particular function. Rhodopsin, the first GPCR to be crystallized, was available in large quantities which made it more amenable to structural and biophysical study (see Palczewski, 2006; Hofmann et al., 2009 for review). These studies indicated that rhodopsin acts like a bimodal switch in that a single photon can activate rhodopsin from a completely inactive state. The ligand for rhodopsin, *cis*-retinal, is always associated with the receptor and acts as an inverse agonist stabilizing it in an inactive form. The transitions induced by light also demonstrated that rhodopsin transits through many different states during its light-induced life cycle, in some cases for extremely short periods of time. Rhodopsin interactions with G proteins are also distinctive, with the system built around the need for signal amplification (Pugh & Lamb, 1993). When comparing the inactive structure with a crystal structure obtained with opsin and a C-terminal fragment of transducin (GαCT), the most notable difference was that the transmembrane helix 6 (TM-VI) had moved substantially outward with respect to the hydrophobic core, thereby creating a binding pocket for the G protein peptide (Scheerer et al., 2008). GαCT binds to a site in opsin which is opened by an outward tilt of TM-VI, a pairing of TM-V and TM-VI, and a restructured TM-VII–helix 8 kink (Scheerer et al., 2008).

Most other GPCRs have some level of constitutive activity and behave as discussed above, more like rheostats than switches. Different ligands can toggle the receptor to different levels of activity and different ligands may drive other GPCRs into a distinct set of conformational states in contrast with the "one" ligand for rhodopsin. Thus the dynamics of receptor activation may be different depending on the particular ligand, G protein, or other signaling partner – that is, molecular context as we discuss below is critical for determining which states a given receptor can occupy. This structural flexibility of other GPCRs as compared with rhodopsin has made it historically difficult to obtain crystal structures (even when it became possible to purify large quantities of receptors). The structures solved have generally involved receptors that have been conformationally silenced by occupancy with strong inverse agonists, by site-directed mutagenesis, or by the addition of large adducts such as antibodies or T4 lysozyme, which stabilized receptor structure. Still, these structures have been highly informative.

Grossly, many similarities existed between the structures of rhodopsin (Palczewski et al., 2000) and other GPCRs that have been crystallized including the β_1AR (Warne et al., 2008), β_2AR (Cherezov et al., 2007; Rasmussen et al., 2007; Rosenbaum et al., 2007), and the adenosine A_{2A}-receptors (Jaakola et al., 2008). Similar overall architecture and the eighth α-helical segment, first identified in rhodopsin, have also been seen in these receptors. Many features of the ligand-binding site predicted from earlier site-directed mutagenesis studies have also been confirmed in the crystal structures (reviewed in Kobilka & Schertler, 2008; Hanson & Stevens, 2009; Lodowski, Angel, & Palczewski, 2009; Rosenbaum, Rasmussen, & Kobilka, 2009). In rhodopsin, the retinal-binding pocket relies mainly on hydrophobic interactions in addition to a covalent linkage with TM VII. β-adrenergic ligands, on the other hand, interact with receptors through two clusters of polar interactions. The first cluster is shown at the tail of the ligand carazolol in cocrystals, where the positively charged secondary amine group and β-OH group participate in polar interactions with a conserved glutamate on TM III and asparagine on TM VII. The second grouping of polar interactions is with the head group of the ligand and a cluster of serine residues on TM V (Cherezov et al., 2007; Rasmussen et al., 2007; Rosenbaum et al., 2007). As discussed, rhodopsin and the β_2AR share overall structural features and a binding pocket for their cognate ligands, at a site located deep within the transmembrane helices. However, the extracellular loops (ECLs) are distinctly structured. In the case of rhodopsin, the N-terminus as well as ECL2 form a lid-like structure that occludes the retinal binding pocket. This structure was not found in βAR and may explain how diffusible ligands gain access to the binding pocket in the β_2AR and other GPCRs. In both the β-adrenergic and adenosine A_{2A}-receptors the extracellular domain is highly constrained and held away from the ligand-binding pocket opening. The adenosine receptor ligand ZM241385 forms mainly polar interactions between a primary amine group and an asparagine residue on TM VI and a glutamate on

ECL2. A π-stacking interaction between the ligand's heterocyclic group and a phenylalanine residue also on ECL2 plays a role in binding affinity (Jaakola et al., 2008). Interestingly, a number of recent studies have demonstrated that the ECL2 may also be a potential target for allosteric receptor modulators (Goupil et al., 2010; Unal, Jagannathan, Bhat, & Karnik, 2010).

However, some surprises relative to rhodopsin were noted in the new structures as well. The role of one highly conserved stretch of residues, the amino acids glutamic acid/aspartic acid–arginine–tyrosine (i.e., the E/DRY motif) in TM III, has received considerable attention with respect to regulating GPCR conformational states (reviewed in Rovati, Capra, & Neubig, 2007). In the consensus view, glutamic acid/aspartic acid maintains the receptor in its ground state, because mutations frequently induce constitutive activity. This hypothesis has been confirmed by the rhodopsin ground-state crystal structure and by computational modeling approaches. However, some class A GPCRs are resistant to mutations that should induce constitutive activity, suggesting alternative roles for the glutamic acid/aspartic acid residue and the E/DRY motif. Of the crystals so far obtained, bovine rhodopsin is the only receptor with an intact ionic lock interaction between the E/DRY motif and glutamate on TM VI. However, in the opsin structures (presumably more reflective of the activated state), the ionic lock is broken and the helical section of TM V is extended considerably relative to the inactive bovine rhodopsin (Scheerer et al., 2008). The human β_2AR has a similar-length TM V as bovine opsin, turkey β_1AR, and human A_{2A}-adenosine receptors, all of which showed a disrupted ionic lock, even in the presence of inactivating mutations and occupation by inverse agonists or antagonists (Cherezov et al., 2007; Rasmussen et al., 2007; Rosenbaum et al., 2007; Jaakola et al., 2008; Warne et al., 2008). With the exception of opsin and rhodopsin, the DRY motif interacts with intracellular loop 2 (ICL2) in the other GPCR structures through a polar interaction between the aspartate residue on the DRY motif and either a serine or tyrosine residue on ICL2.

As discussed above, the inverse agonist carazolol was used to stabilize the β_2AR in order to obtain its structure. The amino acids just below carazolol form a "toggle"' that stabilizes the inactive state of the receptor. Despite superposition of the toggle switch residues of β_2AR with those of the inactive state of rhodopsin, TM-VI of the β_2AR is slightly more tilted, most likely due to the opening of the ionic lock. Other β-adrenergic ligands found to be inverse agonists for AC but agonists for MAPK can be predicted to dock in the β_2AR in a manner highly similar to that of carazolol (Audet & Bouvier, 2008). This suggests that the latter may also activate MAPK. It has been proposed that the "ionic lock opened, toggle switch closed" conformation of the carazolol-bound β_2AR allows β-arrestin-dependent signaling while disfavoring G protein engagement. The implications for these observations for biased agonism are obvious and suggest ways that different ligands might actually stabilize distinct receptor conformations – coupled to specific signaling pathways. We will return to this notion in a broader context at the end of this chapter. Certainly, this will lead to a refined development of more "selective" ligands at the pathway level. We are still avidly awaiting agonist-bound structures of different GPCRs as well as the receptor complexed with G proteins and other interacting partners. Of course there are limitations to static crystal structures in that they only provide a "snapshot" of the receptor conformations possible. More dynamic biophysical techniques are required to examine the changes in these complexes including imaging techniques and NMR.

VI. LARGER RECEPTOR ARRAYS

The most recent crystal structure to appear was that of the CXCR4 chemokine receptor (Wu et al., 2010). In addition to recapitulating many of the features described in the other GPCR structures, this one held a special surprise in that the receptors were crystallized as ligand-bound dimers in five independent structures. The interface between the two monomers included TM V and VI. This result confirms a great deal of work which showed that GPCRs could form both homo- and heterodimeric structures. It is clear from a number of recent studies using reconstitution of GPCRs into proteoliposomes, that these receptors can signal as monomeric proteins (Whorton et al., 2007, 2008). However, it has also become clear in recent years that most if not all GPCRs can form dimers and possibly higher order structures (see Hébert & Bouvier, 1998; Bulenger, Marullo, & Bouvier, 2005; Prinster, Hague, & Hall, 2005; Milligan, 2009 for review). In this section, we focus on the evolving picture of homo- and hetero-dimerization of the different βAR subtypes. Further, we will describe how βARs can be integrated into large signaling arrays involving both their G protein and effector partners, as well as transmembrane receptors from other families. We will also integrate some very recent studies of the organization of GPCRs in the context of receptor oligomers which suggest that signaling may be driven or modulated by different asymmetric arrangements of receptors associated with their signaling partners.

A. βAR Homodimers

Evidence in the literature, based on radiation inactivation experiments and thermodynamic considerations of ligand binding, indicated that GPCRs might actually have been somewhat larger than simple monomers (reviewed in Hébert & Bouvier, 1998). The first direct demonstration of higher order structures for

GPCRs used differential epitope tagging and coimmunoprecipitation to show that the β_2AR was in fact dimeric (Hébert et al., 1996). Interestingly, this study also highlighted the role that TM VI played in dimerization of the receptor using peptides mimicking this transmembrane helix. Direct evidence for β_2AR homodimerization in living cells was provided using resonance energy transfer (RET) experiments (Angers, Salahpour, Joly, Hilairet, Chelsky, & Dennis, 2000; Kuravi, Lan, Barik, & Lambert, 2010). Since then, a combination of RET and copurification approaches has become the gold standard for demonstrating GPCR dimerization (see Pétrin & Hébert, 2010 for review). What has been missing in most studies of receptor homodimerization, or homo-oligomerization for that matter, is an obvious function. It was demonstrated that mutating residues in TM VI, predicted to be important for dimerization of the β_2AR, also reduced the surface trafficking of the receptor suggesting that dimerization was an early event in receptor biosynthesis (Salahpour, Angers, Mercier, Lagacé, Marullo, & Bouvier, 2004). The identification of TM VI being important for receptor dimerization (Hébert et al., 1996; Salahpour et al., 2004) is also in line with the data presented in the recent CXCR4 structures (Wu et al., 2010). Homodimerization of the β_1AR (Mercier, Salahpour, Angers, Breit, & Bouvier, 2002) and β_3AR (Breit, Lagacé, & Bouvier, 2004) has also been demonstrated. We have, until relatively recently, tended to ignore ligand-binding data indicating cooperativity between receptor equivalents as one of the principal manifestations of receptor homo-oligomerization (Ma, Redka, Pisterzi, Angers, & Wells, 2007; Ma, Pawagi, & Wells, 2008), a subject we will return to below.

B. βAR Heterodimerization

It has been somewhat easier to convince skeptics of the potential roles of GPCR heterodimers. This is because a number of studies have demonstrated that heterodimerization can alter signaling profiles or receptor trafficking (reviewed in Terrillon & Bouvier, 2004; Bulenger et al., 2005; Prinster et al., 2005; Milligan, 2009). Not surprisingly, all three βAR subtypes have been shown to form heterodimers with each other. The β_2AR can heterodimerize with both other subtypes (Lavoie et al., 2002; Mercier et al., 2002; Lavoie & Hébert, 2003; Breit et al., 2004) and trafficking was altered in both cases. In the β_1AR/β_2AR heterodimer, the characteristics of the β_1AR predominated, such that the heterodimer trafficked and signaled similar to the β_1AR alone both in HEK 293 cells (Lavoie et al., 2002) and in adult mouse ventricular cardiomyocytes (Zhu et al., 2005). The pharmacology of ligand binding was altered in this pair in that, ligands for both receptors needed to be present to achieve high affinity binding of subtype-selective ligands (Lavoie & Hébert, 2003). In the case of the β$_2$AR/β$_3$AR heterodimer, this pair trafficked similar to the β$_3$AR and was also unable to couple to Gi, unlike the two parent receptors (Breit et al., 2004). To date, no direct demonstration has been provided for interactions between the β$_1$AR and the β$_3$AR, which may be important given their unique subcellular distributions described below.

Without a doubt, the β$_2$AR is the most studied of all the GPCRs with the possible exception of rhodopsin. It has been demonstrated to heterodimerize with a large number of other GPCRs including the α_{1B}AR (Uberti, Hague, Oller, Minneman, & Hall, 2005), 5-HT$_4$R (Berthouze et al., 2005), δ- and κ-opioid receptors (Jordan, Trapaidze, Gomes, Nivarthi, & Devi, 2001; McVey et al., 2001), the EP1 receptor for prostaglandin E$_2$ (McGraw et al., 2006), the bradykinin type 2 receptor (Haack, Tougas, Jones, El-Dahr, Radhakrishna, & McCarty, 2010), angiotensin II type I receptors (Barki-Harrington, Luttrell, L. M., & Rockman, 2003), CXCR4 receptors (Larocca et al., 2010), the cannibinoid CB1 receptor (Hudson, Hébert, & Kelly, 2010; Kuravi et al., 2010), and the olfactory receptors (Hague et al., 2004). A recent study has expanded this list to include the D1 dopamine receptor and the μ-opioid receptor (Kuravi et al., 2010). The α_2AR has been shown to heterodimerize and cointernalize with the β$_1$AR (Junqi et al., 2003) as well as with the β$_2$AR (Kuravi et al., 2010). In some cases, these pairings result in altered trafficking itineraries and in some cases in altered signaling profiles. The majority of these interactions still need to be validated in native tissues, and in many cases, for a clear function to be attributed to heterodimerization.

C. βAR Interactions with Other Receptor Classes

It has become evident in recent years that GPCRs can interact physically and functionally with receptors from other classes including receptor tyrosine kinases (RTKs) and ligand-gated ion channels. In some cases, these interactions manifest through receptor crosstalk, such as in the transactivation of RTKs (Daub, Wallasch, Lankenau, Herrlich, & Ullrich, 1997; Luttrell et al., 1997, reviewed in Luttrell & Luttrell, 2003). In other cases, the GPCR and the RTK are part of larger multiprotein complexes, which can be assembled in response to agonist stimulation and be cotrafficked into endosomes together, as has been demonstrated for the β$_2$AR and the EGFR (Maudsley et al., 2000, reviewed in Rebois & Hébert, 2003). Similar large signaling arrays have been detected for the β$_2$AR and subunits of the GluR1 AMPA type glutamate receptors (Joiner et al., 2010). These interactions may be direct or they may be mediated by shared scaffolding proteins such as AKAPs and PDZ proteins (reviewed in Dai, Hall, & Hell, 2009). The potential implications of direct interactions will be discussed in more detail below.

D. Interactions with G Proteins and Effector Molecules

A number of studies have examined the dynamics of interaction between GPCRs and heterotrimeric G proteins. It has been shown that receptor/G protein complexes are preformed and undergo conformational rearrangement following agonist stimulation (Galés et al., 2005, 2006; Audet et al., 2008). These studies also validated the use of tagged G proteins at different conformational vantage points and showed that agonists could either increase or decrease resonance energy transfer depending on the orientation of the distinct donor/acceptor positions in the same molecules. The latter study showed that complexes containing δ-opioid receptors (DOR) and heterotrimeric G proteins are differentially sensitive to different DOR ligands, highlighting the utility of resonance energy transfer approaches to study and understand efficacy. BRET was also used to demonstrate that the β_2AR forms a complex with heterotrimeric G proteins and effector molecules during biosynthesis, which is subsequently trafficked to the cell surface (Dupré, Robitaille, Ethier, Villeneuve, Mamarbachi, & Hébert, 2006; Rebois et al., 2006).

BRET and FRET have both been used to detect preassembled receptor/G protein complexes and to monitor changes in these interactions in response to ligand stimulation (Galés et al., 2005, 2006; Audet et al., 2008). Although there seems to be a solid case for stability of receptor/G protein interactions in the face of agonist activation, recent data suggest that a spectrum of relative stabilities of the G protein heterotrimer are possible depending on the Gα subunit of the heterotrimeric G protein in question. For example, as described below, it has recently been demonstrated that Go-containing heterotrimers show a markedly increased propensity to dissociate following agonist stimulation than Gs-containing heterotrimers (Digby, Lober, Sethi, & Lambert, 2006; Digby, Sethi, & Lambert, 2008; reviewed in Lambert, 2008).

A number of effectors are also stably associated with both G protein and receptors (in some cases simultaneously) including AC isoforms, L-type calcium channels, calcium-activated potassium channels, and inwardly rectifying potassium channels (Davare et al., 2001; Lavine et al., 2002; Kitano et al., 2003; Liu et al., 2004; Nikolov & Ivanova-Nikolova, 2004; Balijepalli, Foell, Hall, Hell, & Kamp, 2006; Dai et al., 2009). Perhaps more surprisingly, some of these effectors have been shown to be *directly* associated with receptor molecules. For example, it was demonstrated using BRET that β_2AR was associated with both Kir3 ion channels and AC (Lavine et al., 2002; Dupré, Baragli, Rebois, Ethier, & Hébert, 2007), and that D4 dopamine receptors were associated with Kir3 ion channels (Lavine et al., 2002). This latter study also provided the first clues that G$\beta\gamma$ subunits might orchestrate the assembly of receptor/effector complexes. Interfering with G$\beta\gamma$ function (with βARK-CT) but not Gα function (with DN versions of different Gα subunits) prevented D4/Kir3.2 interactions. Protein/

protein interaction assays were also used to show that the opioid-like receptor 1 was physically associated with voltage-gated N-type calcium channels (Beedle et al., 2004; Altier et al., 2006). The existence of stable interactions, independent of receptor activation between G proteins and their effector molecules, including Kir3 ion channels and AC, has also been demonstrated (Dupré et al., 2006; Rebois et al., 2006). A number of recent studies have elegantly demonstrated that G proteins remain associated with Kir3 channels throughout the basic signaling event (Clancy et al., 2005; Lober, Pereira, & Lambert, 2006; Riven, Iwanir, & Reuveny, 2006, reviewed in Zylbergold, Ramakrishnan, & Hébert, 2010). Other key regulatory molecules such as RGS proteins also interact constitutively with receptors, G proteins, and effector molecules (Bernstein, Ramineni, Hague, Cladman, Chidiac, & Levey, 2004; Benians, Nobles, Hosny, & Tinker, 2005; Roy, Baragli, Bernstein, Hepler, Hébert, & Chidiac, 2006). Of course, there are also numerous scaffolding proteins that interact with GPCRs to create an even larger diversity of signaling arrays and signaling outcomes (reviewed in Hall & Lefkowitz, 2002; Bockaert et al., 2004; Daulat et al., 2007; Daulat, Maurice, & Jockers, 2009).

VII. ONTOGENY OF βAR SIGNALING SYSTEMS

It has become clear that transient receptor/G protein/effector (R/G/E) interactions predicted by the standard model of G protein-mediated signal transduction in the mammalian visual system cannot explain the exquisite signaling specificity seen in cells such as cardiomyocytes or neurons. These cells, which may express dozens of possible receptor/G protein heterotrimer/effector combinations, exhibit high signaling fidelity *in vivo* from one receptor activation cycle to the next. *In vitro* studies, where promiscuous coupling is often seen, have not reflected this (reviewed in Gudermann, Kalkbrenner, & Schultz, 1996; Gudermann, Schoneberg, & Schultz, 1997). It has also been shown that particular combinations of heterotrimeric G proteins are responsible for coupling receptors to particular effectors (Kleuss, Hescheler, Ewel, Rosenthal, Schultz, & Wittig, 1991; Kleuss, Scherubl, Hescheler, Schultz, & Wittig, 1992, 1993; 1993; Kalkbrenner et al., 1995; Wang, Mullah, Hansen, Asundi, & Robishaw, 1997; Wang, Mullah, & Robishaw, 1999; Robillard, Ethier, Lachance, & Hébert, 2000; Wang et al., 2001; Albert & Robillard, 2002; Robishaw, Guo, & Wang, 2003; Schwindinger, Betz, Giger, Sabol, Bronson, & Robishaw, 2003). The possibility that receptors (R) and G proteins (G) might be associated prior to receptor activation has been incorporated into models of G protein signaling for some time (Weiss, Morgan, Lutz, & Kenakin, 1996), but experimental evidence that stable "pre-coupled" R–G complexes exist in living cells has been obtained only relatively recently. A large number of studies have demonstrated association, copurification, or coimmunoprecipitation of receptors with G proteins (reviewed in Rebois & Hébert, 2003; Pétrin & Hébert, 2010).

Interestingly, as mentioned above, dimerization has been demonstrated to be required for efficient surface localization of a number of GPCRs including the β_2AR (Salahpour et al., 2004; Dupré et al., 2006) and the α_{1B}AR (Lopez-Gimenez, Canals, Pediani, & Milligan, 2007), and this has been reviewed recently (Milligan, 2010). In fact, significant evidence is accumulating that the assembly of GPCR signaling complexes occurs during their biosynthetic journey, rather than in response to agonist stimulation at the plasma membrane. We have studied the ontogeny of R/G/E complexes, initially focusing on β_1AR and β_2AR (Lavine et al., 2002) as well as AC (Dupré et al., 2007; Baragli, Grieco, Trieu, Villeneuve, & Hébert, 2008) and Kir3 channels (David, Richer, Mamarbachi, Villeneuve, Dupré, & Hébert, 2006; Rebois et al., 2006; Robitaille, Ramakrishnan, Baragli, & Hébert, 2009). Our data suggest that these complexes are formed during biosynthesis rather than through random, agonist-induced interactions at the plasma membrane. First, these interactions occur in the absence of receptor agonists, suggesting that signaling complexes are preassembled (Dupré et al., 2006, 2007; Rebois et al., 2006). Also, these studies demonstrate that many of these proteins interact initially in the endoplasmic reticulum (ER), including monomer equivalents in receptor dimers, receptor and G$\beta\gamma$ subunits as well as effectors such as Kir3 channels and AC with nascent G$\beta\gamma$. These interactions as measured using BRET or coimmunoprecipitation were all insensitive to dominant negative Rab1 or Sar1 (DN Rab1 and Sar1, but not Rabs 2, 6, or 11) constructs (Dupré et al., 2006, 2007), which regulate anterograde receptor trafficking (reviewed in Dupré & Hébert, 2006; Dong, Filipeanu, Duvernay, & Wu, 2007). Rabs and Sar1 are monomeric G proteins that have been demonstrated to be important for vesicular transport to and from different cellular membrane compartments (Zerial & McBride, 2001). Sar1, Rabs1 and 2 are key for trafficking from ER to Golgi, Rabs 6 and 11 for movement from the Golgi to either the nuclear or plasma membrane and Rabs 4, 5, and 7 are important for endosomal targeting. It has recently been demonstrated that different Rab isoforms are important for both the initial membrane targeting of GPCRs (Rab1; Duvernay, Zhou, & Wu, 2004; Filipeanu, Zhou, Claycomb, & Wu, 2004) as well as for their internalization and recycling to the plasma membrane in response to agonist stimulation (Rabs 4, 5, 7, and 11; Seachrist, Anborgh, & Ferguson, 2000; Seachrist et al., 2002; Dale, Seachrist, Babwah, & Ferguson, 2004). We have also demonstrated that Kir3.1/3.4 trafficking is indeed blocked by DN Sar1 and Rab1 as well, where as interactions with G$\beta\gamma$, as measured using BRET or coimmunoprecipitation, were not (Robitaille et al., 2009). However, our data also highlight the fact that the Gα subunit is assembled with nascent receptor/G$\beta\gamma$/effector complexes either in ER export sites or in the Golgi since this interaction *was* blocked by dominant negative Sar1 and Rab 1 (Dupré et al., 2006, 2007). Once the complexes reach the cell surface, the interactions between the β_2AR and the Gs heterotrimer or ACII are sensitive to agonist. However, the affinity of these interactions as measured in BRET saturation experiments did not change (see Rebois et al., 2006). This suggests that conformational changes within the complex occur after agonist stimulation rather than additional recruitment of core interacting partners.

VIII. ASYMMETRIC GPCR COMPLEXES

The notion that GPCRs stably interact with their G protein and effector partners can therefore be suggested as a mechanism to assure rapid and specific signaling. Thus, GPCRs might be viewed as scaffolding proteins for formation of specific hardwired signaling complexes or signaling hubs. These complexes may be distinct for individual receptor monomers, homo- or heterodimers leading to a unique signaling output for each receptor complex. Further, a recent study has shown that the two-receptor equivalents in the context of a D2 dopamine receptor homodimer are organized asymmetrically with respect to their G protein partners (Han, Moreira, Urizar, Weinstein, & Javitch, 2009) such that occupation by ligand of one receptor activates the receptor and occupation of the other modulates signaling allosterically. In the context of a homodimer this may not be so important as either receptor can serve each role and the asymmetry could not be detected (shown schematically in Fig. 2A).

However, a number of recent studies have suggested that GPCRs can form higher order complexes in addition to monomers or simple homo- or heterodimers (Ma et al., 2007, 2008). Protein complementation approaches have now been used to confirm and extend our knowledge regarding dimerization and oligomerization of GPCRs. Reconstitution of split luciferase (*Gaussia* or *Renilla*) and split GFP constructs have shown that dimers of β_2AR (Rebois, Robitaille, Petrin, Zylbergold, Trieu, & Hébert, 2008) and D2 dopamine receptors (Guo et al., 2008) can be detected, complementing immunopurification and RET approaches, and these approaches can be combined to detect and examine larger complexes. A number of investigators have used three partner PCA/RET to show that higher order complexes of GPCRs such as the A_{2A}-adenosine receptor homo- and hetero-oligomers with CB1 cannabinoid/D2 dopamine receptors (Carriba et al., 2008; Gandia et al., 2008; Vidi, Chen, Irudayaraj, & Watts, 2008a; Vidi, Chemel, Hu, & Watts, 2008b) and CXCR4 multimers (Hamatake, Aoki, Futahashi, Urano, Yamamoto, & Komano, 2009) can be detected.

Different FRET approaches have also indicated similar higher order structures for the M2 muscarinic receptor and the β_2AR (Fung et al., 2009; Pisterzi et al., 2010). The latter study provides additional structural details regarding these complexes that impact on the potential asymmetry of GPCR signaling complex

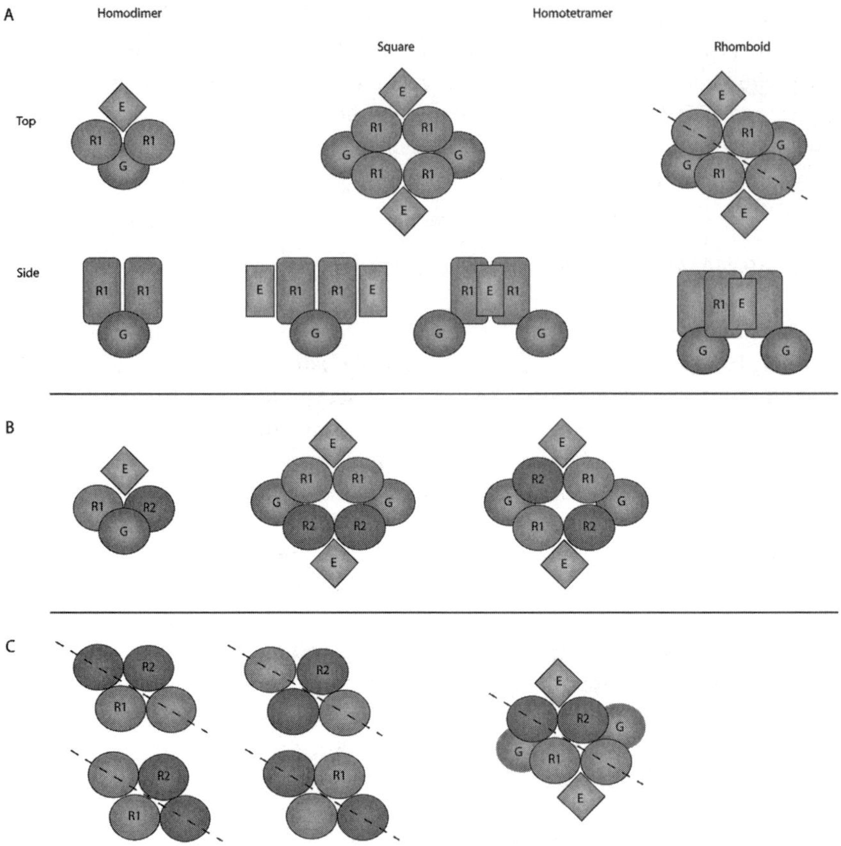

FIGURE 2 Asymmetric organization of receptor homo- and hetero-oligomers. (A) Top and side views of receptor homodimers and homotetramers. Receptor homodimers may be asymmetrically organized with respect to their G protein and effector partners but this is unlikely to have functional consequences *per se* since cooperative effects between the receptor equivalents would be sensed in the same way. However, in the case of receptor homotetramers different shaped complexes are possible. For example, tetramers may have the shape of a square or the shape of a rhomboid, each of which may constrain the organization of interacting proteins such as G proteins and effector molecules. These differential arrangements may be manifested by ligand-binding cooperativity between receptor equivalents and in how this information is transmitted to interacting proteins. (B) The assembly of heterodimers and heterotetramers provides a much larger scope for the assembly of distinctly regulated allosteric signaling machines. At this point we do not know whether two different dimers of homodimers assemble into heterotetramers or whether heterodimers must be formed first. Even in the "square" configuration, a number of asymmetries become possible with respect to how the signaling complex is organized. (C) In the rhomboid configuration, these asymmetries become even more striking. Thus, how receptors are organized and assembled with the interacting proteins might be controlled in the cell to produce distinct signaling architectures.

organization. Using spectral deconvolution and fluorescence lifetime imaging, these authors were also able to show that M2 receptor homotetramers are likely to be in a rhomboid orientation, rather than a simple square array of receptor monomers (Pisterzi et al., 2010, Fig. 2). If receptors form as homodimers or even homotetramers in a square array, the possibilities for asymmetric arrangements are still limited (Fig. 2A). However, in a rhomboid-shaped homotetramer, asymmetries can be introduced with respect to how the entire receptor, G protein, effector complex is arranged (Fig. 2A, right). In a heterodimer, there is already a component of asymmetry added (Fig. 2B, left). Based on the study from the Javitch group (Han et al., 2009), this adds an entirely unappreciated wrinkle to signaling from heterodimers. Using the example of a heterodimer between the β_2AR and the δ-opioid receptor (Jordan et al., 2001; McVey et al., 2001), the asymmetry with respect to how the complex is arranged may mean that in one case, depending on how the complex is formed, we might have a β_2AR modulated by δ-opioid ligands and in another case a δ-opioid receptor modulated by β_2AR ligands. Thus, in one arrangement, protomer A is the signaling receptor and protomer B is the allosteric modulator and the converse is true when the system is organized the other way around. This greatly increases the potential organizational complexity of GPCR signaling and suggests that determinants of signaling complex assembly will be of paramount importance in initially defining signaling specificity in a given tissue, cellular or subcellular compartment (Milligan, 2007, 2009). This has tremendous implications for the formation of receptor heterodimers and hetero-oligomers, in that multiple asymmetrical arrangements are possible depending on the relative orientation of each monomer to the G protein and possibly effector. More diversity is added when we consider heterotetramers which can (1) have variable numbers of each component subunit and (2) different potential arrangements of those subunits (Fig. 2, B, and C). Important questions for us to figure out include how and where heterotetramers can form, in what order subunits are added, in what stoichiometry and how signaling partners are added. As we have seen, receptor complexes can contain multiple receptors, what some authors have termed as receptor mosaics (Agnati et al., 2010). Also, if there are direct interactions between GPCRs and other receptor classes, will these asymmetries be important in their function as well?

IX. BEYOND THE PARADIGM OF A CELL SURFACE RECEPTOR

Receptor internalization, as discussed above, is no longer simply a way of desensitizing receptors. Desensitization, as such, like signaling, must be seen as pathway-specific. Internalization of GPCRs may lead to a switch in signaling pathways by desensitizing the primary, second-messenger-based or cell surface-based pathways while simultaneously activating a second wave of signaling in

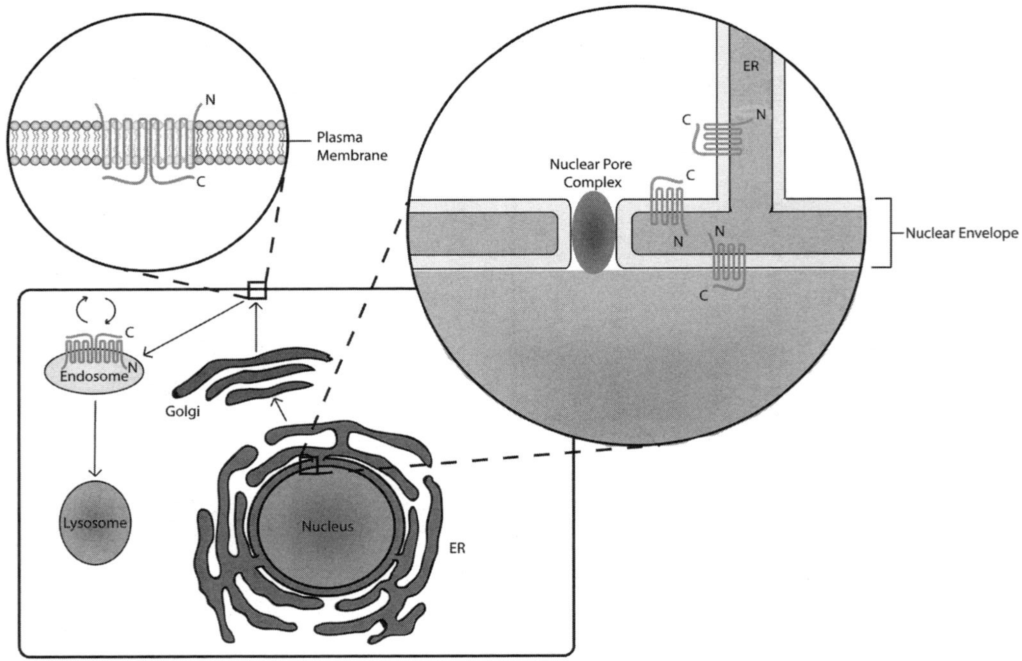

FIGURE 3 Different subcellular sites where βAR signaling complexes have been detected. βAR and other GPCRs signal from the cell surface and also while they are being internalized. However, it has more recently been appreciated that both the β_1AR and the β_3AR, but not the β_2AR are resident on the nuclear membrane, at least in rat and adult mouse ventricular cardiomyocytes (Boivin et al., 2006; Vaniotis et al., 2011). How these receptors are trafficked to distinct endomembrane compartments is not well understood and could either be a result of receptor internalization from the cell surface or via *de novo* delivery from the biosynthetic pathway. The possibility that there may be two distinct orientations for nuclear GPCRs, that is, either capable of delivering signals toward the cytosol or the nucleoplasm, is something that can only be explored in an intact cell context.

endosomes (Fig. 3). The discovery of G protein-independent, or post G protein signaling events still implicates the initial surface targeting of GPCR complexes. In light of our discussion above, it is possible that *de novo* complexes of GPCRs and their signaling partners assembled along the biosynthetic pathway, might be delivered to other endomembrane locations. Here, we focus on data indicating that these signaling systems might be found on the nuclear membrane.

An increasing number of GPCRs have been demonstrated to be targeted at the nuclear membrane, including lysophosphatidic acid receptors (Gobeil et al., 2003), metabotropic glutamate receptors (mGluR5, O'Malley, Jong, Gonchar, Burkhalter, & Romano, 2003; Kumar, Jong, & O'Malley, 2008; Jong, Kumar, & O'Malley, 2009), apelin receptors (Lee et al., 2004), platelet activating factor (PAF) receptors (Marrache et al., 2002), bradykinin B2 receptors (Lee et al., 2004), angiotensin II type I receptors (Lu, Yang, Shaw, & Raizada, 1998; Chen et al., 2000; Zhuo, Imig, Hammond, Orengo, Benes, & Navar, 2002; Lee et al., 2004; Tadevosyan et al., 2010), prostaglandin receptors (Gobeil et al., 2002), endothelin receptors (Boivin, Chevalier, Villeneuve, Rousseau, & Allen, 2003), and α_1-adrenergic receptors (Garcia-Cazarin et al., 2008; Wright et al., 2008, reviewed in Goetzl, 2007; Boivin, Vaniotis, Allen, & Hébert, 2008). Also, mutant V2 vasopressin receptors, which are trapped in intracellular compartments, can signal in response to nonpeptide agonists – indicating that they are in fact functional even when mistrafficked (Robben et al., 2009).

In addition, a large number of signaling proteins, classically associated with receptor-mediated events at the cell surface including heterotrimeric G proteins (Zhang, Barr, Mo, Rojkova, Liu, & Simonds, 2001; Gobeil et al., 2002; Boivin, Villeneuve, Farhat, Chevalier, & Allen, 2005, reviewed in Willard & Crouch, 2000; Dupré & Hébert, 2006; Dupré et al., 2009), AC isoforms (Schulze & Buchwalow, 1998; Yamamoto, Kawamura, & James, 1998), phospholipase A_2 (Schievella, Regier, Smith, & Lin, 1995), phospholipase Cβ (Kim, Park, & Rhee, 1996), and phospholipase D (Freyberg, Sweeney, Siddhanta, Bourgoin, Frohman, & Shields, 2001), RGS proteins (reviewed in Burchett, 2003), β-arrestin1 (Scott et al., 2002; Wang, Wu, Ge, Ma, & Pei, 2003), GRKs (Yi, Gerdes, & Li, 2002; Johnson, Scott, & Pitcher, 2004; Yi, Zhou, Baker, Wang, Gerdes, & Li, 2005), A kinase anchoring proteins (AKAPs), and PKA (Sastri, Barraclough, Carmichael, & Taylor, 2005), among others, have been demonstrated to be trafficked to the nucleus and/or nuclear membrane. Interestingly, enzymes involved in the generation and metabolism of phosphoinositides (Barlow, Laishram, & Anderson, 2010) or processing peptide ligands such as ACE have also been localized to the nuclei of different cell types (Lucero, Kintsurashvili, Marketou, & Gavras, 2010).

Further, these intracrine signaling loops are not restricted to GPCRs and may include a number of other classes of "surface" receptors as well, such as ALK4/ALK5, TGF-β superfamily receptors responsive to activin A (Gressner, Lahme,

Siluschek, Rehbein, Weiskirchen, & Gressner, 2008) and VEGF receptors (Lee et al., 2007, reviewed in Cook & Re, 2007). Although most nuclear GPCRs seem to regulate proximal signaling pathways (i.e., involving generation of second messengers or activation of ERK1/2 and PKB) similar to those seen at the cell surface (reviewed in Boivin et al., 2008), a number of these receptors regulate nuclear events such as DNA synthesis (Watson, Fraher, Natale, Kisiel, Hendy, & Hodsman, 2000), transcription initiation (Boivin et al., 2006; Vaniotis, Del Duca, Trieu, Rohlicek, Hébert, & Allen, 2011), and histone modification (Re et al., 2010).

We have shown that cardiac β_1- and β_3-adrenergic receptors (Boivin et al., 2006; Vaniotis et al., 2011) are targeted at endomembrane locations where they are functional with respect to cellular signaling. Interestingly, subcellular fractionation experiments in adult rat ventricular cardiomyocytes indicated colocalization of βAR with Nup-62, a marker of the nuclear membrane. In order to more carefully characterize the distribution and possible physiological relevance of the three receptor subtypes, we complemented these studies with immunocytochemistry, ligand-binding studies, and functional assays using primary tissue (a key requirement for convincing your peers!). To our surprise, not only were functional β-adrenergic receptors localized to the nuclear membrane but this localization was subtype-specific. Our experiments demonstrated that β_1AR and β_3AR, but not the β_2AR distribute to the nuclear membrane and that the two former βAR isoforms subserve different functions (Boivin et al., 2006). Interestingly, both receptors were differentially coupled to signaling pathways in isolated nuclei. The β_1AR can activate AC, presumably through Gs while the β_3AR activates transcriptional initiation in a PTX-sensitive manner. Further, we showed that both rRNA (18S rRNA) and mRNA (NF-κB and components related to its signaling pathways) levels were modulated by βAR stimulation (Vaniotis et al., 2011). One wonders though if these two receptors can heterodimerize on the nuclear membrane. If so, we will have to re-evaluate the pharmacology of nuclear receptor signaling in that context.

All of the transcriptional events mediated by βAR stimulation in isolated cardiac nuclei were sensitive to inhibitors of ERK1/2, p38, and JNK as well as PKB. Only PKB was activated by nuclear GPCRs showing that other signaling pathways will modulate nuclear βAR signaling via molecular crosstalk (Vaniotis et al., 2011). Perhaps most interesting was the fact that inhibition of PKB switched isoproterenol from an agonist to an inverse agonist with respect to transcriptional initiation. To date, most studies evaluating the signaling downstream of nuclear GPCRs have relied obviously on isolated nuclei. As can be seen in Figure 3, there are two possible orientations for nuclear GPCRs, one with the receptor C-terminus facing the nucleoplasm and the other facing the cytosol. This suggests that accumulation of the ligand into the space between the inner and outer nuclear membrane might result in signals delivered in two directions simultaneously. This could only be studied (and confirmed) in an intact cell context – surely the next challenge facing researchers in this area.

Taken together, these studies suggest that GPCRs do not have to reach the cell surface in order to act as signaling entities as a distinction from receptors that continue to signal (even activating different signaling pathways) after they are internalized.

X. CONCLUSIONS

The holy grail of molecular pharmacologists is to be able to target single pathways associated with a given GPCR. The current focus on pathway-selective, biased ligands is providing optimism that these approaches may actually work. However, until recently, we have focused on the orthosteric ligand-binding site, which cannot provide the necessary level of discrimination possible. Certainly, the focus on allosteric sites for selective pathway modulation provides one way out of this impasse (see Valant, Sexton, & Christopoulos, 2009 for review, and Goupil et al., 2010 for a specific example). We would argue that targeting assembly of signaling complexes might actually provide an even more "selective" set of biased assembly modulators. However, much work remains to identify the molecular determinants of signaling complex assembly before this particular strategy can come to fruition.

Acknowledgments

This work was supported by grants from the Canadian Institutes of Health Research to Terence E. Hébert (MOP-36279) as well as the CIHR Team in GPCR Allosteric Regulation (CTiGAR). Terence E. Hébert is a Chercheur National of the Fonds de la Recherche en Santé du Québec (FRSQ). We thank Vic Rebois (NIH) and the Hébert lab for helpful discussions.

References

Agnati, L. F., Guidolin, D., Albertin, G., Trivello, E., Ciruela, F., & Genedani, S., et al., (2010). An integrated view on the role of receptor mosaics at perisynaptic level: Focus on adenosine A(2A), dopamine D(2), cannabinoid CB(1), and metabotropic glutamate mGlu(5) receptors. *J Recept Signal Transduct Res, 30*, 355–369.

Albert, P. R., & Robillard, L. (2002). G protein specificity: Traffic direction required. *Cell Signal, 14*, 407–418.

Altier, C., Khosravani, H., Evans, R. M., Hameed, S., Peloquin, J. B., & Vartian, B. A., et al., (2006). ORL1 receptor-mediated internalization of N-type calcium channels. *Nat Neurosci, 9*, 31–40.

Angers, S., Salahpour, A., Joly, E., Hilairet, S., Chelsky, D., & Dennis, M. (2000). Detection of β_2-adrenergic receptor dimerization in living cells using bioluminescence resonance energy transfer (BRET). *Proc Natl Acad Sci USA, 97*, 3684–3689.

Audet, M., & Bouvier, M. (2008). Insights into signaling from the β_2-adrenergic receptor structure. *Nat Chem Biol, 4*, 397–403.

Audet, N., Galés, C., Archer-Lahlou, E., Vallieres, M., Schiller, P. W., & Bouvier, M., et al., (2008). Bioluminescence resonance energy transfer assays reveal ligand-specific conformational changes within preformed signaling complexes containing δ-opioid receptors and heterotrimeric G proteins. *J Biol Chem, 283*, 15078–15088.

Azzi, M., Charest, P. G., Angers, S., Rousseau, G., Kohout, T., & Bouvier, M., et al., (2003). β-arrestin-mediated activation of MAPK by inverse agonists reveals distinct active conformations for G protein-coupled receptors. *Proc Natl Acad Sci USA, 100*, 11406–11411.

Balijepalli, R. C., Foell, J. D., Hall, D. D., Hell, J. W., & Kamp, T. J. (2006). Localization of cardiac L-type Ca(2+) channels to a caveolar macromolecular signaling complex is required for β_2-adrenergic regulation. *Proc Natl Acad Sci USA, 103*, 7500–7505.

Baragli, A., Grieco, M. L., Trieu, P., Villeneuve, L. R., & Hébert, T. E. (2008). Heterodimers of adenylyl cyclases 2 and 5 show enhanced functional responses in the presence of Galphas. *Cell Signal, 20*, 480–492.

Barki-Harrington, L., Luttrell, L. M., & Rockman, H. A. (2003). Dual inhibition of β-adrenergic and angiotensin II receptors by a single antagonist: A functional role for receptor–receptor interaction *in vivo*. *Circulation, 108*, 1611–1618.

Barlow, C. A., Laishram, R. S., & Anderson, R. A. (2010). Nuclear phosphoinositides: A signaling enigma wrapped in a compartmental conundrum. *Trends Cell Biol, 20*, 25–35.

Beedle, A. M., McRory, J. E., Poirot, O., Doering, C. J., Altier, C., & Barrere, C., et al., (2004). Agonist-independent modulation of N-type calcium channels by ORL1 receptors. *Nat Neurosci, 7*, 118–125.

Benians, A., Nobles, M., Hosny, S., & Tinker, A. (2005). Regulators of G-protein signaling form a quaternary complex with the agonist, receptor, and G-protein. A novel explanation for the acceleration of signaling activation kinetics. *J Biol Chem, 280*, 13383–13394.

Bernstein, L. S., Ramineni, S., Hague, C., Cladman, W., Chidiac, P., & Levey, A. I. (2004). RGS2 binds directly and selectively to the M1 muscarinic acetylcholine receptor third intracellular loop to modulate Gq/11alpha signaling. *J Biol Chem, 279*, 21248–21256.

Berthouze, M., Ayoub, M., Russo, O., Rivail, L., Sicsic, S., & Fischmeister, R., et al., (2005). Constitutive dimerization of human serotonin 5-HT4 receptors in living cells. *FEBS Lett, 579*, 2973–2980.

Bockaert, J., Roussignol, G., Becamel, C., Gavarini, S., Joubert, L., & Dumuis, A., et al., (2004). GPCR-interacting proteins (GIPs): Nature and functions. *Biochem Soc Trans, 32*, 851–855.

Boivin, B., Chevalier, D., Villeneuve, L. R., Rousseau, E., & Allen, B. G. (2003). Functional endothelin receptors are present on nuclei in cardiac ventricular myocytes. *J Biol Chem, 278*, 29153–29163.

Boivin, B., Lavoie, C., Vaniotis, G., Baragli, A., Villeneuve, L. R., & Ethier, N., et al., (2006). Functional β-adrenergic receptor signaling on nuclear membranes in adult rat and mouse ventricular cardiomyocytes. *Cardiovasc Res, 71*, 69–78.

Boivin, B., Vaniotis, G., Allen, B. G., & Hébert, T. E. (2008). G protein-coupled receptors in and on the cell nucleus: A new signaling paradigm? *J Recept Signal Transduct Res, 28*, 15–28.

Boivin, B., Villeneuve, L. R., Farhat, N., Chevalier, D., & Allen, B. G. (2005). Sub-cellular distribution of endothelin signaling pathway components in ventricular myocytes and heart: Lack of preformed caveolar signalosomes. *J Mol Cell Cardiol, 38*, 665–676.

Breit, A., Lagacé, M., & Bouvier, M. (2004). Hetero-oligomerization between β_2- and β_3-adrenergic receptors generates a β-adrenergic signaling unit with distinct functional properties. *J Biol Chem, 279*, 28756–28765.

Bulenger, S., Marullo, S., & Bouvier, M. (2005). Emerging role of homo- and heterodimerization in G-protein-coupled receptor biosynthesis and maturation. *Trends Pharmacol Sci, 26*, 131–137.

Burchett, S. A. (2003). In through the out door: Nuclear localization of the regulators of G protein signaling. *J Neurochem, 87*, 551–559.

Cao, W., Medvedev, A. V., Daniel, K. W., & Collins, S. (2001). β-adrenergic activation of p38 MAP kinase in adipocytes: cAMP induction of the uncoupling protein 1 (UCP1) gene requires p38 MAP kinase. *J Biol Chem, 276*, 27077–27082.

Carriba, P., Navarro, G., Ciruela, F., Ferre, S., Casado, V., & Agnati, L., et al., (2008). Detection of heteromerization of more than two proteins by sequential BRET-FRET. *Nat Methods, 5*, 727–733.

Chen, R., Mukhin, Y. V., Garnovskaya, M. N., Thielen, T. E., Iijima, Y., & Huang, C., et al., (2000). A functional angiotensin II receptor–GFP fusion protein: Evidence for agonist-dependent nuclear translocation. *Am J Physiol Renal Physiol, 279*, F440–F448.

Cherezov, V., Rosenbaum, D. M., Hanson, M. A., Rasmussen, S. G., Thian, F. S., & Kobilka, T. S., et al., (2007). High-resolution crystal structure of an engineered human β_2-adrenergic G protein-coupled receptor. *Science, 318*, 1258–1265.

Chidiac, P., Hébert, T. E., Valiquette, M., Dennis, M., & Bouvier, M. (1994). Inverse agonist activity of β-adrenergic antagonists. *Mol Pharmacol, 45*, 490–499.

Clancy, S. M., Fowler, C. E., Finley, M., Suen, K. F., Arrabit, C., & Berton, F., et al., (2005). Pertussis-toxin-sensitive Galpha subunits selectively bind to C-terminal domain of neuronal GIRK channels: Evidence for a heterotrimeric G protein–channel complex. *Mol Cell Neurosci, 28*, 375–389.

Cook, J. L., & Re., R. N. (2007). In E. D. Frohlich and R. N. Re (Eds.). *The Local Cardiac Renin–Angiotensin Aldosterone System* (pp. 29–41). Springer.

Dai, S., Hall, D. D., & Hell, J. W. (2009). Supramolecular assemblies and localized regulation of voltage-gated ion channels. *Physiol Rev, 89*, 411–452.

Dale, L. B., Seachrist, J. L., Babwah, A. V., & Ferguson, S. S. (2004). Regulation of angiotensin II type 1A receptor intracellular retention, degradation, and recycling by Rab5, Rab7, and Rab11 GTPases. *J Biol Chem, 279*, 13110–13118.

Daub, H., Wallasch, C., Lankenau, A., Herrlich, A., & Ullrich, A. (1997). Signal characteristics of G protein-transactivated EGF receptor. *EMBO J, 16*, 7032–7044.

Daulat, A. M., Maurice, P., Froment, C., Guillaume, J. L., Broussard, C., & Monsarrat, B., et al., (2007). Purification and identification of G protein-coupled receptor protein complexes under native conditions. *Mol Cell Proteomics, 6*, 835–844.

Daulat, A. M., Maurice, P., & Jockers, R. (2009). Recent methodological advances in the discovery of GPCR-associated protein complexes. *Trends Pharmacol Sci, 30*, 72–78.

Davare, M. A., Avdonin, V., Hall, D. D., Peden, E. M., Burette, A., & Weinberg, R. J., et al., (2001). A β_2-adrenergic receptor signaling complex assembled with the Ca^{2+} channel Cav1.2. *Science, 293*, 98–101.

David, M., Richer, M., Mamarbachi, A. M., Villeneuve, L. R., Dupré, D. J., & Hébert, T. E. (2006). Interactions between GABA-B(1) receptors and Kir 3 inwardly rectifying potassium channels. *Cell Signal, 18*, 2172–2181.

Digby, G. J., Lober, R. M., Sethi, P. R., & Lambert, N. A. (2006). Some G protein heterotrimers physically dissociate in living cells. *Proc Natl Acad Sci USA, 103*, 17789–17794.

Digby, G. J., Sethi, P. R., & Lambert, N. A. (2008). Differential dissociation of G protein heterotrimers. *J Physiol, 586*, 3325–3335.

Dong, C., Filipeanu, C. M., Duvernay, M. T., & Wu, G. (2007). Regulation of G protein-coupled receptor export trafficking. *Biochim Biophys Acta, 1768*, 853–870.

Dupré, D. J., Baragli, A., Rebois, R. V., Ethier, N., & Hébert, T. E. (2007). Signaling complexes associated with adenylyl cyclase II are assembled during their biosynthesis. *Cell Signal, 19*, 481–489.

Dupré, D. J., & Hébert, T. E. (2006). Biosynthesis and trafficking of seven transmembrane receptor signaling complexes. *Cell Signal, 18*, 1549–1559.

Dupré, D. J., Robitaille, M., Ethier, N., Villeneuve, L. R., Mamarbachi, A. M., & Hébert, T. E. (2006). Seven transmembrane receptor core signaling complexes are assembled prior to plasma membrane trafficking. *J Biol Chem, 281*, 34561–34573.

Dupré, D. J., Robitaille, M., Rebois, R. V., & Hébert, T. E. (2009). The role of $G\beta\gamma$ subunits in the organization, assembly, and function of GPCR signaling complexes. *Annu Rev Pharmacol Toxicol, 49*, 31–56.

Duvernay, M. T., Zhou, F., & Wu, G. (2004). A conserved motif for the transport of G protein-coupled receptors from the endoplasmic reticulum to the cell surface. *J Biol Chem, 279*, 30741–30750.

Evans, B. A., Sato, M., Sarwar, M., Hutchinson, D. S., & Summers, R. J. (2010). Ligand-directed signaling at β-adrenoceptors. *Br J Pharmacol, 159*, 1022–1038.

Filipeanu, C. M., Zhou, F., Claycomb, W. C., & Wu, G. (2004). Regulation of the cell-surface expression and function of angiotensin II type 1 receptor by Rab1-mediated ER-to-Golgi transport in cardiac myocytes. *J Biol Chem, 279*, 41077–41084.

Fredriksson, R., Lagerstrom, M. C., Lundin, L. G., & Schioth, H. B. (2003). The G protein-coupled receptors in the human genome form five main families. Phylogenetic analysis, paralogon groups, and fingerprints. *Mol Pharmacol, 63*, 1256–1272.

Freyberg, Z., Sweeney, D., Siddhanta, A., Bourgoin, S., Frohman, M., & Shields, D. (2001). Intracellular localization of phospholipase D1 in mammalian cells. *Mol Biol Cell, 12*, 943–955.

Fung, J. J., Deupi, X., Pardo, L., Yao, X. J., Velez-Ruiz, G. A., & Devree, B. T., et al., (2009). Ligand-regulated oligomerization of β_2-adrenoceptors in a model lipid bilayer. *EMBO J, 28*, 3315–3328.

Galandrin, S., & Bouvier, M. (2006). Distinct signaling profiles of β_1 and β_2 adrenergic receptor ligands toward adenylyl cyclase and mitogen-activated protein kinase reveals the pluridimensionality of efficacy. *Mol Pharmacol, 70*, 1575–1584.

Galandrin, S., Oligny-Longpre, G., & Bouvier, M. (2007). The evasive nature of drug efficacy: Implications for drug discovery. *Trends Pharmacol Sci, 28*, 423–430.

Galés, C., Rebois, R. V., Hogue, M., Trieu, P., Breit, A., & Hébert, T. E., et al., (2005). Real-time monitoring of receptor and G-protein interactions in living cells. *Nat Methods, 2*, 177–184.

Galés, C., Van Durm, J. J., Schaak, S., Pontier, S., Percherancier, Y., & Audet, M., et al., (2006). Probing the activation-promoted structural rearrangements in preassembled receptor–G protein complexes. *Nat Struct Mol Biol, 13*, 778–786.

Gandia, J., Galino, J., Amaral, O. B., Soriano, A., Lluis, C., & Franco, R., et al., (2008). Detection of higher-order G protein-coupled receptor oligomers by a combined BRET–BiFC technique. *FEBS Lett, 582*, 2979–2984.

Garcia-Cazarin, M. L., Smith, J. L., Olszewski, K. A., McCune, D. F., Simmerman, L. A., & Hadley, R. W., et al., (2008). The α_{1D}-adrenergic receptor is expressed intracellularly and coupled to increases in intracellular calcium and reactive oxygen species in human aortic smooth muscle cells. *J Mol Signal, 3*, 6.

Gobeil Jr., F., Bernier, S. G., Vazquez-Tello, A., Brault, S., Beauchamp, M. H., & Quiniou, C., et al., (2003). Modulation of pro-inflammatory gene expression by nuclear lysophosphatidic acid receptor type-1. *J Biol Chem, 278*, 38875–38883.

Gobeil Jr., F., Dumont, I., Marrache, A. M., Vazquez-Tello, A., Bernier, S. G., & Abran, D., et al., (2002). Regulation of eNOS expression in brain endothelial cells by perinuclear EP(3) receptors. *Circ Res, 90*, 682–689.

Goetzl, E. J. (2007). Diverse pathways for nuclear signaling by G protein-coupled receptors and their ligands. *FASEB J, 21*, 638–642.

Goupil, E., Tassy, D., Bourguet, C., Quiniou, C., Wisehart, V., & Pétrin, D., et al., (2010). A novel biased allosteric compound inhibitor of parturition selectively impedes the prostaglandin F2α-mediated Rho/ROCK signaling pathway. *J Biol Chem, 285*, 25624–25636.

Gressner, O. A., Lahme, B., Siluschek, M., Rehbein, K., Weiskirchen, R., & Gressner, A. M. (2008). Intracrine signaling of activin A in hepatocytes upregulates connective tissue growth factor (CTGF/CCN2) expression. *Liver Int, 28*, 1207–1216.

Gudermann, T., Kalkbrenner, F., & Schultz, G. (1996). Diversity and selectivity of receptor–G protein interaction. *Annu Rev Pharmacol Toxicol, 36*, 429–459.

Gudermann, T., Schoneberg, T., & Schultz, G. (1997). Functional and structural complexity of signal transduction via G-protein-coupled receptors. *Annu Rev Neurosci, 20*, 399–427.

Guo, W., Urizar, E., Kralikova, M., Mobarec, J. C., Shi, L., & Filizola, M., et al., (2008). Dopamine D2 receptors form higher order oligomers at physiological expression levels. *EMBO J, 27*, 2293–2304.

Haack, K. K., Tougas, M. R., Jones, K. T., El-Dahr, S. S., Radhakrishna, H., & McCarty, N. A. (2010). A novel bioassay for detecting GPCR heterodimerization: Transactivation of β_2-adrenergic receptor by bradykinin receptor. *J Biomol Screen, 15*, 251–260.

Hague, C., Uberti, M. A., Chen, Z., Bush, C. F., Jones, S. V., & Ressler, K. J., et al., (2004). Olfactory receptor surface expression is driven by association with the β_2-adrenergic receptor. *Proc Natl Acad Sci USA, 101*, 13672–13676.

Hall, R. A., & Lefkowitz, R. J. (2002). Regulation of G protein-coupled receptor signaling by scaffold proteins. *Circ Res, 91*, 672–680.

Hamatake, M., Aoki, T., Futahashi, Y., Urano, E., Yamamoto, N., & Komano, J. (2009). Ligand-independent higher-order multimerization of CXCR4, a G-protein-coupled chemokine receptor involved in targeted metastasis. *Cancer Sci, 100*, 95–102.

Han, Y., Moreira, I. S., Urizar, E., Weinstein, H., & Javitch, J. A. (2009). Allosteric communication between protomers of dopamine class A GPCR dimers modulates activation. *Nat Chem Biol, 5*, 688–695.

Hanson, M. A., & Stevens, R. C. (2009). Discovery of new GPCR biology: One receptor structure at a time. *Structure, 17*, 8–14.

Hébert, T. E., & Bouvier, M. (1998). Structural and functional aspects of G protein-coupled receptor oligomerization. *Biochem Cell Biol, 76*, 1–11.

Hébert, T. E., Moffett, S., Morello, J. P., Loisel, T. P., Bichet, D. G., & Barret, C., et al., (1996). A peptide derived from a β_2-adrenergic receptor transmembrane domain inhibits both receptor dimerization and activation. *J Biol Chem, 271*, 16384–16392.

Hofmann, K. P., Scheerer, P., Hildebrand, P. W., Choe, H. W., Park, J. H., & Heck, M., et al., (2009). A G protein-coupled receptor at work: The rhodopsin model. *Trends Biochem Sci, 34*, 540–552.

Hudson, B. D., Hébert, T. E., & Kelly, M. E. (2010). Physical and functional interaction between CB1 cannabinoid receptors and β_2-adrenoceptors. *Br J Pharmacol, 160*, 627–642.

Jaakola, V. P., Griffith, M. T., Hanson, M. A., Cherezov, V., Chien, E. Y., & Lane, J. R., et al., (2008). The 2.6 angstrom crystal structure of a human A_{2A} adenosine receptor bound to an antagonist. *Science, 322*, 1211–1217.

Johnson, L. R., Scott, M. G., & Pitcher, J. A. (2004). G protein-coupled receptor kinase 5 contains a DNA-binding nuclear localization sequence. *Mol Cell Biol, 24*, 10169–10179.

Joiner, M. L., Lise, M. F., Yuen, E. Y., Kam, A. Y., Zhang, M., & Hall, D. D., et al., (2010). Assembly of a β_2-adrenergic receptor—GluR1 signaling complex for localized cAMP signaling. *EMBO J, 29*, 482–495.

Jong, Y. J., Kumar, V., & O'Malley, K. L. (2009). Intracellular metabotropic glutamate receptor 5 (mGluR5) activates signaling cascades distinct from cell surface counterparts. *J Biol Chem, 284*, 35827–35838.

Jordan, B. A., Trapaidze, N., Gomes, I., Nivarthi, R., & Devi, L. A. (2001). Oligomerization of opioid receptors with β_2-adrenergic receptors: A role in trafficking and mitogen-activated protein kinase activation. *Proc Natl Acad Sci USA, 98*, 343–348.

Kalkbrenner, F., Degtiar, V. E., Schenker, M., Brendel, S., Zobel, A., & Heschler, J., et al., (1995). Subunit composition of G(o) proteins functionally coupling galanin receptors to voltage-gated calcium channels. *EMBO J, 14*, 4728–4737.

Kenakin, T. (2007a). Collateral efficacy in drug discovery: Taking advantage of the good (allosteric) nature of 7TM receptors. *Trends Pharmacol Sci, 28*, 407–415.

Kenakin, T. (2007b). Functional selectivity through protean and biased agonism: Who steers the ship? *Mol Pharmacol, 72*, 1393–1401.

Kenakin, T. (2010). Functional selectivity and biased receptor signaling. *J Pharmacol Exp Ther, 336*, 296–302.

Kim, C. G., Park, D., & Rhee, S. G. (1996). The role of carboxyl-terminal basic amino acids in Gqα-dependent activation, particulate association, and nuclear localization of phospholipase C-β1. *J Biol Chem, 271*, 21187–21192.

Kim, I. M., Tilley, D. G., Chen, J., Salazar, N. C., Whalen, E. J., & Violin, J. D., et al., (2008). β-blockers alprenolol and carvedilol stimulate beta-arrestin-mediated EGFR transactivation. *Proc Natl Acad Sci USA, 105*, 14555–14560.

Kitano, J., Nishida, M., Itsukaichi, Y., Minami, I., Ogawa, M., & Hirano, T., et al., (2003). Direct interaction and functional coupling between metabotropic glutamate receptor subtype 1 and voltage-sensitive Cav2.1 Ca^{2+} channel. *J Biol Chem, 278*, 25101–25108.

Kleuss, C., Hescheler, J., Ewel, C., Rosenthal, W., Schultz, G., & Wittig, B. (1991). Assignment of G protein subtypes to specific receptors inducing inhibition of calcium currents. *Nature, 353*, 43–48.

Kleuss, C., Scherubl, H., Hescheler, J., Schultz, G., & Wittig, B. (1992). Different β-subunits determine G protein interaction with transmembrane receptors. *Nature, 358*, 424–426.

Kleuss, C., Scherubl, H., Hescheler, J., Schultz, G., & Wittig, B. (1993). Selectivity in signal transduction determined by γ subunits of heterotrimeric G proteins. *Science, 259*, 832–834.

Kobilka, B., & Schertler, G. F. (2008). New G-protein-coupled receptor crystal structures: Insights and limitations. *Trends Pharmacol Sci, 29*, 79–83.

Kumar, V., Jong, Y. J., & O'Malley, K. L. (2008). Activated nuclear metabotropic glutamate receptor mGlu5 couples to nuclear Gq/11 proteins to generate inositol 1,4,5-trisphosphate-mediated nuclear Ca^{2+} release. *J Biol Chem, 283*, 14072–14083.

Kuravi, S., Lan, T. H., Barik, A., & Lambert, N. A. (2010). Third-party bioluminescence resonance energy transfer indicates constitutive association of membrane proteins: Application to class A G protein-coupled receptors and G proteins. *Biophys J, 98*, 2391–2399.

Lambert, N. A. (2008). Dissociation of heterotrimeric G proteins in cells. *Sci Signal, 1*, re5.

Larocca, T. J., Schwarzkopf, M., Altman, P., Zhang, S., Gupta, A., & Gomes, I., et al., (2010). β_2-adrenergic receptor signaling in the cardiac myocyte is modulated by interactions with CXCR4. *J Cardiovasc Pharmacol, 56*, 548–559.

Lavine, N., Ethier, N., Oak, J. N., Pei, L., Liu, F., & Trieu, P., et al., (2002). G protein-coupled receptors form stable complexes with inwardly rectifying potassium channels and adenylyl cyclase. *J Biol Chem, 277*, 46010–46019.

Lavoie, C., & Hébert, T. E. (2003). Pharmacological characterization of putative β_1-β_2-adrenergic receptor heterodimers. *Can J Physiol Pharmacol, 81*, 186–195.

Lavoie, C., Mercier, J. F., Salahpour, A., Umapathy, D., Breit, A., & Villeneuve, L. R., et al., (2002). β_1/β_2-adrenergic receptor heterodimerization regulates β_2-adrenergic receptor internalization and ERK signaling efficacy. *J Biol Chem, 277*, 35402–35410.

Lee, A., Rana, B. K., Schiffer, H. H., Schork, N. J., Brann, M. R., & Insel, P. A., et al., (2003). Distribution analysis of nonsynonymous polymorphisms within the G-protein-coupled receptor gene family. *Genomics, 81*, 245–248.

Lee, D. K., Lanca, A. J., Cheng, R., Nguyen, T., Ji, X. D., & Gobeil Jr., F., et al., (2004). Agonist-independent nuclear localization of the Apelin, angiotensin AT1, and bradykinin B2 receptors. *J Biol Chem, 279*, 7901–7908.

Lee, T. H., Seng, S., Sekine, M., Hinton, C., Fu, Y., & Avraham, H. K., et al., (2007). Vascular endothelial growth factor mediates intracrine survival in human breast carcinoma cells through internally expressed VEGFR1/FLT1. *PLoS Med, 4*, e186.

Liu, G., Shi, J., Yang, L., Cao, L., Park, S. M., & Cui, J., et al., (2004). Assembly of a Ca(^{2} $^+$)-dependent BK channel signaling complex by binding to β_2 adrenergic receptor. *EMBO J, 23*, 2196–2205.

Lober, R. M., Pereira, M. A., & Lambert, N. A. (2006). Rapid activation of inwardly rectifying potassium channels by immobile G protein-coupled receptors. *J Neurosci, 26*, 12602–12608.

Lodowski, D. T., Angel, T. E., & Palczewski, K. (2009). Comparative analysis of GPCR crystal structures. *Photochem Photobiol, 85*, 425–430.

Lopez-Gimenez, J. F., Canals, M., Pediani, J. D., & Milligan, G. (2007). The α_{1b}-adrenoceptor exists as a higher-order oligomer: Effective oligomerization is required for receptor maturation, surface delivery, and function. *Mol Pharmacol, 71*, 1015–1029.

Lu, D., Yang, H., Shaw, G., & Raizada, M. K. (1998). Angiotensin II-induced nuclear targeting of the angiotensin type 1 (AT1) receptor in brain neurons. *Endocrinology, 139*, 365–375.

Lucero, H. A., Kintsurashvili, E., Marketou, M. E., & Gavras, H. (2010). Cell signaling, internalization, and nuclear localization of the angiotensin converting enzyme in smooth muscle and endothelial cells. *J Biol Chem, 285*, 5555–5568.

Luttrell, D. K., & Luttrell, L. M. (2003). Signaling in time and space: G protein-coupled receptors and mitogen-activated protein kinases. *Assay Drug Dev Technol, 1*, 327–338.

Luttrell, L. M., & Gesty-Palmer, D. (2010). Beyond desensitization: Physiological relevance of arrestin-dependent signaling. *Pharmacol Rev, 62*, 305–330.

Luttrell, L. M., van Biesen, T., Hawes, B. E., Koch, W. J., Krueger, K. M., & Touhara, K., et al., (1997). G protein-coupled receptors and their regulation: Activation of the MAP kinase signaling pathway by G protein-coupled receptors. *Adv Second Messenger Phosphoprotein Res, 31*, 263–277.

Ma, A. W., Pawagi, A. B., & Wells, J. W. (2008). Heterooligomers of the muscarinic receptor and G proteins purified from porcine atria. *Biochem Biophys Res Commun, 374*, 128–133.

Ma, A. W., Redka, D. S., Pisterzi, L. F., Angers, S., & Wells, J. W. (2007). Recovery of oligomers and cooperativity when monomers of the M2 muscarinic cholinergic receptor are reconstituted into phospholipid vesicles. *Biochemistry, 46*, 7907–7927.

Marrache, A. M., Gobeil Jr., F., Bernier, S. G., Stankova, J., Rola-Pleszczynski, M., & Choufani, S., et al., (2002). Proinflammatory gene induction by platelet-activating factor mediated via its cognate nuclear receptor. *J Immunol, 169*, 6474–6481.

Maudsley, S., Pierce, K. L., Zamah, A. M., Miller, W. E., Ahn, S., & Daaka, Y., et al., (2000). The β_2-adrenergic receptor mediates extracellular signal-regulated kinase activation via assembly of a multi-receptor complex with the epidermal growth factor receptor. *J Biol Chem, 275*, 9572–9580.

McGraw, D. W., Mihlbachler, K. A., Schwarb, M. R., Rahman, F. F., Small, K. M., & Almoosa, K. F., et al., (2006). Airway smooth muscle prostaglandin-EP1 receptors directly modulate β_2-adrenergic receptors within a unique heterodimeric complex. *J Clin Invest, 116*, 1400–1409.

McVey, M., Ramsay, D., Kellett, E., Rees, S., Wilson, S., & Pope, A. J., et al., (2001). Monitoring receptor oligomerization using time-resolved fluorescence resonance energy transfer and bioluminescence resonance energy transfer. The human δ-opioid receptor displays constitutive oligomerization at the cell surface, which is not regulated by receptor occupancy. *J Biol Chem, 276*, 14092–14099.

Mercier, J. -F., Salahpour, A., Angers, S., Breit, A., & Bouvier, M. (2002). Quantitative assessment of β_1- and β_2-adrenergic receptor homo- and heterodimerization by bioluminescence resonance energy transfer. *J Biol Chem, 277*, 44925–44931.

Milligan, G. (2007). G protein-coupled receptor dimerisation: Molecular basis and relevance to function. *Biochim Biophys Acta, 1768*, 825–835.

Milligan, G. (2009). G protein-coupled receptor hetero-dimerization: Contribution to pharmacology and function. *Br J Pharmacol, 158*, 5–14.

Milligan, G. (2010). The role of dimerisation in the cellular trafficking of G-protein-coupled receptors. *Curr Opin Pharmacol, 10*, 23–29.

Nikolov, E. N., & Ivanova-Nikolova, T. T. (2004). Coordination of membrane excitability through a GIRK1 signaling complex in the atria. *J Biol Chem, 279*, 23630–23636.

Noma, T., Lemaire, A., Naga Prasad, S. V., Barki-Harrington, L., Tilley, D. G., & Chen, J., et al., (2007). β-arrestin-mediated β_1-adrenergic receptor transactivation of the EGFR confers cardioprotection. *J Clin Invest, 117*, 2445–2458.

O'Malley, K. L., Jong, Y. J., Gonchar, Y., Burkhalter, A., & Romano, C. (2003). Activation of metabotropic glutamate receptor mGlu5 on nuclear membranes mediates intranuclear Ca^{2+} changes in heterologous cell types and neurons. *J Biol Chem, 278*, 28210–28219.

Palczewski, K. (2006). G protein-coupled receptor rhodopsin. *Annu Rev Biochem, 75*, 743–767.

Palczewski, K., Kumasaka, T., Hori, T., Behnke, C. A., Motoshima, H., & Fox, B. A., et al., (2000). Crystal structure of rhodopsin: A G protein-coupled receptor. *Science, 289*, 739–745.

Patel, C. B., Noor, N., & Rockman, H. A. (2010). Functional selectivity in adrenergic and angiotensin signaling systems. *Mol Pharmacol, 78*, 983–992.

Pétrin, D., & Hébert, T. E. (2010). Imaging-based approaches to understanding G protein-coupled receptor signaling complexes. *Methods Molec Biol,* (in press).

Pisterzi, L. F., Jansma, D. B., Georgiou, J., Woodside, M. J., Chou, J. T., & Angers, S., et al., (2010). Oligomeric size of the m2 muscarinic receptor in live cells as determined by quantitative fluorescence resonance energy transfer. *J Biol Chem, 285*, 16723–16738.

Prinster, S. C., Hague, C., & Hall, R. A. (2005). Heterodimerization of G protein-coupled receptors: Specificity and functional significance. *Pharmacol Rev, 57*, 289–298.

Pugh Jr., E. N., & Lamb, T. D. (1993). Amplification and kinetics of the activation steps in phototransduction. *Biochim Biophys Acta, 1141*, 111–149.

Rajagopal, K., Lefkowitz, R. J., & Rockman, H. A. (2005). When 7 transmembrane receptors are not G protein-coupled receptors. *J Clin Invest, 115*, 2971–2974.

Rasmussen, S. G., Choi, H. J., Rosenbaum, D. M., Kobilka, T. S., & Thian, F. S., et al., (2007). Crystal structure of the human β_2-adrenergic G-protein-coupled receptor. *Nature, 450*, 383–387.

Re, M., Pampillo, M., Savard, M., Dubuc, C., & McArdle, C. A., et al., (2010). The human gonadotropin releasing hormone type I receptor is a functional intracellular GPCR expressed on the nuclear membrane. *PLoS One, 5*, e11489.

Rebois, R. V., & Hébert, T. E. (2003). Protein complexes involved in heptahelical receptor-mediated signal transduction. *Receptors Channels, 9*, 169–194.

Rebois, R. V., Robitaille, M., Galés, C., Dupré, D. J., & Baragli, A., et al., (2006). Heterotrimeric G proteins form stable complexes with adenylyl cyclase and Kir3.1 channels in living cells. *J Cell Sci, 119*, 2807–2818.

Rebois, R. V., Robitaille, M., Petrin, D., Zylbergold, P., Trieu, P., & Hébert, T. E. (2008). Combining protein complementation assays with resonance energy transfer to detect multipartner protein complexes in living cells. *Methods, 45*, 214–218.

Riven, I., Iwanir, S., & Reuveny, E. (2006). GIRK channel activation involves a local rearrangement of a preformed G protein channel complex. *Neuron, 51*, 561–573.

Robben, J. H., Kortenoeven, M. L., Sze, M., Yae, C., Milligan, G., & Oorschot, V. M., et al., (2009). Intracellular activation of vasopressin V2 receptor mutants in nephrogenic diabetes insipidus by nonpeptide agonists. *Proc Natl Acad Sci USA, 106*, 12195–12200.

Robillard, L., Ethier, N., Lachance, M., & Hébert, T. E. (2000). Gβγ subunit combinations differentially modulate receptor and effector coupling *in vivo*. *Cell Signal, 12*, 673–682.

Robishaw, J. D., Guo, Z. P., & Wang, Q. (2003). Ribozymes as tools for suppression of G protein γ subunits. *Methods Mol Biol, 237*, 169–180.

Robitaille, M., Ramakrishnan, N., Baragli, A., & Hébert, T. E. (2009). Intracellular trafficking and assembly of specific Kir3 channel/G protein complexes. *Cell Signal, 21*, 488–501.

Rosenbaum, D. M., Cherezov, V., Hanson, M. A., Rasmussen, S. G., Thian, F. S., & Kobilka, T. S., et al., (2007). GPCR engineering yields high-resolution structural insights into β_2-adrenergic receptor function. *Science, 318*, 1266–1273.

Rosenbaum, D. M., Rasmussen, S. G., & Kobilka, B. K. (2009). The structure and function of G-protein-coupled receptors. *Nature, 459*, 356–363.

Rovati, G. E., Capra, V., & Neubig, R. R. (2007). The highly conserved DRY motif of class A G protein-coupled receptors: Beyond the ground state. *Mol Pharmacol, 71*, 959–964.

Roy, A. A., Baragli, A., Bernstein, L. S., Hepler, J. R., Hébert, T. E., & Chidiac, P. (2006). RGS2 interacts with Gs and adenylyl cyclase in living cells. *Cell Signal, 18*, 336–348.

Salahpour, A., Angers, S., Mercier, J. F., Lagacé, M., Marullo, S., & Bouvier, M. (2004). Homodimerization of the β_2-adrenergic receptor as a prerequisite for cell surface targeting. *J Biol Chem, 279*, 33390–33397.

Samama, P., Pei, G., Costa, T., Cotecchia, S., & Lefkowitz, R. J. (1994). Negative antagonists promote an inactive conformation of the β_2-adrenergic receptor. *Mol Pharmacol, 45*, 390–394.

Sastri, M., Barraclough, D. M., Carmichael, P. T., & Taylor, S. S. (2005). A-kinase-interacting protein localizes protein kinase A in the nucleus. *Proc Natl Acad Sci USA, 102*, 349–354.

Sato, M., Horinouchi, T., Hutchinson, D. S., Evans, B. A., & Summers, R. J. (2007). Ligand-directed signaling at the β_3-adrenoceptor produced by 3-(2-Ethylphenoxy)-1-[(1,S)-1,2,3,4-tetrahydronapth-1-ylamino]-2*S*-2-propanol oxalate (SR59230A) relative to receptor agonists. *Mol Pharmacol, 72*, 1359–1368.

Sato, M., Hutchinson, D. S., Evans, B. A., & Summers, R. J. (2008). The beta3-adrenoceptor agonist 4-[[(Hexylamino)carbonyl]amino]-*N*-[4-[2-[[(2*S*)-2-hydroxy-3-(4-hydroxyphenoxy) propyl]amino]ethyl]-phenyl]-benzenesulfonamide (L755507) and antagonist (*S*)-*N*-[4-[2-[[3-[3-(acetamidomethyl)phenoxy]-2-hydroxypropyl]amino]-ethyl] phenyl]benzenesulfonamide (L748337) activate different signaling pathways in Chinese hamster ovary-K1 cells stably expressing the human β_3-adrenoceptor. *Mol Pharmacol, 74*, 1417–1428.

Scheerer, P., Park, J. H., Hildebrand, P. W., Kim, Y. J., Krauss, N., & Choe, H. W., et al., (2008). Crystal structure of opsin in its G-protein-interacting conformation. *Nature, 455*, 497–502.

Schievella, A. R., Regier, M. K., Smith, W. L., & Lin, L. L. (1995). Calcium-mediated translocation of cytosolic phospholipase A2 to the nuclear envelope and endoplasmic reticulum. *J Biol Chem, 270*, 30749–30754.

Schulze, W., & Buchwalow, I. B. (1998). Adenylyl cyclase in the heart: an enzymocytochemical and immunocytochemical approach. *Microsc Res Tech, 40*, 473–478.

Schwindinger, W. F., Betz, K. S., Giger, K. E., Sabol, A., Bronson, S. K., & Robishaw, J. D. (2003). Loss of G protein γ7 alters behavior and reduces striatal alpha(olf) level and cAMP production. *J Biol Chem, 278*, 6575–6579.

Scott, M. G., Le Rouzic, E., Perianin, A., Pierotti, V., Enslen, H., & Benichou, S., et al., (2002). Differential nucleocytoplasmic shuttling of β-arrestins. Characterization of a leucine-rich nuclear export signal in β-arrestin2. *J Biol Chem, 277*, 37693–37701.

Seachrist, J. L., Anborgh, P. H., & Ferguson, S. S. (2000). β_2-adrenergic receptor internalization, endosomal sorting, and plasma membrane recycling are regulated by rab GTPases. *J Biol Chem, 275*, 27221–27228.

Seachrist, J. L., Laporte, S. A., Dale, L. B., Babwah, A. V., Caron, M. G., & Anborgh, P. H., et al., (2002). Rab5 association with the angiotensin II type 1A receptor promotes Rab5 GTP binding and vesicular fusion. *J Biol Chem, 277*, 679–685.

Shiina, T., Kawasaki, A., Nagao, T., & Kurose, H. (2000). Interaction with beta-arrestin determines the difference in internalization behaviour between β_1- and β_2-adrenergic receptors. *J Biol Chem, 275*, 29082–29090.

Shiina, T., Nagao, T., & Kurose, H. (2001). Low affinity of β_1-adrenergic receptor for β-arrestins explains the resistance to agonist-induced internalization. *Life Sci, 68*, 2251–2257.

Suzuki, T., Nguyen, C. T., Nantel, F., Bonin, H., Valiquette, M., & Frielle, T., et al., (1992). Distinct regulation of β_1- and β_2-adrenergic receptors in Chinese hamster fibroblasts. *Mol Pharmacol, 41*, 542–548.

Tadevosyan, A., Maguy, A., Villeneuve, L. R., Babin, J., Bonnefoy, A., & Allen, B. G., et al., (2010). Nuclear-delimited angiotensin receptor-mediated signaling regulates cardiomyocyte gene expression. *J Biol Chem, 285*, 22338–22349.

Terrillon, S., & Bouvier, M. (2004). Roles of G protein-coupled receptor dimerization. *EMBO Rep, 5*, 30–34.

Tilley, D. G., Kim, I. M., Patel, P. A., Violin, J. D., & Rockman, H. A. (2009). β-arrestin mediates β_1-adrenergic receptor-epidermal growth factor receptor interaction and downstream signaling. *J Biol Chem, 284*, 20375–20386.

Uberti, M. A., Hague, C., Oller, H., Minneman, K. P., & Hall, R. A. (2005). Heterodimerization with β_2-adrenergic receptors promotes surface expression and functional activity of α_{1D}-adrenergic receptors. *J Pharmacol Exp Ther, 313*, 16–23.

Unal, H., Jagannathan, R., Bhat, M. B., & Karnik, S. S. (2010). Ligand-specific conformation of extracellular loop-2 in the angiotensin II type 1 receptor. *J Biol Chem, 285*, 16341–16350.

Vaidehi, N., & Kenakin, T. (2010). The role of conformational ensembles of seven transmembrane receptors in functional selectivity. *Curr Opin Pharmacol, 10*, 775–781.

Valant, C., Sexton, P. M., & Christopoulos, A. (2009). Orthosteric/allosteric bitopic ligands: Going hybrid at GPCRs. *Mol Interv, 9*, 125–135.

Vaniotis, G., Del Duca, D., Trieu, P., Rohlicek, C. V., Hébert, T. E., & Allen, B. G. (2011). Nuclear β-adrenergic receptors modulate gene expression in adult rat heart. *Cell Signal, 23*, 89–98.

Vidi, P. A., Chen, J., Irudayaraj, J. M., & Watts, V. J. (2008a). Adenosine A(2A) receptors assemble into higher-order oligomers at the plasma membrane. *FEBS Lett, 582*, 3985–3990.

Vidi, P. A., Chemel, B. R., Hu, C. D., & Watts, V. J. (2008b). Ligand-dependent oligomerization of dopamine D(2) and adenosine A(2A) receptors in living neuronal cells. *Mol Pharmacol, 74*, 544–551.

Wang, P., Wu, Y., Ge, X., Ma, L., & Pei, G. (2003). Subcellular localization of β-arrestins is determined by their intact N domain and the nuclear export signal at the C terminus. *J Biol Chem, 278*, 11648–11653.

Wang, Q., Jolly, J. P., Surmeier, J. D., Mullah, B. M., Lidow, M. S., & Bergson, C. M., et al., (2001). Differential dependence of the D1 and D5 dopamine receptors on the G protein γ7 subunit for activation of adenylylcyclase. *J Biol Chem, 276*, 39386–39393.

Wang, Q., Mullah, B., Hansen, C., Asundi, J., & Robishaw, J. D. (1997). Ribozyme-mediated suppression of the G protein γ7 subunit suggests a role in hormone regulation of adenylylcyclase activity. *J Biol Chem, 272*, 26040–26048.

Wang, Q., Mullah, B. K., & Robishaw, J. D. (1999). Ribozyme approach identifies a functional association between the G protein β1γ7 subunits in the β-adrenergic receptor signaling pathway. *J Biol Chem, 274*, 17365–17371.

Warne, T., Serrano-Vega, M. J., Baker, J. G., Moukhametzianov, R., Edwards, P. C., & Henderson, R., et al., (2008). Structure of a β_1-adrenergic G-protein-coupled receptor. *Nature, 454*, 486–491.

Watson, P. H., Fraher, L. J., Natale, B. V., Kisiel, M., Hendy, G. N., & Hodsman, A. B. (2000). Nuclear localization of the type 1 parathyroid hormone/parathyroid hormone-related peptide receptor in MC3T3-E1 cells: Association with serum-induced cell proliferation. *Bone, 26*, 221–225.

Weiss, J. M., Morgan, P. H., Lutz, M. W., & Kenakin, T. P. (1996). The cubic ternary complex receptor-occupancy model. I. Model description. *J Theor Biol, 178*, 151–167.

Whorton, M. R., Bokoch, M. P., Rasmussen, S. G., Huang, B., Zare, R. N., & Kobilka, B., et al., (2007). A monomeric G protein-coupled receptor isolated in a high-density lipoprotein particle efficiently activates its G protein. *Proc Natl Acad Sci USA, 104*, 7682–7687.

Whorton, M. R., Jastrzebska, B., Park, P. S., Fotiadis, D., Engel, A., & Palczewski, K., et al., (2008). Efficient coupling of transducin to monomeric rhodopsin in a phospholipid bilayer. *J Biol Chem, 283*, 4387–4394.

Willard, F. S., & Crouch, M. F. (2000). Nuclear and cytoskeletal translocation and localization of heterotrimeric G-proteins. *Immunol Cell Biol, 78*, 387–394.

Wright, C. D., Chen, Q., Baye, N. L., Huang, Y., Healy, C. L., & Kasinathan, S., et al., (2008). Nuclear α_1-adrenergic receptors signal activated ERK localization to caveolae in adult cardiac myocytes. *Circ Res, 103*, 992–1000.

Wu, B., Chien, E. Y., Mol, C. D., Fenalti, G., Liu, W., & Katritch, V., et al., (2010). Structures of the CXCR4 chemokine GPCR with small-molecule and cyclic peptide antagonists. *Science, 330*, 1066–1071.

Yamamoto, S., Kawamura, K., & James, T. N. (1998). Intracellular distribution of adenylate cyclase in human cardiocytes determined by electron microscopic cytochemistry. *Microsc Res Tech, 40*, 479–487.

Yi, X. P., Gerdes, A. M., & Li, F. (2002). Myocyte redistribution of GRK2 and GRK5 in hypertensive, heart-failure-prone rats. *Hypertension, 39*, 1058–1063.

Yi, X. P., Zhou, J., Baker, J., Wang, X., Gerdes, A. M., & Li, F. (2005). Myocardial expression and redistribution of GRKs in hypertensive hypertrophy and failure. *Anat Rec A Discov Mol Cell Evol Biol, 282*, 13–23.

Zerial, M., & McBride, H. (2001). Rab proteins as membrane organizers. *Nat Rev Mol Cell Biol, 2*, 107–117.

Zhang, J. H., Barr, V. A., Mo, Y., Rojkova, A. M., Liu, S., & Simonds, W. F. (2001). Nuclear localization of G protein β5 and regulator of G protein signaling 7 in neurons and brain. *J Biol Chem, 276*, 10284–10289.

Zhu, W. Z., Chakir, K., Zhang, S., Yang, D., Lavoie, C., & Bouvier, M., et al., (2005). Heterodimerization of β_1- and β_2-adrenergic receptor subtypes optimizes β-adrenergic modulation of cardiac contractility. *Circ Res, 97*, 244–251.

Zhuo, J. L., Imig, J. D., Hammond, T. G., Orengo, S., Benes, E., & Navar, L. G. (2002). Ang II accumulation in rat renal endosomes during Ang II-induced hypertension: Role of AT(1) receptor. *Hypertension, 39*, 116–121.

Zylbergold, P., Ramakrishnan, N., & Hébert, T. E. (2010). The role of G proteins in assembly and function of Kir 3 inwardly rectifying potassium channels. *Channels, 4*, 411–421.

CHAPTER 3

β-Arrestin-Biased Signaling by the β-Adrenergic Receptors

Sudha K. Shenoy

Departments of Medicine and Cell Biology, Duke University Medical Center, Durham, North Carolina, USA

I. Overview
II. Introduction
III. β-Arrestin-Biased Signaling by The β_2AR
  A. Ligand Bias
  B. Receptor Bias
IV. β-Arrestin-Biased Signaling by The β_1AR
  A. Ligand Bias
  B. Receptor Bias
V. Factors that Define β-Arrestin-Dependent Signaling
  A. Temporal and Spatial Features
  B. βAR Phosphorylation by GRKs
  C. Receptor Endocytosis
  D. Scaffolding Properties of β-arrestin
  E. β-arrestin Modifications
  F. Conformational Changes in β-arrestin
VI. Inhibitors of β-Arrestin-Dependent Signaling
  A. Spinophilin
  B. NHERF 1 and 2
  C. Ubiquitin Specific Protease 33 (USP33)
VII. Future Perspectives
  References

I. OVERVIEW

Physiological effects of endogenous catecholamines, epinephrine and norepinephrine are mediated by the β-adrenergic receptors (βARs), which are members of the large family of seven-transmembrane receptors (7TMRs, aka G protein-coupled receptors). Upon agonist stimulation, βARs couple to the heterotrimeric Gs and increase intracellular cAMP by activating adenylyl

Current Topics in Membranes, Volume 67

0065-230X/10 $35.00
DOI: 10.1016/B978-0-12-384921-2.00003-3

cyclase. Cessation of agonist-activated βAR-Gs-mediated signaling occurs upon recruitment of adaptor proteins, β-arrestins (β-arrestin1 and β-arrestin2), to the cytoplasmic surface of the receptor. This process is enhanced by receptor phosphorylation by G protein-coupled receptor kinases (GRKs). Until recently, G protein-mediated signaling was considered as the only mode of signaling responsible for the physiological effects of βAR stimulation. This long-standing view is rapidly changing due to the discovery that β-arrestins not only desensitize G protein signaling but also have multifaceted roles in βAR regulation and can actually mediate cell signaling. Thus, β-arrestins first compete and block G protein signaling, next divert receptors through the endocytic pathway, and further initiate signal transduction by activating various kinase cascades. Molecular changes involving protein conformation as well as posttranslational modifications of β-arrestins could form the basis of their dynamic interactions during receptor trafficking and β-arrestin-dependent signaling. Studies have demonstrated that β-arrestin-biased signaling plays critical roles in physiological processes such as cardiovascular regulation and bone remodeling. An in-depth understanding of β-arrestin-biased signaling at both biochemical and physiological levels is required to successfully exploit this signaling pathway (s) for medicinal therapeutics and to develop pathway-selective drugs with minimal side effects.

II. INTRODUCTION

β-Adrenergic receptors (βARs) are considered as prototypic members of the super-family of cell-surface receptors known as seven-transmembrane receptors (7TMRs aka G protein-coupled receptors or GPCRs), which are represented by about a thousand genes in the Human Genome (Lagerstrom & Schioth, 2008; Pierce, Premont, & Lefkowitz, 2002). 7TMRs are signal transducers for a wide range of extracellular stimuli that include hormones, neurotransmitters, lipids, peptides, ions, and sensory stimuli. Their clinical importance is evident from the fact that about 50% of prescription drugs target members of this family.

7TMRs have a basic molecular architecture of seven transmembrane helices that are connected by three intra- and three extracellular loops. The N-terminal region of the receptor protein is exposed to extracellular milieu and the carboxyl tail is cytoplasmic. Upon agonist stimulation, conformational changes occur in the transmembrane domains of the 7TMR and expose cytoplasmic domains for heterotrimeric G protein binding and subsequent GDP–GTP exchange. This leads to dissociation of activated subunits of Gα and Gβγ followed by an acute modulation of levels of second messengers and activities of various effector enzymes in the cell (Fig. 1) (Neves, Ram, & Iyengar, 2002). Agonist-activated 7TMRs are rapidly phosphorylated on the cytoplasmic domains by

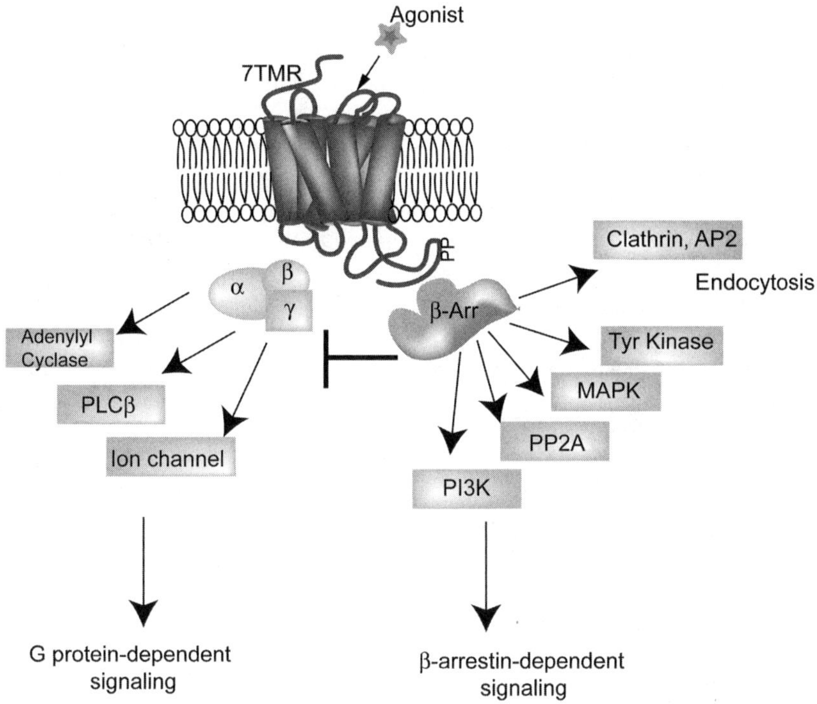

FIGURE 1 7TMR signal transduction occurs via two independent pathways. Agonist stimulation of cell-surface 7TMRs leads to coupling and activation of heterotrimeric G proteins and signal transduction via second messenger dependent pathways and effectors. Agonist occupied receptors are phosphorylated in the cytoplasmic domains by GRKs. Phosphorylated receptors display high affinity interaction with β-arrestins. β-arrestin recruitment leads to an immediate blocking of G protein coupling and signal transduction (desensitization). β-arrestin recruits endocytic proteins such as clathrin and adaptin protein2 (AP2) and facilitates receptor internalization. Additionally β-arrestin can function as a signal transducer by recruiting and activating a variety of kinases.

the seryl-threonyl kinases called G protein-coupled receptor kinases (GRKs) (Pitcher, Freedman, & Lefkowitz, 1998; Premont & Gainetdinov, 2007). Seven GRKs (GRK1–7) are expressed in mammalian cells; GRK1 and 7 are confined to visual tissue and phosphorylate the visual 7TMRs, rhodopsin and cone opsin; GRK4 has restricted tissue distribution; and GRKs 2, 3, 5, and 6 are ubiquitously expressed and regulate most nonvisual 7TMRs. Phosphorylated receptors present a high-affinity binding interface for recruiting the cytosolic adaptor proteins β-arrestins at the cytoplasmic domains. β-arrestin binding competitively blocks G protein coupling and leads to desensitization of 7TMR signaling (Fig. 1) (Lefkowitz & Shenoy, 2005).

The arrestin family has four members: arrestin1 that binds rhodopsin, β-arrestin1 (aka arrestin2) and β-arrestin2 (aka arrestin3) that bind most non-visual 7TMRs and arrestin4 that binds the cone opsins (DeWire, Ahn, Lefkowitz, & Shenoy, 2007; Lefkowitz & Shenoy, 2005). Although β-arrestins 1 and 2 were discovered as signal blockers, research through the past decade has expanded their roles to involve initiation of novel signaling via the same receptors where they block G protein coupling (Fig. 1). This β-arrestin-mediated signaling has been demonstrated for an expanding list of 7TMRs and in many cases shown to have important physiological roles (DeWire, Ahn, Lefkowitz, & Shenoy, 2007). This chapter will describe β-arrestin-mediated signaling and regulation of the two well-characterized βAR subtypes: β_1 and β_2 ARs. The third subtype, β_3AR, is neither phosphorylated nor internalized and does not robustly recruit β-arrestins (Breit, Lagace, & Bouvier, 2004; Liggett, Freedman, Schwinn, & Lefkowitz, 1993; Nantel et al., 1993). On the other hand, recent studies have shown that both β_1 and β_2 ARs engage novel β-arrestin-mediated signaling (Noma et al., 2007; Patel, Noor, & Rockman, 2010; Shenoy et al., 2006; Violin & Lefkowitz, 2007).

III. β-ARRESTIN-BIASED SIGNALING BY THE β_2AR

Almost all human cells express β_2ARs: smooth muscle cells of various organs (lungs, intestine, blood vessels, and uterus), cardiomyocytes, skeletal muscles neurons, etc. The main physiological functions of β_2ARs include relaxation of smooth muscles and regulation of cardiac contractility (Lohse, Engelhardt, & Eschenhagen, 2003; Rockman, Koch, & Lefkowitz, 2002). When stimulated by agonists, β_2ARs activate the heterotrimeric Gs proteins and increase cellular cAMP level through the activation of adenylyl cyclase. This increase in cAMP then primarily augments the enzymatic activity of protein kinase A (PKA). PKA-mediated phosphorylation of a variety of PKA substrates that include ion channels, cytoskeletal proteins, and other effector enzymes culminates in a biological response (e.g., increase in contractility in the heart, or increase in relaxation of smooth muscles). Intriguingly, the β_2AR itself is a PKA substrate and upon phosphorylation of serines in the third intracellular loop and carboxyl tail, the receptor effectively uncouples from Gs and couples to the inhibitory G protein, Gi (Daaka, Luttrell, & Lefkowitz, 1997; Zamah, Delahunty, Luttrell, & Lefkowitz, 2002). Such G protein switching, not only ensures a feedback loop for blocking excess adenylyl cyclase activity, but also leads to additional signaling mediated by the pertussis toxin sensitive Gi proteins (Neves, Ram, & Iyengar, 2002). In addition, subsequent studies have also demonstrated that in the presence of inhibitors that block PKA and Gi activity, β_2AR stimulation can nonetheless evoke MAP Kinase signaling via the extracellular signal regulated kinases 1 and 2 (ERK1/2) (Azzi et al., 2003; Shenoy et al., 2006). The ERK activation that results within the first 5 min after β2AR stimulation is attenuated by inhibitors of G protein signaling (PKA inhibitor – H89 and Gi inhibitor – pertussis toxin), but the signals beyond 5 min of agonist treatment are unaffected (Shenoy et al., 2006). This G protein-independent signaling is ablated when either β-arrestin1 or 2 levels are silenced by siRNA targeting these isoforms, suggesting that the β_2AR can mediate G protein independent, β-arrestin-dependent ERK activation (Shenoy et al., 2006). Accordingly, the G protein-dependent and the β-arrestin-dependent pools of activated ERK are temporally separated: the initial acute phase promoted by G protein coupling and the later sustained phase mediated by β-arrestins (Fig. 2A). This bimodal pattern of ERK activation was also shown for other 7TMRs, but the involvement of β-arrestin isoforms was found to differ in specific cases (DeWire et al., 2007). Thus, some 7TMRs such as the β_2AR engage both β-arrestins, whereas in some of the isoforms display reciprocal effects with respect to the other: for example, Angiotensin-stimulated β-arrestin-dependent ERK is inhibited by β-arrestin1, but promoted by β-arrestin2, whereas Protease Activated receptor 1 stimulated ERK is promoted by β-arrestin1 and blocked by β-arrestin2 (Ahn, Shenoy, Wei, & Lefkowitz, 2004; Kuo, Lu, & Fu, 2006). The reasons as to why some receptors require both β-arrestins and some depend on one specific isoform is currently not understood. It is likely that some receptors adopt a conformation more favorable to recruit heterodimers of β-arrestins 1 and 2 and some preferentially activate homodimers or monomers of one isoform of β-arrestin. Hetero and homo oligomerization of overexpressed β-arrestins have been demonstrated by coimmunoprecipitation and resonance energy transfer based assays and appear to regulate subcellular distribution of β-arrestins as well as binding with inositol hexakisphosphate (IP6) (Milano, Kim, Stefano, Benovic, & Brenner, 2006; Storez et al., 2005).

A. Ligand Bias

Discovery of β-arrestin-mediated signaling has also revealed that ligands can preferentially activate G proteins versus β-arrestins or vice versa, leading to a behavior termed as "biased agonism," also called "ligand-directed trafficking," "protean agonism," "pleuridimensional efficacy," and "collateral efficacy" (Galandrin, Oligny-Longpre, & Bouvier, 2007; Kenakin & Miller, 2010; Vaidehi & Kenakin, 2010; Violin & Lefkowitz, 2007). Accordingly, a biased ligand or a biased receptor selectively activates one pathway unlike the unbiased or balanced ligand that activates multiple pathways with equal efficacy (Fig. 2B). Unlike the long-standing view that receptors exist either in one active and one inactive conformation, recent studies also indicate that

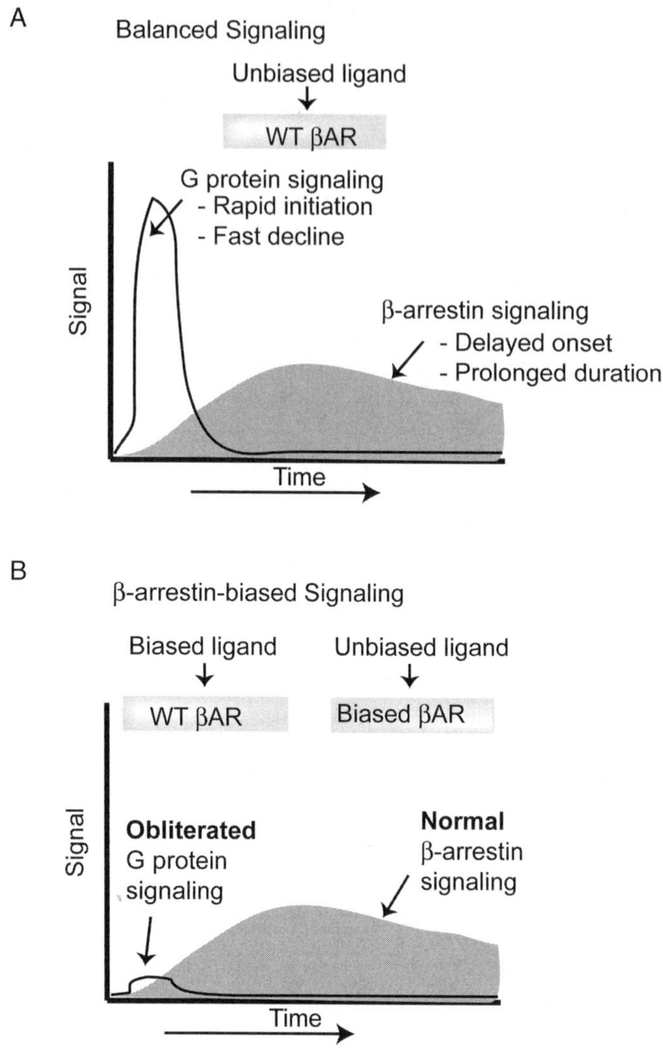

FIGURE 2 Balanced and biased signaling via the βARs. (A) When wild type βARs are stimulated with a balanced or unbiased agonist (e.g., isoproterenol), both G protein and β-arrestin dependent signaling to the MAP Kinase ERK are elicited. The G protein signals are rapid and decline within a few minutes. β-arrestin signaling is delayed and is initiated after a few minutes and is sustained for 20–30 minutes. For some 7TMRs such as the angiotensin II 1a receptor this can be prolonged for 1–2 h. (B) When wild type β_2AR is stimulated with a β-arrestin-biased ligand (e.g., carvedilol) or a β-arrestin-biased βAR (e.g., β_2AR-TYY) is stimulated with an unbiased agonist (e.g., isoproterenol), β-arrestin-dependent signals are elicited normally whereas G protein signaling is obliterated.

receptors not only exist in more than two conformations, but present multiple "active" conformations within the defined "G protein-binding" and "β-arrestin-binding" conformations.

In HEK-293 and mouse embryonic fibroblasts, the β_2AR antagonists propranolol and ICI-118551 were shown to activate ERK only in the presence of β-arrestins, although these compounds do not stimulate cAMP production (Azzi et al., 2003). Moreover, various βAR ligands that activate adenylyl cyclase and MAPK signaling pathways at the β_2AR were demonstrated to have ligand-dependent activation profile differences (Baker, 2010; Galandrin & Bouvier, 2006). Studies also indicate that the majority of known β_2AR agonists exhibit relative efficacies for β-arrestin-associated activities (β-arrestin membrane translocation and β_2AR internalization) identical to their relative efficacies for G protein-dependent signaling (cyclic AMP generation) (Drake, Violin, Whalen, Wisler, Shenoy, & Lefkowitz, 2008). Remarkably, these analyses also discovered three βAR ligands to have a marked bias toward β-arrestin signaling, since these ligands stimulated β-arrestin-dependent receptor functions to a greater magnitude, but displayed moderate efficacy for G protein signaling (Drake et al., 2008). Very strikingly, structural comparison of these biased ligands revealed that all three are catecholamines containing an ethyl substitution on the α-carbon, a motif absent on all of the other unbiased ligands that were tested (Drake et al., 2008). Other studies have revealed that alteration of the stereochemistry of a β_2AR ligand, fenoterol changed the selectivity for G protein coupling such that the *R,R* isomer coupled only to Gs whereas *R,S* isomer coupled to both Gs and pertussis toxin sensitive Gi (Woo et al., 2009). Accordingly, small structural alterations of ligands could provide a means to convert an unbiased ligand to become biased for a particular mode of signaling.

When a panel of 16 βAR antagonists were compared for their blocking effects utilizing read outs based on fluorescence resonance energy transfer (FRET) biosensors for cAMP increase and biochemical assays for MAP Kinase activation, a diverse spectrum of efficacies was observed for both Gs-dependent and β-arrestin-dependent signaling (Wisler et al., 2007). Additionally, one compound, carvedilol, possessed unique signaling profile of negative efficacy for Gs-dependent adenylyl cyclase activation but positive efficacy for β-arrestin-dependent ERK activation. Intriguingly, carvedilol stimulated phosphorylation of the β_2AR, β-arrestin translocation to the receptor, and receptor internalization, all of which are characteristic of β-arrestin-mediated cellular processes. Thus, carvedilol acts as a biased ligand at the β_2AR and promotes signaling via β-arrestin-dependent ERK activation in the absence of G protein activation (Wisler et al., 2007). The unique effects produced by carvedilol on β-arrestin-dependent signaling may correlate with the clinical efficacy and survival advantages provided by this beta-blocker. On the other hand, the β-arrestin-dependent ERK activation promoted by carvedilol is not robust when compared with the unbiased agonist isoproterenol and further modification of its chemical structure or identification of a novel compound with better potency toward β-arrestin-dependent signaling could have increased therapeutic advantage.

The two endogenous ligands that bind the β_2AR, namely, epinephrine and norepinephrine have also been characterized for various biological effects. By transfecting Flag-β_2AR into neonatal cardiomyocytes isolated from β_1/β_2AR-KO mice, effects specific for this receptor subtype were studied (Wang, De Arcangelis, Gao, Ramani, Jung, & Xiang, 2008). It was demonstrated that while epinephrine induced a rapid contractile response, GRK phosphorylation, receptor recycling, and Gs/Gi switching properties at the β_2AR, norepinephrine induced slower contractile response, slow kinetics of receptor phosphorylation, retarded recycling, and no Gs/Gi switching (Wang et al., 2008). Additionally when a modified β_2AR (truncated at residue 369) with engineered FRET sensors, namely YFP at the carboxyl tail and CFP at the third intracellular loop, was stimulated with epinephrine or norepinephrine, the cAMP responses were similar, but norepinephrine produced only 50% of the conformational responses (change in intramolecular FRET) as that induced by epinephrine (Reiner, Ambrosio, Hoffmann, & Lohse, 2010). Furthermore, norepinephrine produced modest β-arrestin recruitment and receptor internalization, suggesting that its partial agonism may in fact correspond to a particular active conformation at the β_2AR than epinephrine. Surprisingly, however, in this system, both epinephrine and norepinephrine induced a similar extent of receptor phosphorylation. Because the amount of receptor phosphorylation determines the amount of β-arrestin recruitment, and because norepinephrine recruits less β-arrestin, predictably, norepinephrine should have led to lesser phosphorylation of the β_2AR than epinephrine stimulation. On the other hand, it is likely that these two ligands lead to the same amount of phosphorylation at discrete sites on the β_2AR, out of which one affects β-arrestin binding.

B. Receptor Bias

Earlier studies showed that deletion of regions within the third intracellular loop of the β_2AR could impair G protein coupling (Barber, Ganz, Bongiorno, & Strader, 1992; Cheung, Huang, & Strader, 1992; O'Dowd, Hnatowich, Regan, Leader, Caron, & Lefkowitz, 1988). The idea that alteration of key residues for G protein interaction might yield a receptor that would still recruit β-arrestin was tested by generating a β_2AR mutant with alterations in three residues Thr-68, Tyr-132, and Tyr-219 within the human β_2AR (Shenoy et al., 2006). This mutant denoted as β_2AR-TYY, serves as an example of β-arrestin-biased

7TMR because it does not induce cAMP production upon agonist-stimulation, nonetheless recruits β-arrestin, internalizes normally and activates ERK in a β-arrestin-dependent manner, (Shenoy et al., 2006).

IV. β-ARRESTIN-BIASED SIGNALING BY THE β_1AR

β_1ARs are expressed mainly in cardiomyocytes and kidney juxtaglomerular cells, and regulate cardiac output and renin release. Similar to the β_2AR, the β_1 subtype is also expressed in most tissues albeit at low levels. Traditional signaling via the β_1AR follows a similar paradigm as the β_2AR, by leading to cAMP production via activation of Gs, phosphorylation by PKA, recruitment and seryl-threonyl phosphorylation by GRKs and β-arrestin-dependent desensitization and internalization (Freedman, Liggett, Drachman, Pei, Caron, & Lefkowitz, 1995; Frielle, Collins, Daniel, Caron, Lefkowitz, & Kobilka, 1987; Martin, Whalen, Zamah, Pierce, & Lefkowitz, 2004; Rockman, Koch, & Lefkowitz, 2002; Shiina, Kawasaki, Nagao, & Kurose, 2000). Depending on the cell type, expression levels and co-expression of the tyrosine kinase receptor EGFR, β_1AR signaling to ERK is either β-arrestin dependent or independent (Galandrin, Oligny-Longpre, Bonin, Ogawa, Gales, & Bouvier, 2008; Noma et al., 2007). The β-arrestin-dependent pathway involves transactivation of EGFR and is cardioprotective since ablation of the β-arrestin-dependent EGFR transactivation results in increased apoptosis and deterioration of cardiac function (Noma et al., 2007). Cellular studies indicate that the β_1AR and EGFR form a complex at the plasma membrane, and upon stimulation with a β_1AR agonist, dobutamine, EGFR is trans-activated in a β-arrestin-dependent manner, leading to accumulation of phosphorylated ERK in the cytoplasm (Tilley, Kim, Patel, Violin, & Rockman, 2009). EGFR inhibitor AG1478 or β-arrestin knockdown blocks this activity. On the other hand, EGFR activation induced by EGF is unaffected by β-arrestin expression, but completely blocked by AG1478. The immediate downstream effectors of the induced cytoplasmic ERK activity by the β_1AR–EGFR complexes are currently unknown.

A. Ligand Bias

The β_1AR also exhibits complex ligand efficacy profiles: when a variety of traditional ligands were tested for the effects on adenylyl cyclase activity versus MAPK activation, compounds that had a negative efficacy for cAMP production nonetheless activated MAP Kinase, ERK (Galandrin & Bouvier, 2006). Furthermore, compounds that were agonists for adenylyl cyclase were either

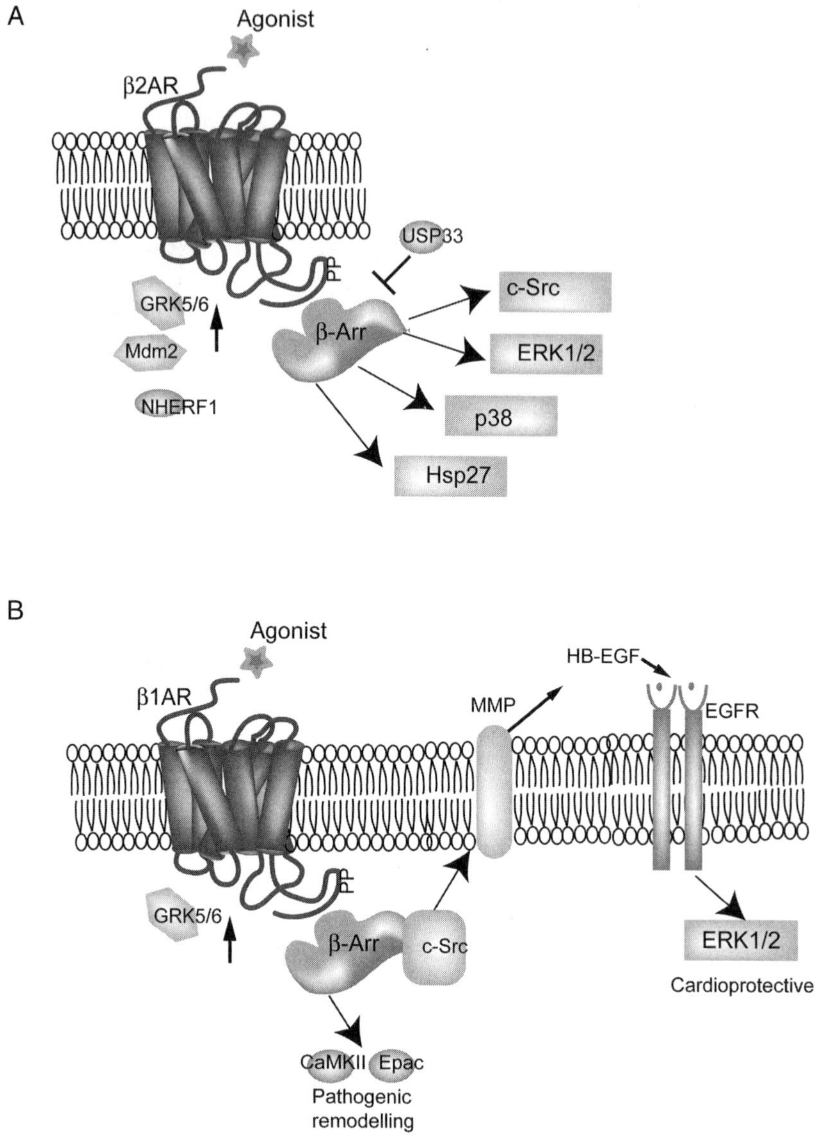

FIGURE 3 β-arrestin-biased signaling by the βARs. (A) Upon stimulation of the β_2AR, β-arrestin-mediated signals are evoked as described in Figure 1. The main effector pathways include coupling to nonreceptor tyrosine kinases, scaffolding and activation of MAP Kinases (p38 and ERK1/2), and antiapoptotic signaling via Hsp27. GRK5/6 phosphorylation at the β_2AR augments β-arrestin-dependent signaling. Mdm2-β-arrestin binding and subsequent ubiquitination is required for β-arrestin-dependent ERK activation. Binding of USP33 and deubiquitination

antagonists or agonists for MAPK. In another study that tested 20 beta-blockers for their ability to induce β-arrestin-dependent signaling by EGFR transactivation, only alprenolol and carvedilol were found to act as β-arrestin-biased agonists, since these compounds though ineffective for activating G protein signaling at the β_1AR nevertheless promoted EGFR activation (Kim et al., 2008).

B. Receptor Bias

So far, a modified β_1AR that is biased toward β-arrestin-dependent signaling has not been characterized. Previous studies have however, implicated that there might be unique regions in the transmembrane domains required for receptor activation and constitutive activity (Zeitoun, Santos, Gardner, White, & Bahouth, 2006). Future studies involving mutagenesis as well as characterization of β-arrestin-dependent outcomes should illuminate us on a receptor conformation that could be biased toward β-arrestin-dependent signaling.

Our current understanding about β-arrestin-biased signaling via the β_1 and β_2 ARs and effector pathways involved are summarized in Figure 3. Additionally, the sections that follow will describe the salient features of β-arrestin-mediated signaling and how it might be regulated.

V. FACTORS THAT DEFINE β-ARRESTIN-DEPENDENT SIGNALING

A. Temporal and Spatial Features

Studies conducted with inhibitors of G proteins and gene silencing of β-arrestin isoforms have shown that β-arrestin-dependent signaling involving ERK1/2 is delayed at its onset and in most cases, the response lasts longer than that effected by G proteins (Fig. 2) (DeWire et al., 2007). For the β_2AR, G protein-dependent ERK peaks within 2–5 minutes after agonist stimulation and β-arrestin-dependent ERK peaks between 5 and 10–60 minutes. Thus, traditional unbiased agonists such as isoproterenol engage two modes of signaling at

Figure 3 (Continued). antagonizes β-arrestin-dependent signaling to ERK. (B) Stimulation of the β_1AR leads to the recruitment of β-arrestin to GRK5/6-phosphorylated receptors. Two modes of β-arrestin-dependent signaling are elicited by the β_1AR. β-arrestin–c-Src complexes activate a matrix-metalloproteinase (MMP) that cleaves and releases the heparin-binding epidermal growth factor (HB-EGF) into the extracellular milieu, thus leading to the transactivation and tyrosine autophosphorylation of EGFR and subsequent endocytosis and ERK activation. In a second signaling pathway, upon binding to the carboxyl tail of the β_1AR, β-arrestins recruit Epac and CaMKII leading to signaling via CaMKII.

the β_2AR, an initial acute phase through G proteins and subsequent sustained phase mediated by β-arrestins. The temporal nature of the two pathways might become slightly shifted when either the receptor or ligand is totally biased. If a receptor is coupled to G protein, but does not engage β-arrestins, the G protein signaling may last longer due to the absence of efficient desensitization by β-arrestin–receptor interaction. It is also plausible that a β-arrestin-biased ligand or receptor induces slightly earlier responses than an unbiased ligand or receptor, suggesting that the G proteins themselves or associated components inhibit β-arrestin activity during the rapid signaling phase induced by unbiased agonists. The temporal pattern of MAP Kinase activation could also present differences from the above general characteristics described for ERK1/2. The family of MAPKs includes ERK1 ($p44^{MAPK}$), ERK2 ($p42^{MAPK}$), ERK5 (BMK), c-Jun NH2-terminal kinases (JNKs 1-3, also called Stress activated protein kinases SAPKs), and p38 MAPKs (α, β, γ, δ isoforms). In HEK-293 cells, a PKA-independent, β-arrestin1-dependent activation of p38α and β that lasts up to 60 min after isoproterenol stimulation was detected (Gong, Li, Xu, Du, Lv, & Zhang, 2008). Surprisingly, activation of these p38 isoforms reinitiated after 60 min and lasted for 6 h and this second phase of activation was PKA-dependent and β-arrestin independent (Gong et al., 2008).

The temporal characteristic of G protein and β-arrestin-dependent pathways induced by 7TMRs allows segregation of downstream signals in discrete locations. Most often, the G protein induced ERK activity is translocated into the nucleus whereas β-arrestin-dependent ERK is localized in cytoplasm or associated with endosomal vesicles (Ahn, Shenoy, Wei, & Lefkowitz, 2004). Thus, it is likely that each pathway would target some unique downstream effectors. Bioinformatics and proteomic analyses carried out suggest that a distinction between the two pathways could also result from differences in the magnitude of activation of kinases common to both pathways (Xiao et al., 2010).

B. βAR *Phosphorylation by* GRKs

In general, GRK-mediated phosphorylation of both β_1 and β_2 ARs is critical for β-arrestin recruitment, since mutation of all the phosphorylation sites in the carboxyl tail attenuates β-arrestin binding (DeWire et al., 2007; Noma et al., 2007). The extent of phosphorylation on receptor C-tail and the number of seryl-threonyl clusters also influence whether a 7TMR–β arrestin complex is formed transiently or stably at the plasma membrane (Oakley, Laporte, Holt, Barak, & Caron, 2001). The βARs lack a serine-rich cluster unlike receptors such as the AT_1aR and V2 vasopressin receptor (V2R) and hence bind β-arrestin only transiently at the plasma membrane. When β-arrestin–7TMR complexes are stable, they internalize together and localize on endocytic vesicles (Oakley,

Laporte, Holt, Caron, & Barak, 2000; Shenoy & Lefkowitz, 2003). More often, these internalized complexes are bound to active MAP kinases and are termed as signalosomes (Lefkowitz & Shenoy, 2005; Luttrell et al., 2001).

In addition to increasing the affinity for β-arrestin binding, GRK-mediated phosphorylation of the βARs and other 7TMRs also influence β-arrestin-dependent signaling (Reiter & Lefkowitz, 2006). However, phosphorylation by different GRKs (among the ubiquitously expressed GRK 2, 3, 5, and 6) may have distinct effects on β-arrestin. Thus, while GRK5/6 mediated phosphorylation promotes β-arrestin-dependent ERK activation, GRK2/3 mediated receptor phosphorylation prevents engagement of β-arrestin's signaling function (Reiter & Lefkowitz, 2006; Shenoy et al., 2006). Nonetheless, GRK2-mediated phosphorylation accounts for most of the total phosphorylation signals on activated β_2AR (Shenoy et al., 2006). The β-arrestin-dependent transactivation of EGFR elicited by the β_1AR also requires GRK5/6 activity both in HEK-293 cells and in mice heart (Noma et al., 2007). Although previous *in vitro* studies have attempted elucidation of specific phosphorylation sites on the βARs that are targeted by distinct GRKs, this challenging question remains to be addressed *in vivo* (Fredericks, Pitcher, & Lefkowitz, 1996).

In the case of the β-arrestin-biased receptor, β_2AR-TYY, GRK2-mediated phosphorylation is not detected unless a membrane-targeted form of GRK2 is coexpressed in cells (Shenoy et al., 2006). This is because β_2AR-TYY activation does not release $G\beta\gamma$ to bind GRK2 and facilitate its membrane localization (Inglese, Koch, Caron, & Lefkowitz, 1992; Pitcher et al., 1992). Furthermore, phosphorylation of β_2AR-TYY occurs at a level of 20% of the signals detected for the wild type β_2AR and is mainly detected at serines 355 and 356 on the carboxyl tail (Shenoy et al., 2006). Notably, β-arrestin recruitment and β-arrestin-dependent internalization are detected at the β_2AR-TYY suggesting that β-arrestin binding to 7TMRs might depend on phosphorylation occurring at specific sites and not on the extent of phosphorylation in general. In addition, overexpression of GRKs 5 and 6 promote phosphorylation of these residues, allowing the formation of a stable β_2AR–β-arrestin complex and augmentation of ERK signaling, whereas GRK2 does not (Shenoy et al., 2006). These findings further suggest that β_2AR, when phosphorylated by GRK5/6 presents an activated conformation that is conducive for β-arrestin-dependent signaling.

C. Receptor Endocytosis

In response to agonist-stimulation, both β_1 and β_2ARs generally internalize via clathrin-coated vesicles within a few minutes. This internalization mechanism was originally considered to be responsible for turning off signaling, since activated receptors are sequestered away from the cell-surface and cannot bind extracellular ligands (Chuang & Costa, 1979; Stadel et al., 1983). However, not only are β-arrestins required for mediated βAR internalization, but there are further indications that β-arrestin-dependent signals might be intimately linked to the internalization process (DeFea, Zalevsky, Thoma, Dery, Mullins, & Bunnett, 2000; Ferguson, Downey, Colapietro, Barak, Menard, & Caron, 1996; Lefkowitz & Shenoy, 2005). The initiation of β-arrestin-dependent signaling coincides with the time frame of early stages of receptor endocytosis. The dual role of β-arrestin in facilitating both receptor endocytosis and signaling poses a difficulty to address its signaling role independent of its endocytic role. Hence, whether β-arrestin-dependent signaling initiates endocytosis or if β-arrestin-dependent endocytosis instigates signaling remains an unsettled issue.

D. Scaffolding Properties of β-arrestin

β-arrestins 1 and 2 bind a variety of kinases and regulatory proteins and in many cases, β-arrestin functions as a receptor-activated scaffold to converge upstream and downstream components of a particular signal transduction pathway (Lefkowitz & Shenoy, 2005; Miller & Lefkowitz, 2001). For example, β-arrestin binds cRAF-1 (upstream MAPK kinase kinase) and ERK2 (downstream MAPK) and recruits MEK1 (MAPK kinase), thus converging core components for activating ERK2 (DeFea et al., 2000; Luttrell et al., 2001). Furthermore, such a scaffold assembly is often potentiated by 7TMR activation, thus providing a stimulus-dependent signal transduction event to direct the cell's biological response. Yet another feature of such signaling scaffolds is that upon high-affinity interactions of β-arrestin and 7TMR, they become localized on endosomes, often termed as signalosomes, which bestows compartmentalization of 7TMR signaling. Thus, β-arrestin-dependent signals are not only temporally distinct, but also spatially segregated from the initial second-messenger responses generated by G protein signaling. Spatial segregation of cAMP signals via the β_1 and β_2ARs have also been reported to occur in rat cardiomyocytes, and is attributed to differential distribution of the two receptor subtypes, that is β_1AR at the cell crest and β_2AR in the transverse tubule regions of the myocyte (Nikolaev et al., 2010).

Both β_1 and β_2ARs form complexes with the EGFR, and in both cases agonist stimulation of the βAR evokes EGFR transactivation (Maudsley et al., 2000; Noma et al., 2007). These EGFR activities are regulated by β-arrestins since β-arrestin mutants defective in receptor interaction or depletion of β-arrestin isoforms attenuates EGFR-mediated responses (Maudsley et al., 2000; Noma et al., 2007). β-arrestins also recruit the nonreceptor tyrosine kinase c-Src to activated β_2ARs, and this was one of the early demonstration that β-arrestin can initiate their own signaling via a 7TMR (Luttrell et al., 1999). β_1AR-mediated EGFR transactivation involves recruitment of a Src-dependent matrix-metalloprotease activity that cleaves and releases HB-EGF, which stimulates EGFR signaling. Upon β_1AR activation, β-arrestins also scaffold Ca^{2+}/calmodulin kinase II (CaMKII) and the cAMP-dependent guanine-nucleotide exchange factor (Epac), thus promoting CaMKII signaling (Mangmool, Shukla, & Rockman, 2010). While the β_1AR-mediated β-arrestin-dependent EGFR transactivation is cardioprotective, that via CaMKII might promote a pathological process resulting in adverse cardiac remodeling. This pathway is specifically dependent on the interaction of β-arrestin with the carboxyl tail of the β_1AR, since replacing this region with the β_2AR carboxyl tail prevents β-arrestin-dependent CaMKII signaling (Mangmool, Shukla, & Rockman, 2010). β_2AR stimulation, on the other hand, engages a β-arrestin/Hsp27 complex that protects against apoptosis induced by staurosporine treatment (Rojanathammanee et al., 2009). Thus, emerging evidence suggests that β-arrestin-dependent signaling displays a variety of protein interactions and regulates diverse cellular pathways via both β_1 and β_2 ARs (Fig. 3).

E. β-arrestin Modifications

β-arrestins undergo posttranslational modifications most often in response to 7TMR stimulation and as discussed below these molecular changes in β-arrestin conformation affect receptor endocytosis, and signaling.

1. Phosphorylation of β-arrestin

β-arrestin 1 and 2 are cytosolic phosphoproteins and upon agonist-stimulated recruitment to the β_2AR, they become rapidly dephosphorylated (Lin, Chen, Shenoy, Cong, Exum, & Lefkowitz, 2002; Lin et al., 1997; Lin, Miller, Luttrell, & Lefkowitz, 1999). In β-arrestin1, the phosphorylation site is identified as serine 412, which is phosphorylated by ERK1/2 when β_2ARs are activated and by GRK5 when 5-HT4 receptors are activated (Barthet et al., 2009; Lin et al., 1997). Threonine 383 in rat β-arrestin2 (which is threonine 382 of bovine β-arrestin2) is phosphorylated by casein kinase II (Kim, Barak, Caron, & Benovic, 2002; Lin et al., 2002). Studies conducted with both phosphomimetic (serine/threonine mutated to aspartate) and phosphorylation impaired (serine/threonine mutated to alanine) mutants have indicated that dephosphorylation of β-arrestin1 and 2 are important for clathrin interaction and clathrin-dependent-receptor internalization. Such mutations of serine 412 or threonine 383 respectively in β-arrestin1 or 2 do not alter their properties with respect to receptor binding and desensitization. However, when β-arrestin1 is phosphorylated by

GRK5, it triggers a unique sequence of events in which the G protein- independent signaling of 5-HT4 receptor to c-Src/ERK is inhibited (Barthet et al., 2009).

A tyrosine (Y54) residue found only in the β-arrestin1 isoform is phosphorylated by c-Src and this lessens interaction with mu subunit of adaptin2, thus lowering its ability to augment β_2AR internalization (Marion, Fralish, Laporte, Caron, & Barak, 2007). β-arrestin2 has a natural substitution at this residue (phenylalanine 54) and has a more productive interaction with AP2 than β-arrestin1. Phosphorylation of β-arrestin thus appears to be a mechanism of feed-back regulation for specific protein interactions during endocytosis in general and may also regulate signaling events for certain 7TMRs.

2. Ubiquitination of β-arrestin

Ubiquitin (Ub) is a small, ubiquitous, highly conserved protein of 76 residues. Posttranslational attachment of Ub, known as ubiquitination, is a highly regulated process wherein the C-terminal glycine of Ub becomes covalently attached to the epsilon amino group of a lysine residue in a substrate (Hershko & Ciechanover, 1998). The process requires three distinct enzyme activities: E1-Ub activating enzyme, E2-Ub carrier enzyme, and E3-Ub ligating enzyme.

Ubiquitination of β-arrestin is mediated by the E3 ligase Mdm2 in a transient manner upon β_2AR activation, and this process has several key consequences (Shenoy, McDonald, Kohout, & Lefkowitz, 2001). β_2AR-dependent β-arrestin ubiquitination is inhibited by Mdm2 mutants, which lack the RING domain but retain the β-arrestin binding region. Isoproterenol-stimulated β-arrestin ubiquitination is not detectable in Mdm2/p53 double knock out mouse embryonic fibroblasts. Quite surprisingly under both these conditions, which impair β-arrestin ubiquitination, β_2AR internalization, which normally occurs as a rapid agonist-promoted response, is ablated (Shenoy, McDonald, Kohout, & Lefkowitz, 2001). β-arrestin ubiquitination is also important for forming a high affinity complex with the β_2AR (Shenoy et al., 2007; Shenoy & Lefkowitz, 2003). Overexpression of Mdm2 augments β-arrestin ubiquitination as well as stabilization of β-arrestin–β_2AR binding promoting their cointernalization and colocalization on endosomes (Shenoy et al., 2009). Comparison of two modified β-arrestins, one which lacks all ubiquitin acceptor sites in β-arrestin and the other where stable ubiquitination is conferred by fusing ubiquitin at the carboxyl terminus of β-arrestin, with wild type β-arrestin provided insights as to how β-arrestin ubiquitin might regulate both β_2AR endocytosis and ERK signaling (Shenoy et al., 2007). The stably ubiquitinated β-arrestin2-Ub fusion protein, bound activated β_2AR, clathrin and ERK2 with higher affinity than wild type β-arrestin and formed stable signalosomes, where as the nonubiquitinated β-arrestin showed only weak or highly transient binding with these partners (Shenoy et al., 2007). Ubiquitination is also targeted at specific domains of β-arrestin2 depending on the 7TMR that is being activated (Shenoy & Lefkowitz, 2005).

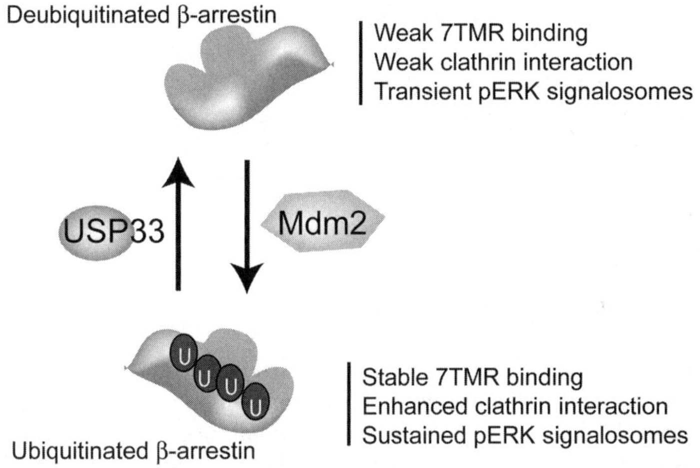

FIGURE 4 Reciprocal regulation of β-arrestin-biased signaling by ubiquitination and deubiquitination. Agonist-stimulation of the β_2AR leads to ubiquitination of β-arrestin2, which is mediated by the E3 ubiquitin ligase Mdm2. Ubiquitinated β-arrestin interacts with the receptor, endocytic protein clathrin, and the signaling kinase ERK much more robustly than deubiquitinated β-arrestin2. The deubiquitinating enzyme USP33 catalyzes reversal of β-arrestin ubiquitination, which decreases receptor interaction, retards internalization, and disassembles β-arrestin–ERK complexes.

The role of β-arrestin ubiquitination in receptor binding, endocytosis, and signaling is further illuminated by the discovery that it can be reversed by a specific deubiquitinating enzyme, USP33 (Shenoy et al., 2009). Upon β2AR activation, the recruited β-arrestins are rapidly ubiquitinated and deubiquitinated and this is facilitated by a high affinity interaction between β-arrestin2 and USP33. This interaction leads to deubiquitination of β-arrestin and such a loss of ubiquitin moieties leads to its dissociation from the internalizing β_2AR and promote only transient binding with activated ERK. Indeed, depletion of USP33 not only stabilizes β-arrestin ubiquitination, but also promotes its stable interaction with the internalized β_2AR on endosomes and also increases the magnitude of β-arrestin-dependent ERK activity (increase at later time points of agonist treatment) (Shenoy et al., 2009). In contrast, depletion of the E3 ubiquitin ligase Mdm2, ablates both β-arrestin ubiquitination and β-arrestin-dependent ERK activity (Shenoy et al., 2009). Thus, the kinetics of β-arrestin ubiquitination and deubiquitination are tightly regulated by these cellular enzymes ensuring the appropriate duration and magnitude of β-arrestin-biased signaling (Fig. 4).

3. S-Nitrosylation of β-arrestin2

β-arrestin2, interacts with and is S-nitrosylated at a single cysteine (C-terminal residue 410) by endothelial NO synthase (eNOS), and S-nitrosylation of β-arrestin2 is promoted by endogenous S-nitrosogluthathione (Ozawa et al., 2008). Interestingly, S-nitrosylation of cysteine 410 potentiates the interaction of β-arrestin2 with clathrin and β-adaptin, in striking contrast to the inhibitory effect of S-nitrosylation on the interaction between β-arrestin 2 and eNOS (Ozawa et al., 2008). Although S-nitrosylation of β-arrestin2 was reported to increase β_2AR internalization, its effect on β-arrestin-dependent signaling is unknown.

F. Conformational Changes in β-arrestin

β-arrestins 1 and 2 function as binding partners for numerous receptors and nonreceptor proteins. In many cases, the interactions are primarily induced after 7TMR stimulation, indicating that the molecular structure or presentation of β-arrestin after being recruited to the activated receptor is altered due to a conformational change (Gurevich & Gurevich, 2004; Lefkowitz & Shenoy, 2005; Palczewski, Pulvermuller, Buczylko, & Hofmann, 1991). The X-ray structures of bovine visual arrestin and β-arrestin1 in the basal inactive state indicate that arrestin is an elongated molecule with two domains (N- and C-domain), connected through a 12-residue linker region (Han, Gurevich, Vishnivetskiy, Sigler, & Schubert, 2001; Hirsch, Schubert, Gurevich, & Sigler, 1999). A notable feature is that of a hydrogen-bonded polar core, embedded between the N- and C-domains at the fulcrum of the β-arrestin molecule. Disruption of the polar core by the phosphate moieties on the activated receptors and the resulting rearrangement of the "three element interface" is suggested to promote β-arrestin activation and conformational change (Vishnivetskiy, Schubert, Climaco, Gurevich, Velez, & Gurevich, 2000).

Additionally, conformational changes have been demonstrated for β-arrestins 1 and 2 to occur in the presence of a phosphopeptide that mimics the GRK phosphorylated carboxyl tail of the V2R (Nobles, Guan, Xiao, Oas, & Lefkowitz, 2007; Xiao, Shenoy, Nobles, & Lefkowitz, 2004). Essentially, addition of the phosphopeptide led to the exposure of a buried tryptic cleavage site (arginine 393 in β-arrestin1 and arginine 394 in β-arrestin2) as well as the release of buried carboxyl terminus of β-arrestin, containing the previously mapped sites for clathrin interaction (Krupnick, Goodman, Keen, & Benovic, 1997). In this "activated" conformation induced by the phosphopeptide, β-arrestin binding to clathrin was much more robust than in the presence of a nonphosphorylated peptide. Conformational changes have also been measured in the β-arrestin molecule sandwiched between luciferase and YFP by a process called intramolecular BRET (Charest, Terrillon, & Bouvier, 2005). In this biosensor, alterations due to structural rearrangements in the β-arrestin molecule produce a change in BRET, which could be a measure of conformational change. In these measurements, two BRET signals are detected, one that immediately follows agonist stimulation corresponding to β-arrestin conformational change upon β-arrestin receptor interaction, and a second wave of conformational change that occurs upon β-arrestin's interaction with partners that are recruited to the activated receptor complex. On the other hand, absence of an increase or decrease in BRET does not rule out the possibility of a conformational change because structural rearrangements might occur, but the distance between luciferase and YFP remain constant (Salahpour & Masri, 2007). Furthermore, studies with the above BRET biosensor have also indicated that β-arrestin undergoes distinct conformational changes when the β-arrestin-biased β_2AR-TYY is stimulated with isoproterenol, than the wild type β_2AR (Shukla, Violin, Whalen, Gesty-Palmer, Shenoy, & Lefkowitz, 2008). Future studies are necessary to determine whether the β-arrestin conformational changes induced by 7TMR binding are merely a means to facilitate binding of activated nonreceptor partners, or a process that initiates β-arrestin-biased signaling.

VI. INHIBITORS OF β-ARRESTIN-DEPENDENT SIGNALING

Unlike the G protein-dependent signaling that involves catalytic GTP–GDP exchange reactions, β-arrestin-dependent signaling relies on the formation of a molecular complex with the activated 7TMR, which then acts as a signal transducer by recruiting and activating cellular kinases. Currently this concept is based on the theory that 7TMR–β-arrestin complexes mimic an agonist-activated ternary complex. However, it is unknown as to what triggers and defines the magnitude of β-arrestin signaling and if there are mechanisms to regulate or "desensitize" β-arrestin-dependent signaling. By nature's definition, cells need a "turn off" mechanism to downregulate signaling and prevent overstimulation and abnormal growth. As mentioned above, interaction with Mdm2 and receptor phosphorylation by GRK5/6 enzymes promote a high affinity interaction of β-arrestin and the β_2AR, thus activating β-arrestin-biased signaling whereas interaction with USP33 has reciprocal effects (Shenoy et al., 2006, 2009). In addition to USP33 other inhibitors of 7TMR signaling and β-arrestin binding have been identified and are described in this section.

A. Spinophilin

Spinophilin is a ubiquitously expressed protein containing an F-actin–binding domain, a phosphatase 1 (PP1) binding and regulatory domain, a protein-interaction PDZ domain, and C-terminal coiled-coil domains and interacts with at least two subfamilies of GPCRs, the α_2AR subtypes and the D_2 dopamine receptor (Smith, Oxford, & Milgram, 1999; Wang & Limbird, 2002; Wang et al., 2004). Epinephrine treatment leads to an enhancement of overexpressed Spinophilin recruitment to the third intracellular loop of α_2AR, which is attenuated upon pertussis toxin treatment. Spinophilin also blocks GRK2 recruitment and prevents a stable phosphorylation of the receptor and intriguingly, this attenuation requires β-arrestin expression. Furthermore, spinophilin antagonizes some of β-arrestin's functions, namely arrestin-dependent stabilization of α_2AR phosphorylation, clathrin-dependent endocytosis, recycling, and resensitization of the receptor (Wang & Limbird, 2007). In this system, arrestin-dependent signals are observed as an augmentation of ERK activity due to cell-surface replenishment of recycled receptors. Additionally, arrestin expression diverts the α_2AR signaling to ERK to proceed through recruitment of c-Src (Wang, Lu, Zhao, & Limbird, 2006). Spinophilin binds to α_2AR–Gβγ complex and preferentially binds activated receptors that have not been phosphorylated by GRK2. Spinophilin binding blocks GRK2 phosphorylation as well as subsequent β-arrestin recruitment, and antagonizes β-arrestin-dependent trafficking and signaling (Wang & Limbird, 2007). Future studies should reveal if Spinophilin regulation can extend to most 7TMRs and whether, like β-arrestin, it has the potential to interact with a majority of 7TMRs.

B. NHERF 1 *and* 2

The β_2AR contains a PDZ binding motif at its carboxyl terminus X-Serine–X-Leucine which serves as the binding site for the PDZ proteins Na^+–H^+ exchange regulatory factor 1 (NHERF1; also known as EBP50 and SlC9A3R1) and closely related NHERF2 (Hall et al., 1998a; Hall et al., 1998b). Agonist stimulation of the β_2AR leads to the recruitment of NHERF1 and a modulation of Na^+–H^+ exchanger type 3 (NHE3; also known as SlC9A3) activity. Disruption of β_2AR PDZ domain ablates NHERF1 binding and the effect of Na/H exchange, without affecting β_2AR coupling to Gs. Overexpression of GRK5 inhibits NHERF binding to the β_2AR whereas GRK6A phosphorylates and regulates NHERF1 (Cao, Deacon, Reczek, Bretscher, & von Zastrow, 1999; Hall et al., 1999). Disruption of the β_2AR PDZ motif inhibits receptor recycling after isoproterenol-induced internalization and inhibits the receptor coupling to Gi in neonatal cardiomyocytes (Xiang & Kobilka, 2003). Recent studies have shown that NHERF1 and β-arrestin2 interact and further form a ternary complex with the 7TMR PTH1R (Klenk et al., 2010). However, whether NHERF1 affects the β-arrestin-biased signaling via the PTH1R that was recently reported to play a role in anabolic bone formation

(Gesty-Palmer et al., 2010) or modulates β-arrestin-biased signaling via the βARs remain to be seen.

C. *Ubiquitin Specific Protease 33* (USP33)

As aforementioned, in addition to receptor phosphorylation, agonist-stimulated ubiquitination of β-arrestins also governs the stability of receptor–β-arrestin interactions. Earlier studies showed that β-arrestin ubiquitination is crucial for both its endocytic and signaling functions. Importantly, the kinetics of β-arrestin deubiquitination correlates with the dissociation of β-arrestins from activated receptors, suggesting that ubiquitin specific protease(s) (USP(s)) that reverse protein ubiquitination might play specific regulatory roles in 7TMR endocytosis and signal transduction. Recently the reciprocal roles of ubiquitination and deubiquitination of β-arrestin2 in regulating 7TMR trafficking and signaling have further become evident upon the discovery of a novel β-arrestin interaction with USP33. Stimulation of the β_2AR (that binds β-arrestin transiently at the plasma membrane) induces transient ubiquitination of β-arrestin2 mediated by Mdm2 and subsequently promotes association of β-arrestin with the deubiquitinase, USP33 (Fig. 4). This interaction facilitates the deubiquitination of β-arrestin leading to its dissociation from the β_2AR. In contrast, stable β-arrestin-binders such as the V2R (that forms stable complexes with β-arrestin and localize on endosomes) promote a β-arrestin conformation that does not favor the association of USP33 with β-arrestin. Depletion of USP33 prolongs the interaction between β-arrestin and the β_2AR allowing sustained signaling (Shenoy et al., 2009). This implies that during β_2AR signaling, USP33 plays a regulatory role by dissolving the receptor–arrestin signalosome by promoting deubiquitination and dissociation of β-arrestin2. In this context, USP33 functions much like an "antagonist" to inhibit β-arrestin-dependent ERK activation and regulates the extent of downstream signaling (Fig. 4).

Quite interestingly, USP33 and its homolog USP20 also interact with the β_2AR at the cytoplasmic domains and reverse receptor ubiquitination, which directs β_2AR trafficking and lysosomal degradation (Berthouze, Venkataramanan, Li, & Shenoy, 2009; Shenoy et al., 2001). β_2AR binds these USPs constitutively and agonist activation diminishes the interaction, whereas a simultaneous increase in β-arrestin–USP33 binding is observed. Thus, agonist-stimulation induces a reciprocal pattern of USP33 interaction with the β_2AR and β-arrestin2: dissociation of USP33 from the β_2AR and association of USP33 with β-arrestin2. The net result is initiation of β_2AR ubiquitination and concomitant reversal of β-arrestin ubiquitination leading to dissociation of β-arrestin from the internalized ubiquitinated β_2AR. Thus, USP20 and 33 function as in a tag team to separate internalizing β_2AR and β-arrestin leading to a tight regulation and balancing of signaling and internalization processes.

VII. FUTURE PERSPECTIVES

The multifunctional β-arrestins have emerged as important signal transducers for various 7TMRs including the βARs. β-arrestin-dependent signaling can be evoked at 7TMRs independent of the canonical G protein signaling, and it appears that each of these signaling modes elicits distinct biochemical and physiological effects. The discovery that ligands at the 7TMRs could be biased toward either the canonical G protein or novel β-arrestin signaling pathways, each having its own functional consequences has important implications for developing drugs targeting 7TMRs and pharmaceutical discovery platforms are being modified to fit these new paradigms. Additionally, recent breakthroughs in solving crystal structures of the βARs, will also influence structure-based drug design, which until recently were dependent on homology models based on the visual rhodopsin structure (Palczewski et al., 2000; Rasmussen et al., 2007; Warne et al., 2008).

The cardioprotective effects of β-arrestin-dependent signaling mediated by the β_1AR–EGFR complex and the unique signaling induced by carvedilol via β-arrestin, suggest the exciting possibility that compounds that block the deleterious G protein signaling in the heart and promote β-arrestin-dependent signaling at the βARs could prove to be a new class of drugs to treat heart failure. G protein biased drugs could also have their own advantages: for example, treatment of asthma could be more effective with βAR agonists that do not engage GRKs and β-arrestins, thus avoiding tachyphylaxis. A few β-arrestin-biased ligands for other 7TMRs have been identified and shown to have beneficial effects in animal models. On the other hand, our understanding about the exact molecular mechanisms governing β-arrestin-mediated signaling and the physiological roles of its multitude of protein partners is currently limited. Future studies geared at an in depth understanding of the various molecular interactions and their roles in β-arrestin-biased signaling combined with physiological studies in animal models will help to precisely target 7TMRs and develop new drugs with minimal side effects.

References

Ahn, S., Shenoy, S. K., Wei, H., & Lefkowitz, R. J. (2004). Differential kinetic and spatial patterns of beta-arrestin and G protein-mediated ERK activation by the angiotensin II receptor. *J Biol Chem, 279*(34), 35518–35525.

Azzi, M., Charest, P. G., Angers, S., Rousseau, G., Kohout, T., & Bouvier, M., et al. (2003). Beta-arrestin-mediated activation of MAPK by inverse agonists reveals distinct active conformations for G protein-coupled receptors. *Proc Natl Acad Sci USA, 100*(20), 11406–11411.

Baker, J. G. (2010). The selectivity of beta-adrenoceptor agonists at human beta1-, beta2- and beta3-adrenoceptors. *Br J Pharmacol, 160*(5), 1048–1061.

Barber, D. L., Ganz, M. B., Bongiorno, P. B., & Strader, C. D. (1992). Mutant constructs of the beta-adrenergic receptor that are uncoupled from adenylyl cyclase retain functional activation of Na–H exchange. *Mol Pharmacol, 41*(6), 1056–1060.

Barthet, G., Carrat, G., Cassier, E., Barker, B., Gaven, F., & Pillot, M., et al. (2009). Beta-arrestin1 phosphorylation by GRK5 regulates G protein-independent 5-HT4 receptor signalling. *Embo J, 28*(18), 2706–2718.

Berthouze, M., Venkataramanan, V., Li, Y., & Shenoy, S. K. (2009). The deubiquitinases USP33 and USP20 coordinate beta2 adrenergic receptor recycling and resensitization. *Embo J, 28*(12), 1684–1696.

Breit, A., Lagace, M., & Bouvier, M. (2004). Hetero-oligomerization between beta2- and beta3-adrenergic receptors generates a beta-adrenergic signaling unit with distinct functional properties. *J Biol Chem, 279*(27), 28756–28765.

Cao, T. T., Deacon, H. W., Reczek, D., Bretscher, A., & von Zastrow, M. (1999). A kinase-regulated PDZ-domain interaction controls endocytic sorting of the beta2-adrenergic receptor. *Nature, 401* (6750), 286–290.

Charest, P. G., Terrillon, S., & Bouvier, M. (2005). Monitoring agonist-promoted conformational changes of beta-arrestin in living cells by intramolecular BRET. *EMBO Rep, 6*(4), 334–340.

Cheung, A. H., Huang, R. R., & Strader, C. D. (1992). Involvement of specific hydrophobic, but not hydrophilic, amino acids in the third intracellular loop of the beta-adrenergic receptor in the activation of Gs. *Mol Pharmacol, 41*(6), 1061–1065.

Chuang, D. M., & Costa, E. (1979). Evidence for internalization of the recognition site of beta-adrenergic receptors during receptor subsensitivity induced by (–)-isoproterenol. *Proc Natl Acad Sci USA, 76*(6), 3024–3028.

Daaka, Y., Luttrell, L. M., & Lefkowitz, R. J. (1997). Switching of the coupling of the beta2-adrenergic receptor to different G proteins by protein kinase A. *Nature, 390*(6655), 88–91.

DeFea, K. A., Zalevsky, J., Thoma, M. S., Dery, O., Mullins, R. D., & Bunnett, N. W. (2000). beta-arrestin-dependent endocytosis of proteinase-activated receptor 2 is required for intracellular targeting of activated ERK1/2. *J Cell Biol, 148*(6), 1267–1281.

DeWire, S. M., Ahn, S., Lefkowitz, R. J., & Shenoy, S. K. (2007). Beta-arrestins and cell signaling. *Annu Rev Physiol, 69*, 483–510.

Drake, M. T., Violin, J. D., Whalen, E. J., Wisler, J. W., Shenoy, S. K., & Lefkowitz, R. J. (2008). beta-arrestin-biased agonism at the beta2-adrenergic receptor. *J Biol Chem, 283*(9), 5669–5676.

Ferguson, S. S., Downey IIIrd, W. E., Colapietro, A. M., Barak, L. S., Menard, L., & Caron, M. G. (1996). Role of beta-arrestin in mediating agonist-promoted G protein-coupled receptor internalization. *Science, 271*(5247), 363–366.

Fredericks, Z. L., Pitcher, J. A., & Lefkowitz, R. J. (1996). Identification of the G protein-coupled receptor kinase phosphorylation sites in the human beta2-adrenergic receptor. *J Biol Chem, 271* (23), 13796–13803.

Freedman, N. J., Liggett, S. B., Drachman, D. E., Pei, G., Caron, M. G., & Lefkowitz, R. J. (1995). Phosphorylation and desensitization of the human beta 1-adrenergic receptor. Involvement of G protein-coupled receptor kinases and cAMP-dependent protein kinase. *J Biol Chem, 270*(30), 17953–17961.

Frielle, T., Collins, S., Daniel, K. W., Caron, M. G., Lefkowitz, R. J., & Kobilka, B. K. (1987). Cloning of the cDNA for the human beta 1-adrenergic receptor. *Proc Natl Acad Sci USA, 84*(22), 7920–7924.

Galandrin, S., & Bouvier, M. (2006). Distinct signaling profiles of beta1 and beta2 adrenergic receptor ligands toward adenylyl cyclase and mitogen-activated protein kinase reveals the pluridimensionality of efficacy. *Mol Pharmacol, 70*(5), 1575–1584.

Galandrin, S., Oligny-Longpre, G., Bonin, H., Ogawa, K., Gales, C., & Bouvier, M. (2008). Conformational rearrangements and signaling cascades involved in ligand-biased mitogen-activated protein kinase signaling through the beta1-adrenergic receptor. *Mol Pharmacol, 74* (1), 162–172.

Galandrin, S., Oligny-Longpre, G., & Bouvier, M. (2007). The evasive nature of drug efficacy: Implications for drug discovery. *Trends Pharmacol Sci, 28*(8), 423–430.

Gesty-Palmer, D., Flannery, P., Yuan, L., Corsino, L., Spurney, R., & Lefkowitz, R. J., et al. (2010). A beta-arrestin-biased agonist of the parathyroid hormone receptor (PTH1R) promotes bone formation independent of G protein activation. *Sci Transl Med, 1*(1), 1ra1.

Gong, K., Li, Z., Xu, M., Du, J., Lv, Z., & Zhang, Y. (2008). A novel protein kinase A-independent, beta-arrestin-1-dependent signaling pathway for p38 mitogen-activated protein kinase activation by beta2-adrenergic receptors. *J Biol Chem, 283*(43), 29028–29036.

Gurevich, V. V., & Gurevich, E. V. (2004). The molecular acrobatics of arrestin activation. *Trends Pharmacol Sci, 25*(2), 105–111.

Hall, R. A., Ostedgaard, L. S., Premont, R. T., Blitzer, J. T., Rahman, N., & Welsh, M. J., et al. (1998a). A C-terminal motif found in the beta2-adrenergic receptor, P2Y1 receptor and cystic fibrosis transmembrane conductance regulator determines binding to the Na^+/H^+ exchanger regulatory factor family of PDZ proteins. *Proc Natl Acad Sci USA, 95*(15), 8496–8501.

Hall, R. A., Premont, R. T., Chow, C. W., Blitzer, J. T., Pitcher, J. A., & Claing, A., et al. (1998b). The beta2-adrenergic receptor interacts with the Na^+/H^+-exchanger regulatory factor to control Na^+/H^+ exchange. *Nature, 392*(6676), 626–630.

Hall, R. A., Spurney, R. F., Premont, R. T., Rahman, N., Blitzer, J. T., & Pitcher, J. A., et al. (1999). G protein-coupled receptor kinase 6A phosphorylates the Na(+)/H(+) exchanger regulatory factor via a PDZ domain-mediated interaction. *J Biol Chem, 274*(34), 24328–24334.

Han, M., Gurevich, V. V., Vishnivetskiy, S. A., Sigler, P. B., & Schubert, C. (2001). Crystal structure of beta-arrestin at 1.9 A: Possible mechanism of receptor binding and membrane Translocation. *Structure, 9*(9), 869–880.

Hershko, A., & Ciechanover, A. (1998). The ubiquitin system. *Annu Rev Biochem, 67*, 425–479.

Hirsch, J. A., Schubert, C., Gurevich, V. V., & Sigler, P. B. (1999). The 2.8 A crystal structure of visual arrestin: A model for arrestin's regulation. *Cell, 97*(2), 257–269.

Inglese, J., Koch, W. J., Caron, M. G., & Lefkowitz, R. J. (1992). Isoprenylation in regulation of signal transduction by G-protein-coupled receptor kinases. *Nature, 359*(6391), 147–150.

Kenakin, T., & Miller, L. J. (2010). Seven transmembrane receptors as shape shifting proteins: The impact of allosteric modulation and functional selectivity on new drug discovery. *Pharmacol Rev, 62*(2), 265–304.

Kim, I. M., Tilley, D. G., Chen, J., Salazar, N. C., Whalen, E. J., & Violin, J. D., et al. (2008). Beta-blockers alprenolol and carvedilol stimulate beta-arrestin-mediated EGFR transactivation. *Proc Natl Acad Sci USA, 105*(38), 14555–14560.

Kim, Y. M., Barak, L. S., Caron, M. G., & Benovic, J. L. (2002). Regulation of arrestin-3 phosphorylation by casein kinase II. *J Biol Chem, 277*(19), 16837–16846.

Klenk, C., Vetter, T., Zurn, A., Vilardaga, J. P., Friedman, P. A., & Wang, B., et al. (2010). Formation of a ternary complex among NHERF1, beta-arrestin, and parathyroid hormone receptor. *J Biol Chem, 285*(39), 30355–30362.

Krupnick, J. G., Goodman Jr., O. B., Keen, J. H., & Benovic, J. L. (1997). Arrestin/clathrin interaction. Localization of the clathrin binding domain of nonvisual arrestins to the carboxy terminus. *J Biol Chem, 272*(23), 15011–15016.

Kuo, F. T., Lu, T. L., & Fu, H. W. (2006). Opposing effects of beta-arrestin1 and beta-arrestin2 on activation and degradation of Src induced by protease-activated receptor 1. *Cell Signal, 18*(11), 1914–1923.

Lagerstrom, M. C., & Schioth, H. B. (2008). Structural diversity of G protein-coupled receptors and significance for drug discovery. *Nat Rev Drug Discov, 7*(4), 339–357.

Lefkowitz, R. J., & Shenoy, S. K. (2005). Transduction of receptor signals by beta-arrestins. *Science, 308*(5721), 512–517.

Liggett, S. B., Freedman, N. J., Schwinn, D. A., & Lefkowitz, R. J. (1993). Structural basis for receptor subtype-specific regulation revealed by a chimeric beta 3/beta 2-adrenergic receptor. *Proc Natl Acad Sci USA, 90*(8), 3665–3669.

Lin, F. T., Chen, W., Shenoy, S., Cong, M., Exum, S. T., & Lefkowitz, R. J. (2002). Phosphorylation of beta-arrestin2 regulates its function in internalization of beta(2)-adrenergic receptors. *Biochemistry, 41*(34), 10692–10699.

Lin, F. T., Krueger, K. M., Kendall, H. E., Daaka, Y., Fredericks, Z. L., & Pitcher, J. A., et al. (1997). Clathrin-mediated endocytosis of the beta-adrenergic receptor is regulated by phosphorylation/dephosphorylation of beta-arrestin1. *J Biol Chem, 272*(49), 31051–31057.

Lin, F. T., Miller, W. E., Luttrell, L. M., & Lefkowitz, R. J. (1999). Feedback regulation of beta-arrestin1 function by extracellular signal-regulated kinases. *J Biol Chem, 274*(23), 15971–15974.

Lohse, M. J., Engelhardt, S., & Eschenhagen, T. (2003). What is the role of beta-adrenergic signaling in heart failure? *Circ Res, 93*(10), 896–906.

Luttrell, L. M., Ferguson, S. S., Daaka, Y., Miller, W. E., Maudsley, S., & Della Rocca, G. J., et al. (1999). Beta-arrestin-dependent formation of beta2 adrenergic receptor–Src protein kinase complexes. *Science, 283*(5402), 655–661.

Luttrell, L. M., Roudabush, F. L., Choy, E. W., Miller, W. E., Field, M. E., & Pierce, K. L., et al. (2001). Activation and targeting of extracellular signal-regulated kinases by beta-arrestin scaffolds. *Proc Natl Acad Sci USA, 98*(5), 2449–2454.

Mangmool, S., Shukla, A. K., & Rockman, H. A. (2010). beta-Arrestin-dependent activation of Ca(2+)/calmodulin kinase II after beta(1)-adrenergic receptor stimulation. *J Cell Biol, 189*(3), 573–587.

Marion, S., Fralish, G. B., Laporte, S., Caron, M. G., & Barak, L. S. (2007). N-terminal tyrosine modulation of the endocytic adaptor function of the beta-arrestins. *J Biol Chem, 282*(26), 18937–18944.

Martin, N. P., Whalen, E. J., Zamah, M. A., Pierce, K. L., & Lefkowitz, R. J. (2004). PKA-mediated phosphorylation of the beta1-adrenergic receptor promotes Gs/Gi switching. *Cell Signal, 16*(12), 1397–1403.

Maudsley, S., Pierce, K. L., Zamah, A. M., Miller, W. E., Ahn, S., & Daaka, Y., et al. (2000). The beta (2)-adrenergic receptor mediates extracellular signal-regulated kinase activation via assembly of a multi-receptor complex with the epidermal growth factor receptor. *J Biol Chem, 275*(13), 9572–9580.

Milano, S. K., Kim, Y. M., Stefano, F. P., Benovic, J. L., & Brenner, C. (2006). Nonvisual arrestin oligomerization and cellular localization are regulated by inositol hexakisphosphate binding. *J Biol Chem, 281*(14), 9812–9823.

Miller, W. E., & Lefkowitz, R. J. (2001). Expanding roles for beta-arrestins as scaffolds and adapters in GPCR signaling and trafficking. *Curr Opin Cell Biol, 13*(2), 139–145.

Nantel, F., Bonin, H., Emorine, L. J., Zilberfarb, V., Strosberg, A. D., & Bouvier, M., et al. (1993). The human beta 3-adrenergic receptor is resistant to short term agonist-promoted desensitization. *Mol Pharmacol, 43*(4), 548–555.

Neves, S. R., Ram, P. T., & Iyengar, R. (2002). G protein pathways. *Science, 296*(5573), 1636–1639.

Nikolaev, V. O., Moshkov, A., Lyon, A. R., Miragoli, M., Novak, P., & Paur, H., et al. (2010). Beta2-adrenergic receptor redistribution in heart failure changes cAMP compartmentation. *Science, 327*(5973), 1653–1657.

Nobles, K. N., Guan, Z., Xiao, K., Oas, T. G., & Lefkowitz, R. J. (2007). The active conformation of beta-arrestin1: direct evidence for the phosphate sensor in the N-domain and conformational differences in the active states of beta-arrestins1 and -2. *J Biol Chem, 282*(29), 21370–21381.

Noma, T., Lemaire, A., Naga Prasad, S. V., Barki-Harrington, L., Tilley, D. G., & Chen, J., et al. (2007). Beta-arrestin-mediated beta1-adrenergic receptor transactivation of the EGFR confers cardioprotection. *J Clin Invest, 117*(9), 2445–2458.

O'Dowd, B. F., Hnatowich, M., Regan, J. W., Leader, W. M., Caron, M. G., & Lefkowitz, R. J. (1988). Site-directed mutagenesis of the cytoplasmic domains of the human beta 2-adrenergic receptor. Localization of regions involved in G protein-receptor coupling. *J Biol Chem, 263*(31), 15985–15992.

Oakley, R. H., Laporte, S. A., Holt, J. A., Barak, L. S., & Caron, M. G. (2001). Molecular determinants underlying the formation of stable intracellular G protein-coupled receptor-beta-arrestin complexes after receptor endocytosis. *J Biol Chem, 276*(22), 19452–19460.

Oakley, R. H., Laporte, S. A., Holt, J. A., Caron, M. G., & Barak, L. S. (2000). Differential affinities of visual arrestin, beta arrestin1, and beta arrestin2 for G protein-coupled receptors delineate two major classes of receptors. *J Biol Chem, 275*(22), 17201–17210.

Ozawa, K., Whalen, E. J., Nelson, C. D., Mu, Y., Hess, D. T., & Lefkowitz, R. J., et al. (2008). S-nitrosylation of beta-arrestin regulates beta-adrenergic receptor trafficking. *Mol Cell, 31*(3), 395–405.

Palczewski, K., Kumasaka, T., Hori, T., Behnke, C. A., Motoshima, H., & Fox, B. A., et al. (2000). Crystal structure of rhodopsin: A G protein-coupled receptor. *Science, 289*(5480), 739–745.

Palczewski, K., Pulvermuller, A., Buczylko, J., & Hofmann, K. P. (1991). Phosphorylated rhodopsin and heparin induce similar conformational changes in arrestin. *J Biol Chem, 266*(28), 18649–18654.

Patel, C. B., Noor, N., & Rockman, H. A. (2010). Functional selectivity in adrenergic and angiotensin signaling systems. *Mol Pharmacol, 78*(6), 983–992.

Pierce, K. L., Premont, R. T., & Lefkowitz, R. J. (2002). Seven-transmembrane receptors. *Nat Rev Mol Cell Biol, 3*(9), 639–650.

Pitcher, J. A., Freedman, N. J., & Lefkowitz, R. J. (1998). G protein-coupled receptor kinases. *Annu Rev Biochem, 67*, 653–692.

Pitcher, J. A., Inglese, J., Higgins, J. B., Arriza, J. L., Casey, P. J., & Kim, C., et al. (1992). Role of beta gamma subunits of G proteins in targeting the beta-adrenergic receptor kinase to membrane-bound receptors. *Science, 257*(5074), 1264–1267.

Premont, R. T., & Gainetdinov, R. R. (2007). Physiological roles of G protein-coupled receptor kinases and arrestins. *Annu Rev Physiol, 69*, 511–534.

Rasmussen, S. G., Choi, H. J., Rosenbaum, D. M., Kobilka, T. S., Thian, F. S., & Edwards, P. C., et al. (2007). Crystal structure of the human beta2 adrenergic G-protein-coupled receptor. *Nature, 450*(7168), 383–387.

Reiner, S., Ambrosio, M., Hoffmann, C., & Lohse, M. J. (2010). Differential signaling of the endogenous agonists at the {beta}2-adrenergic receptor. *J Biol Chem, 285*(46), 36188–36198.

Reiter, E., & Lefkowitz, R. J. (2006). GRKs and beta-arrestins: Roles in receptor silencing, trafficking and signaling. *Trends Endocrinol Metab, 17*(4), 159–165.

Rockman, H. A., Koch, W. J., & Lefkowitz, R. J. (2002). Seven-transmembrane-spanning receptors and heart function. *Nature, 415*(6868), 206–212.

Rojanathammanee, L., Harmon, E. B., Grisanti, L. A., Govitrapong, P., Ebadi, M., & Grove, B. D., et al. (2009). The 27-kDa heat shock protein confers cytoprotective effects through a beta 2-adrenergic receptor agonist-initiated complex with beta-arrestin. *Mol Pharmacol, 75*(4), 855–865.

Salahpour, A., & Masri, B. (2007). Experimental challenge to a 'rigorous' BRET analysis of GPCR oligomerization. *Nat Methods, 4*(8), 599–600 author reply 601.

Shenoy, S. K., Barak, L. S., Xiao, K., Ahn, S., Berthouze, M., & Shukla, A. K., et al. (2007). Ubiquitination of beta-arrestin links seven-transmembrane receptor endocytosis and ERK activation. *J Biol Chem, 282*(40), 29549–29562.

Shenoy, S. K., Drake, M. T., Nelson, C. D., Houtz, D. A., Xiao, K., & Madabushi, S., et al. (2006). beta-arrestin-dependent, G protein-independent ERK1/2 activation by the beta2 adrenergic receptor. *J Biol Chem, 281*(2), 1261–1273.

Shenoy, S. K., & Lefkowitz, R. J. (2003). Trafficking patterns of beta-arrestin and G protein-coupled receptors determined by the kinetics of beta-arrestin deubiquitination. *J Biol Chem, 278*(16), 14498–14506.

Shenoy, S. K., & Lefkowitz, R. J. (2005). Receptor-specific ubiquitination of beta-arrestin directs assembly and targeting of seven-transmembrane receptor signalosomes. *J Biol Chem, 280*(15), 15315–15324.

Shenoy, S. K., McDonald, P. H., Kohout, T. A., & Lefkowitz, R. J. (2001). Regulation of receptor fate by ubiquitination of activated beta 2-adrenergic receptor and beta-arrestin. *Science, 294*(5545), 1307–1313.

Shenoy, S. K., Modi, A. S., Shukla, A. K., Xiao, K., Berthouze, M., & Ahn, S., et al. (2009). Beta-arrestin-dependent signaling and trafficking of 7-transmembrane receptors is reciprocally regulated by the deubiquitinase USP33 and the E3 ligase Mdm2. *Proc Natl Acad Sci USA, 106*(16), 6650–6655.

Shiina, T., Kawasaki, A., Nagao, T., & Kurose, H. (2000). Interaction with beta-arrestin determines the difference in internalization behavior between beta1- and beta2-adrenergic receptors. *J Biol Chem, 275*(37), 29082–29090.

Shukla, A. K., Violin, J. D., Whalen, E. J., Gesty-Palmer, D., Shenoy, S. K., & Lefkowitz, R. J. (2008). Distinct conformational changes in beta-arrestin report biased agonism at seven-transmembrane receptors. *Proc Natl Acad Sci USA, 105*(29), 9988–9993.

Smith, F. D., Oxford, G. S., & Milgram, S. L. (1999). Association of the D2 dopamine receptor third cytoplasmic loop with spinophilin, a protein phosphatase-1-interacting protein. *J Biol Chem, 274*(28), 19894–19900.

Stadel, J. M., Strulovici, B., Nambi, P., Lavin, T. N., Briggs, M. M., & Caron, M. G., et al. (1983). Desensitization of the beta-adrenergic receptor of frog erythrocytes. Recovery and characterization of the down-regulated receptors in sequestered vesicles. *J Biol Chem, 258*(5), 3032–3038.

Storez, H., Scott, M. G., Issafras, H., Burtey, A., Benmerah, A., & Muntaner, O., et al. (2005). Homo- and hetero-oligomerization of beta-arrestins in living cells. *J Biol Chem, 280*(48), 40210–40215.

Tilley, D. G., Kim, I. M., Patel, P. A., Violin, J. D., & Rockman, H. A. (2009). beta-Arrestin mediates beta1-adrenergic receptor-epidermal growth factor receptor interaction and downstream signaling. *J Biol Chem, 284*(30), 20375–20386.

Vaidehi, N., & Kenakin, T. (2010). The role of conformational ensembles of seven transmembrane receptors in functional selectivity. *Curr Opin Pharmacol, 10*(6), 775–781.

Violin, J. D., & Lefkowitz, R. J. (2007). Beta-arrestin-biased ligands at seven-transmembrane receptors. *Trends Pharmacol Sci, 28*(8), 416–422.

Vishnivetskiy, S. A., Schubert, C., Climaco, G. C., Gurevich, Y. V., Velez, M. G., & Gurevich, V. V. (2000). An additional phosphate-binding element in arrestin molecule. Implications for the mechanism of arrestin activation. *J Biol Chem, 275*(52), 41049–41057.

Wang, Q., & Limbird, L. E. (2002). Regulated interactions of the alpha 2A adrenergic receptor with spinophilin, 14-3-3zeta, and arrestin 3. *J Biol Chem, 277*(52), 50589–50596.

Wang, Q., & Limbird, L. E. (2007). Regulation of alpha2AR trafficking and signaling by interacting proteins. *Biochem Pharmacol, 73*(8), 1135–1145.

Wang, Q., Lu, R., Zhao, J., & Limbird, L. E. (2006). Arrestin serves as a molecular switch, linking endogenous alpha2-adrenergic receptor to SRC-dependent, but not SRC-independent, ERK activation. *J Biol Chem, 281*(36), 25948–25955.

Wang, Q., Zhao, J., Brady, A. E., Feng, J., Allen, P. B., & Lefkowitz, R. J., et al. (2004). Spinophilin blocks arrestin actions *in vitro* and in vivo at G protein-coupled receptors. *Science, 304*(5679), 1940–1944.

Wang, Y., De Arcangelis, V., Gao, X., Ramani, B., Jung, Y. S., & Xiang, Y. (2008). Norepinephrine- and epinephrine-induced distinct beta2-adrenoceptor signaling is dictated by GRK2 phosphorylation in cardiomyocytes. *J Biol Chem, 283*(4), 1799–1807.

Warne, T., Serrano-Vega, M. J., Baker, J. G., Moukhametzianov, R., Edwards, P. C., & Henderson, R., et al. (2008). Structure of a beta1-adrenergic G-protein-coupled receptor. *Nature, 454*(7203), 486–491.

Wisler, J. W., DeWire, S. M., Whalen, E. J., Violin, J. D., Drake, M. T., & Ahn, S., et al. (2007). A unique mechanism of beta-blocker action: Carvedilol stimulates beta-arrestin signaling. *Proc Natl Acad Sci USA, 104*(42), 16657–16662.

Woo, A. Y., Wang, T. B., Zeng, X., Zhu, W., Abernethy, D. R., & Wainer, I. W., et al. (2009). Stereochemistry of an agonist determines coupling preference of beta2-adrenoceptor to different G proteins in cardiomyocytes. *Mol Pharmacol, 75*(1), 158–165.

Xiang, Y., & Kobilka, B. (2003). The PDZ-binding motif of the beta2-adrenoceptor is essential for physiologic signaling and trafficking in cardiac myocytes. *Proc Natl Acad Sci USA, 100*(19), 10776–10781.

Xiao, K., Shenoy, S. K., Nobles, K., & Lefkowitz, R. J. (2004). Activation-dependent conformational changes in {beta}-arrestin 2. *J Biol Chem, 279*(53), 55744–55753.

Xiao, K., Sun, J., Kim, J., Rajagopal, S., Zhai, B., & Villen, J., et al. (2010). Global phosphorylation analysis of beta-arrestin-mediated signaling downstream of a seven transmembrane receptor (7TMR). *Proc Natl Acad Sci USA, 107*(34), 15299–15304.

Zamah, A. M., Delahunty, M., Luttrell, L. M., & Lefkowitz, R. J. (2002). Protein kinase A-mediated phosphorylation of the beta 2-adrenergic receptor regulates its coupling to Gs and Gi. Demonstration in a reconstituted system. *J Biol Chem, 277*(34), 31249–31256.

Zeitoun, O., Santos, N. M., Gardner, L. A., White, S. W., & Bahouth, S. W. (2006). Mutagenesis within helix 6 of the human beta1-adrenergic receptor identifies Lysine324 as a residue involved in imparting the high-affinity binding state of agonists. *Mol Pharmacol, 70*(3), 838–850.

CHAPTER 4

Anterograde Trafficking of Nascent α_{2B}-Adrenergic Receptor: Structural Basis and Roles of Small GTPases

Guangyu Wu

Department of Pharmacology and Toxicology, Georgia Health Sciences University, Augusta, GA, USA

I. Overview
II. Introduction
III. The Structural Basis of α_{2B}-AR Export from the ER and the Golgi
 A. The C-terminal $F(x)_6IL$ Motif in α_{2B}-AR Export from the ER
 B. The L48 Residue in α_{2B}-AR Exit from the ER
 C. The N-terminal Y12/S13 Motif in α_{2B}-AR Export from the Golgi
 D. The ICL3 in the Basolateral Targeting of Three α_2-ARs
IV. The Role of Small GTPases in the Export Trafficking of α_{2B}-AR
 A. Sar1 in α_{2B}-AR Exit from the ER
 B. ARF GTPases in α_{2B}-AR Exit from the ER and the Golgi
 C. Rab GTPases in the ER–Golgi-Cell Surface Transport of α_{2B}-AR
V. Conclusions and Perspectives
 Acknowledgment
 References

I. OVERVIEW

Similar to many other G protein-coupled receptors (GPCRs), the functionality of α_{2B}-adrenergic receptor (α_{2B}-AR) is dependent on its proper transport to the cell surface. However, compared with the well-understood endocytic and recycling pathways, the molecular mechanism underlying the anterograde trafficking of newly synthesized α_{2B}-AR from the endoplasmic reticulum (ER) through the Golgi to the plasma membrane remains poorly elucidated. Recent studies have revealed that α_{2B}-AR targeting to the cell surface is a highly regulated process, which is coordinated by many intrinsic determinants and regulatory proteins. This chapter will review the roles of recently identified motifs and the Sar1/ARF and Rab GTPases in α_{2B}-AR exit from intracellular organelles and transport from the ER to the cell surface.

Current Topics in Membranes, Volume 67 0065-230X/10 $35.00

DOI: 10.1016/B978-0-12-384921-2.0005

II. INTRODUCTION

As the largest superfamily of cell surface receptors, G protein-coupled receptors (GPCRs) regulate a variety of cellular functions through coupling to heterotrimeric G proteins, which in turn modulate the activity of downstream effectors, including adenylyl cyclases, phospholipases, protein kinases, and ion channels (Hanyaloglu & von Zastrow, 2008; Pierce, Premont, & Lefkowitz, 2002; Rosenbaum, Rasmussen, & Kobilka, 2009). The life of GPCRs begins in the endoplasmic reticulum (ER) where they are synthesized. Once correctly folded and properly assembled, GPCRs are able to pass the ER quality control system and exit from the ER, beginning the journey of intracellular trafficking (Dong, Filipeanu, Duvernay, & Wu, 2007). The nascent receptors then pass many intracellular compartments, including the ER-Golgi intermediate compartment (ERGIC), the Golgi, and the trans-Golgi network (TGN), *en route* to the cell surface, which is the functional destination for most GPCRs. An important feature for the cell surface GPCRs is that they may undergo internalization in response to sustained agonist stimulation during which the receptors are transported from the plasma membrane to endosomes. The internalized receptors in endosomes may be sorted to different destinations, including the recycling pathway for return to the cell surface, the lysosomal compartment for degradation, and the Golgi for retrograde transport. Therefore, the balance of these dynamic intracellular trafficking events dictates the amount of the receptors at the plasma membrane, which in turn controls the magnitude of cellular response to a given extracellular signal. Over the past decades, most studies on the intracellular trafficking of GPCRs have focused on the endocytosis and recycling processes. These studies have not only greatly advanced our knowledge about the mechanisms of GPCR trafficking but also revealed physiological functions for the trafficking in regulating receptor signal propagation and in the pathogenesis of human diseases (Hanyaloglu & von Zastrow, 2008; Marchese, Chen, Kim, & Benovic, 2003; Moore, Milano, & Benovic, 2007; Tan, Brady, Nickols, Wang, & Limbird, 2004; Wu, Benovic, Hildebrandt, & Lanier, 1998; Wu, Krupnick, Benovic, & Lanier, 1997; Xia, Gray, Compton-Toth, & Roth, 2003). In contrast, the molecular mechanism underlying anterograde transport of nascent GPCRs from the ER through the Golgi apparatus to the cell surface and the role of export traffic in the functional regulation of the receptors have just begun to be elucidated.

The progress achieved over the past few years indicates that, similar to the endocytic and recycling pathways, the ER-to-cell surface movement of GPCRs is a highly regulated, dynamic process, which is orchestrated by structural features of the receptors and many regulatory proteins. First, it has been demonstrated that ER export is a rate-limiting step for the cell surface transport of GPCRs (Petaja-Repo, Hogue, Laperriere, Walker, & Bouvier, 2000). Second, a number of studies have recently identified highly conserved hydrophobic sequences, which are required for GPCR export from the ER (Bermak, Li, Bullock, & Zhou, 2001; Carrel, Hamon, & Darmon, 2006; Robert, Clauser, Petit, & Ventura, 2005; Schulein et al., 1998). These studies suggest that, similar to many other plasma membrane proteins, GPCR exit from the ER may be dictated by specific export motifs. Third, cell surface transport of GPCRs is modulated by direct interactions with multiple regulatory proteins such as the receptor activity modifying proteins (RAMPs), the ER chaperone proteins, and accessory proteins which may behave as chaperones/escort proteins, stabilizing receptor conformation and promoting their delivery to the plasma membrane (Dong et al., 2007). Fourth, dimerization (homo- and hetero-dimerization) may also participate in the regulation of GPCR export to the cell surface, likely through influencing their correct folding or assembly in the ER (Bouvier, 2001; Salahpour, Angers, Mercier, Lagace, Marullo, & Bouvier, 2004; Zhang et al., 2009; Zhou, Filipeanu, Duvernay, & Wu, 2006). Finally, GPCR transport from the ER through the Golgi to the cell surface is mediated through distinct pathways, in which the Ras-like Rab GTPases play a crucial role (Dong & Wu, 2007; Filipeanu, Zhou, Claycomb, & Wu, 2004; Filipeanu, Zhou, Fugetta, & Wu, 2006; Wu, Zhao, & He, 2003).

My laboratory has used adrenergic and angiotensin II receptors as representatives to search for the players that control the cell surface targeting of the receptors by addressing two important questions: Are there conserved structural elements in GPCRs which function as motifs dictating their exit from intracellular compartments? And could the export trafficking of GPCRs be selectively regulated by well-defined transport regulators? Over the past several years, we have identified several highly conserved residues essential for the receptors to exit from the ER and the Golgi apparatus (Dong & Wu, 2006; Duvernay et al., 2009a, 2009b; Duvernay, Zhou, & Wu, 2004; Zhou et al., 2006). We have also demonstrated that small GTPases, specifically the Rab and Sar1/ARF subfamilies, may selectively or differentially modulate the anterograde traffic of GPCRs along the secretory pathway (Dong & Wu, 2007; Dong et al., 2010a, 2010b; Dong, Zhou, Fugetta, Filipeanu, & Wu, 2008; Filipeanu et al., 2004, 2006; Wu et al., 2003; Zhang et al., 2009).

In this chapter, we will review the role of structural determinants and small GTPases, specifically the Sar1/ARF and Rab subfamilies, in the regulation of α_{2B}-AR exit from intracellular compartments and transport from the ER to the cell surface. There are three α_2-AR subtypes, designated as α_{2A}-AR, α_{2B}-AR, and α_{2C}-AR, all of which play an important role in regulating sympathetic nervous system, both peripherally and centrally. All three α_2-ARs have similar structural features: whereas the third intracellular loop (ICL3) is quite large with more than 170 amino acid residues, other loops and the termini are relatively short with less than 25 residues.

III. THE STRUCTURAL BASIS OF α_{2B}-AR EXPORT FROM THE ER AND THE GOLGI

Although all three α_2-ARs have a strong similarity in their structures and functions, they are markedly different in their abilities to move to the cell surface. In particular, α_{2C}-AR transports to the cell surface in a cell type- and temperature-dependent fashion. For example, α_{2C}-AR is able to efficiently move to the plasma membrane in some neuroendocrine cells, such as PC12 and AtT20 cells, in a temperature-independent manner, whereas the majority of α_{2C}-AR is arrested in the intracellular compartments including the ER and the Golgi, unable to transport to the cell surface at 37°C in fibroblasts and vascular smooth muscle cells, and reducing temperature may facilitate the cell surface transport of the intracellularly accumulated receptors (Bailey, Eid, Mitra, Flavahan, & Flavahan, 2004; Daunt, Hurt, Hein, Kallio, Feng, & Kobilka, 1997; Jeyaraj, Chotani, Mitra, Gregg, Flavahan, & Morrison, 2001). Such an effect of lowering temperature on α_{2C}-AR translocation may contribute to Raynaud syndrome which is characterized by enhanced peripheral vasoconstriction during cold exposure or emotional stress and can be ameliorated by using α_2-AR antagonists. Interestingly, it has been demonstrated that the intracellular accumulation of α_{2C}-AR may be under the control of multiple arginine residues in the C-terminus and hydrophobic residues in the N-terminus which may function as ER retention motifs trapping the receptor in the ER (Angelotti, Daunt, Shcherbakova, Kobilka, & Hurt, 2010; Ma et al., 2001).

In contrast to α_{2C}-AR, both α_{2A}-AR and α_{2B}-AR are normally expressed at the cell surface and recent studies have demonstrated that their transport from the ER to the cell surface is controlled by multiple highly conserved specific motifs. Specifically, the F436, I443, and L444 residues [F(x)$_6$IL motif] in the C-terminus and a single L48 residue in the first intracellular loop (ICL1) are required for α_{2B}-AR to exit from the ER (Duvernay et al., 2004, 2009b, 2009b), whereas the Y12/S13 motif located in the N-terminus is crucial for α_{2B}-AR export from the Golgi (Dong & Wu, 2006). In addition, the ICL3 may possess signals for the retention of the receptor in the basolateral subdomain in polarized cells (Brady, Wang, Colbran, Allen, Greengard, & Limbird, 2003; Edwards & Limbird, 1999; Keefer, Kennedy, & Limbird, 1994; Keefer & Limbird, 1993; Prezeau, Richman, Edwards, & Limbird, 1999; Saunders, Keefer, Bonner, & Limbird, 1998; Saunders & Limbird, 2000; Wozniak & Limbird, 1996).

A. *The C-terminal F(x)$_6$IL Motif in α_{2B}-AR Export from the ER*

Protein export from the ER is a selective process that may be dictated by short, linear sequences called ER export motifs (Kappeler, Klopfenstein, Foguet,

Paccaud, & Hauri, 1997; Nishimura & Balch, 1997; Nishimura et al., 1999; Nishimura, Plutner, Hahn, & Balch, 2002; Nufer et al., 2002; Nufer, Kappeler, Guldbrandsen, & Hauri, 2003; Votsmeier & Gallwitz, 2001; Wendeler, Paccaud, & Hauri, 2007). Of various ER export motifs identified, the diacidic motifs have been found in the cytoplasmic C-termini of several membrane proteins such as vesicular stomatitis virus glycoprotein (VSVG), cystic fibrosis transmembrane conductance regulator, and potassium channels (KAT1, TASK-3, and Kir2.1) (Ma et al., 2001; Nishimura & Balch, 1997; Nishimura et al., 1999; Sevier, Weisz, Davis, & Machamer, 2000; Wang et al., 2004b; Zuzarte, Rinne, Schlichthorl, Schubert, Daut, & Preisig-Muller, 2007) and demonstrated to function as ER export motifs. Interestingly, export function of the di-acidic motifs is mediated through their interaction with components of COPII transport vesicles, particularly Sec24 subunits. This interaction results in the concentration of cargo in ER exit sites and facilitates cargo recruitment onto the vesicles (Farhan, Reiterer, Korkhov, Schmid, Freissmuth, & Sitte, 2007).

The C-terminal tails of GPCRs consist of a putative amphipathic 8th α-helix in the membrane-proximal region and a nonstructural membrane-distal region. The function of the C-terminus, particularly the membrane-proximal 8th α-helix portion, in regulating cell surface transport of the receptors has been described for a number of GPCRs including angiotensin II type 1 receptor (AT1R), rhodopsin, vasopressin V2 receptor, dopamine D1 receptor, adenosine A1 receptor, melanin-concentrating hormone receptor 1, and luteinizing hormone/choriogonadotropin receptor (Duvernay et al., 2004; Gaborik, Mihalik, Jayadev, Jagadeesh, Catt, & Hunyady, 1998; Heymann & Subramaniam, 1997; Pankevych, Korkhov, Freissmuth, & Nanoff, 2003; Rodriguez, Xie, Wang, Collison, & Segaloff, 1992; Tetsuka, Saito, Imai, Doi, & Maruyama, 2004). We first demonstrated that deletion of the entire C-terminus almost abolished the cell surface expression of α_{2B}-AR and subsequent mutagenesis of individual residues in the C-terminus revealed F436 and I443/L444 residues in the membrane-proximal portion essential for α_{2B}-AR transport to the cell surface (Duvernay et al., 2004) (Fig. 1A). Consistent with the lack of cell surface expression, the mutated receptor lacking the F436 and I443/L444 was unable to initiate downstream signaling, such as activation of ERK1/2 (Duvernay et al., 2004). Further subcellular distribution analysis showed that the mutated receptors were strongly accumulated in the ER, suggestive of defective ER export. Interestingly, the function of F436 and I443/L444 in mediating α_{2B}-AR export cannot be fully substituted by any other hydrophobic residues (Duvernay et al., 2009b). These data indicate that the $F(x)_6IL$ motif modulates α_{2B}-AR export at the level of the ER and this function is mediated by its unique properties.

Consistent with the role of the $F(x)_6IL$ motif in α_{2B}-AR transport, several similar motifs, such as the $E(x)_3LL$, $FN(x)_2LL(x)_3L$, and $F(x)_3F(x)_3F$ motifs,

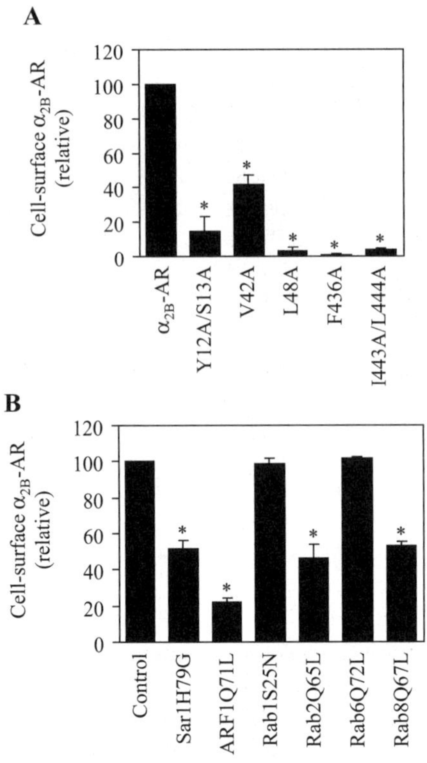

FIGURE 1 Effect of mutating specific residues (A) and expressing small GTPase mutants (B) on the cell surface expression of α_{2B}-AR as measured by intact cell ligand binding. (A) α_{2B}-AR and its mutants in which specific residues were mutated to alanines were transiently transfected into HEK293 cells. (B) α_{2B}-AR was transfected with or without individual small GTPase mutants into HEK293 cells. The cell surface expression of α_{2B}-AR was measured by intact cell ligand binding by using [^{3}H]-RX821002 at a concentration of 20 nM. *, $p < 0.05$ versus wild type α_{2B}-AR (A) or control (B). (The data are adapted from the references Dong & Wu, 2006; Duvernay et al., 2009a, 2009b).

have been identified to control the ER-to-cell surface transport of other GPCRs (Bermak et al., 2001; Robert et al., 2005; Schulein et al., 1998). Importantly, the $F(x)_6LL$ motif (where x can be any residues and L leucine or isoleucine) is highly conserved in the membrane-proximal C-termini of many family A GPCRs (Duvernay et al., 2004) and indeed, this motif is also required for ER export of several other GPCRs, including α_{1B}-AR, β_2-AR, and AT1R (Duvernay et al., 2009b).

To further provide insights into how the $F(x)_6IL$ motif controls α_{2B}-AR transport, we analyzed the structural features of the motif by homology modeling based on the newly published crystal structure of β_2-AR. F436 residue is buried within the hydrophobic core of the receptor and in close proximity to V42 in the first transmembrane domain and mutation of V42 also significantly impairs α_{2B}-AR export to the cell surface (Fig. 1A). Furthermore, the defect in the transport of the F436A mutant can be partially rescued by a number of treatments, such as chemical chaperones and lowing temperature, and the mutant has enhanced abilities to bind to the chaperone proteins calnexin and calreticulin. These data suggest that the F436 residue is likely involved in the regulation of proper α_{2B}-AR folding in the ER, which is mediated through intramolecular interactions with other hydrophobic residues, such as V42 in the first transmembrane domain, enabling the receptor to pass the ER quality control and to export from the ER.

How I443/L444 residues influence α_{2B}-AR export from the ER remains unknown. The dileucine-based motifs have been demonstrated to be involved in both endocytosis and basolateral delivery. The fact that the branched carbon side chains of the I443/L444 residues are exposed to the cytosolic space suggests that they are capable of providing a docking site for other proteins (Duvernay et al., 2009b). Indeed, our recent studies have demonstrated that Rab8 GTPase modulates β_2-AR transport from the TGN, which is likely mediated through its physical association with the C-terminal dileucine motif of the receptor. However, mutation of the I443/L444 residues did not alter α_{2B}-AR interaction with Rab8 (Dong et al., 2010a). Therefore, to search for proteins interacting with the dileucine motif in the cytoplasm, particularly components of transport machinery or other trafficking-related regulatory proteins, will help to elucidate the mechanism of the I443/L444 motif in α_{2B}-AR export from the ER.

B. The L48 Residue in α_{2B}-AR Exit from the ER

The ICL1 of α_{2B}-AR is very short, composed of only 12 amino acid residues. Similar to the C-terminus, the ICL1 is absolutely necessary for proper transport of α_{2B}-AR to the cell surface, as the ICL1-deleted receptor was accumulated in intracellular compartments and unable to transport to the cell surface (Duvernay et al., 2009a). Mutagenesis studies identified a single L48 residue essential for the cell surface transport of α_{2B}-AR (Duvernay et al., 2009a) (Fig. 1A) and the mutated receptor was very well co-localized with the ER marker DsRed2-ER (Fig. 2A), suggesting that L48 residue is involved in the regulation of α_{2B}-AR exit from the ER.

An isolated leucine residue in the center of the ICL1 is remarkably conserved among the class A GPCRs. About 85% of the family A GPCRs in human and 83% in all species contain a leucine residue in the center of ICL1

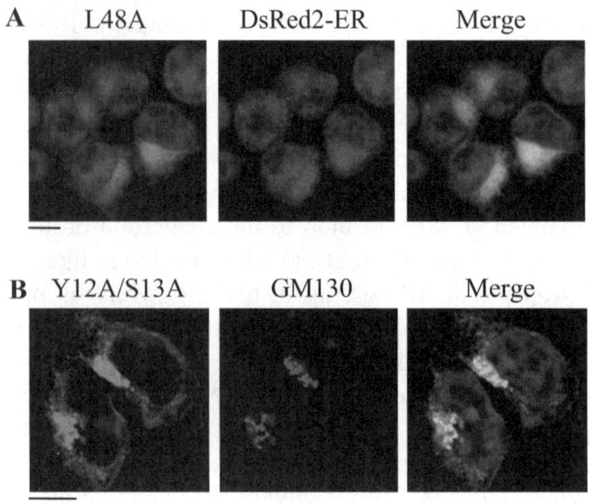

FIGURE 2 Colocalization of the α_{2B}-AR mutants L48A and Y12A/S13A with ER and Golgi markers, respectively. (A) Colocalization of the α_{2B}-AR mutant L48A with the ER marker DsRed2-ER. HEK293 cells were transfected with the GFP-tagged L48A mutant together with pDsRed2-ER and the subcellular distribution and colocalization of the receptor with DsRed2-ER were revealed by fluorescence microscopy. (B) Colocalization of the α_{2B}-AR Y12A/S13A mutant with the *cis*-Golgi marker GM130. HEK293 cells were transfected with Y12A/S13A and its co-localization with GM130 was revealed by fluorescence microscopy following staining with antibodies against GM130 at 1:50 dilution. Scale bars, 10 μm. (The data are adapted from the references Dong & Wu, 2006; Duvernay et al., 2009a).

(Duvernay et al., 2009a). Mutation of this conserved residue also significantly attenuated the cell surface expression of several other GPCRs, including β_1-AR, AT1R, and α_{1B}-AR (Duvernay et al., 2009a). These data suggest that the single leucine residue in the ICL1 may be a common signal mediating the ER export of a number of GPCRs.

C. The N-terminal Y12/S13 Motif in α_{2B}-AR Export from the Golgi

Recent studies have demonstrated that, similar to exit from the ER, protein export from the Golgi/TGN is a selective process that may be dictated by specific export motifs. Newly synthesized proteins are sorted at the Golgi/TGN to be delivered to final cellular destinations, such as endosomes, lysosomes, and the plasma membrane. There are several well-defined endosomal sorting signals including tyrosine-based motifs (NPxY and YxxØ, where x can be any residue and Ø is a hydrophobic residue) and dileucine-based motifs ([D/E]xxxL[L/I] and DxxLL). Whereas YxxØ and [D/E]xxxL[L/I] motifs are recognized by the adaptor protein complexes, DxxLL is recognized by Golgi-localized γ-ear-containing ARF1-binding proteins (GGAs) (Hirst, Lui, Bright, Totty, Seaman, & Robinson, 2000; Puertollano, Aguilar, Gorshkova, Crouch, & Bonifacino, 2001; Puertollano, Randazzo, Presley, Hartnell, & Bonifacino, 2001). These motifs function to sort protein transport from the TGN to the endosomal compartment (Boucher, Larkin, Brodeur, Gagnon, Theriault, & Lavoie, 2008; Chen, Yuan, & Lobel, 1997; Hou, Suzuki, Pessin, & Watson, 2006; Johnson & Kornfeld, 1992; Lori, Florencia, & Frederick, 2007).

The fact that G protein-coupled olfactory and chemokine receptors as well as the opsin mutant E150K are released from the ER, but accumulated in the Golgi (Gimelbrant, Haley, & McClintock, 2001; Venkatesan, Petrovic, Van Ryk, Locati, Weissman, & Murphy, 2002; Zhu et al., 2006) suggests that GPCR export from the Golgi and transport from the Golgi to the cell surface is a regulated process. We found that the N-terminus, specifically Y12 and S13 residues in the membrane-proximal N-terminal region, is absolutely required for the transport of α_{2B}-AR to the cell surface. Single and double substitution of the Y12/S13 motif significantly reduced the cell surface expression of α_{2B}-AR (Dong & Wu, 2006) (Fig. 1A). However, unlike the $F(x)_6IL$ and L48 mutants that were accumulated in the ER, the Y12/S13 motif mutants were retained in the Golgi apparatus (Dong & Wu, 2006) (Fig. 2B), suggesting that the Y12/S13 motif mediates α_{2B}-AR export at the level of the Golgi. The YS motif only exists in the membrane proximal N-termini of three α_2-AR family members and indeed, it exerts a similar function on α_{2A}-AR trafficking (Dong & Wu, 2006). Therefore, the YS motif may function as an export signal specifically modulating the Golgi export of the members of α_2-AR subfamily.

In addition to α_{2B}-AR, an important role for the N-terminus in the intracellular trafficking of GPCRs has been described for other GPCRs. For example, the deletion of the N-termini facilitates the cell surface export of α_{1D}-AR and α_{2C}-AR, suggesting that the N-termini may contain signals retaining the receptors in the ER (Angelotti et al., 2010; Hague, Chen, Pupo, Schulte, Toews, & Minneman, 2004). Taken together, these studies demonstrate that, similar to the C-termini, the N-termini may also contain signals modulating the export of GPCRs from intracellular compartments.

The Y12/S13 motif represents the first Golgi export motif identified in the GPCR superfamily. As the N-terminus is positioned towards the lumen of ER and Golgi during the export process, the YS motif is not able to directly interact with components of transport machinery in the cytoplasm. Furthermore, the fact that YS mutant receptors are able to exit from the ER to reach the Golgi compartment suggests that they are properly folded. Therefore, the defective transport is unlikely caused by misfolding. Further investigation is needed to clarify the molecular mechanism underlying the function of YS motif in the regulation of receptor export from the Golgi.

D. The ICL3 in the Basolateral Targeting of Three α_2-ARs

It has been well demonstrated that the ICL3 is involved in the regulation of receptor coupling to G proteins, phosphorylation, internalization, and signal termination (DeGraff, Gurevich, & Benovic, 2002; Jewell-Motz, Small, Theiss, & Liggett, 2000; Pao & Benovic, 2005; Small, Brown, Forbes, & Liggett, 2001; Wade, Lim, Lan, Chung, Nanamori, & Neubig, 1999; Wade, Scribner, Dalman, Taylor, & Neubig, 1996; Wang & Limbird, 2002; Wang et al., 2004a; Wu et al., 1997, 1998; Wu, Bogatkevich, Mukhin, Benovic, Hildebrandt, & Lanier, 2000). The role of the ICL3 in the localization and trafficking of three α_2-ARs have been extensively studied in polarized Madin–Darby canine kidney II (MDCKII) cells in the laboratory of Dr. Lee Limbird (Brady et al., 2003; Edwards & Limbird, 1999; Keefer et al., 1994; Prezeau et al., 1999; Saunders et al., 1998; Saunders & Limbird, 2000; Wozniak & Limbird, 1996). It has been demonstrated that three newly synthesized α_2-ARs use different pathways to target to the basolateral domain and have distinct retention profiles in MDCKII cells. Consistent with different transport abilities of the three α_2-ARs in some cell types, at steady state, both α_{2A}-AR and α_{2B}-ARs are almost exclusively located at the basolateral surface, while about half of α_{2C}-AR is localized at the basolateral membrane and another half in the intracellular compartments. More interestingly, it appears that α_{2A}-AR and α_{2B}-AR utilize different paths for their basolateral targeting. α_{2B}-AR is first *randomly* transported to both the apical and basolateral surfaces and then selectively retained at the basolateral domain, whereas α_{2A}-AR is directly delivered to the basolateral membrane. Despite the remarkable differences in basolateral targeting, three α_2-ARs exhibit comparable half-life of about 10–12 h at the basolateral domain (Wozniak & Limbird, 1996).

The ICL3 and the C-terminus are not involved in the regulation of direct basolateral delivery of α_{2A}-AR and indeed, the basolateral targeting information for α_{2A}-AR is identified in the membrane-embedded regions (Keefer et al., 1994; Keefer & Limbird, 1993; Saunders et al., 1998). However, removal of the ICL3 significantly facilitates the turnover of the cell surface α_{2A}-AR, shortening its half-life to about 4 h (Edwards & Limbird, 1999). This function of the ICL3 in stabilizing α_{2A}-AR and α_{2B}-AR at the basolateral surface is directly linked to its ability to physically associate with spinophilin (Brady et al., 2003; Richman, Brady, Wang, Hensel, Colbran, & Limbird, 2001). Taken together, these data suggest that the stabilization/retention of α_{2A}-AR and α_{2B}-AR at specific membrane domains is most likely mediated through ICL3 interactions with other proteins.

IV. THE ROLE OF SMALL GTPASE IN THE EXPORT TRAFFICKING OF α_{2B}-AR

The Ras-like small GTPase superfamily consists of more than 150 members and can be divided into Ras, Rho/Rac/Cdc42, Ran, Sar1/ARF and Rab subfamilies. The small GTPases in the Ras and Rho/Rac/Cdc42 subfamilies have been well documented to function as signaling proteins modulating gene expression, cell division, and cytoskeletal reorganization, the small GTPases in the Rab and Sar1/Arf subfamilies regulate vesicle trafficking, and the Ran GTPases regulate nucleocytoplasmic transport (Takai, Sasaki, & Matozaki, 2001). The roles of the small GTPases in the transport of newly synthesized GPCRs from the ER to the cell surface have been recently studied. Through manipulating the function of endogenous small GTPases by expressing their GDP- and GTP-bound mutants and siRNA targeting to specific GTPases, we and others have recently demonstrated that multiple small GTPases in the Sar1/ARF and Rab subfamilies modulate GPCR cell surface transport *en route* from the ER and the Golgi/TGN.

A. Sar1 in α_{2B}-AR Exit from the ER

The small GTPase Sar1 and the heterodimeric Sec23/24 and Sec13/31 complexes are the components of COPII-coated transport vesicles, which exclusively mediate export of newly synthesized cargo from the ER. It has been well demonstrated that GDP/GTP exchange and GTP hydrolysis by Sar1 GTPase play a crucial role in the formation and budding of COPII-coated vesicles on the ER membrane. Assembly of the COPII coat takes place on the ER membrane at discrete locations called ER exit sites and is initiated by the exchange of GDP for GTP on Sar1 GTPase. GTP activation of Sar1 GTPase recruits the Sec23/24 and Sec13/31 complexes onto the ER membrane forming the COPII-coated vesicles. Hydrolysis of GTP to GDP by Sar1 GTPase results in the dissociation of Sar1 GTPase from the ER membrane and the release of the COPII vesicles (Gurkan, Stagg, Lapointe, & Balch, 2006; Pucadyil & Schmid, 2009).

As a first study to define the role of the ER-derived COPII transport vesicles in GPCR export from the ER, we determined the effect of transient expression of the GTP-restricted mutant Sar1H79G, which presumably blocks the release of the COPII vesicles from the ER membrane, on the cell surface expression and subcellular distribution of α_{2B}-AR (Dong et al., 2008). Expression of Sar1H79G significantly attenuated the cell surface expression of α_{2B}-AR (Fig. 1B) and arrested α_{2B}-AR in ER exit sites (Dong et al., 2008). These data indicate that α_{2B}-AR export from the ER and transport to the cell surface is dependent on the normal function of the small GTPase Sar1. These data also suggest that, similar to many other proteins, α_{2B}-AR exit from the ER is mediated through the Sar1-dependent COPII-coated vesicles. Similar to α_{2B}-AR, the cell surface expression of β_2-AR, AT1R, and human calcium receptor (hCaR) was attenuated by Sar1H79G mutant and siRNA-mediated knockdown of Sar1 (Dong et al., 2008; Zhuang, Chowdhury, Northup, & Ray, 2010), further confirming a general role for Sar1 GTPase in the cell surface transport of the GPCR superfamily.

B. ARF GTPases in α_{2B}-AR Exit from the ER and the Golgi

Of the five ARF GTPases (ARF1, 3, 4, 5, and 6) identified in humans, ARF1 and ARF6 are the best characterized and well understood members. ARF6 primarily engages in the regulation of endocytic trafficking and cytoskeleton remodeling, whereas ARF1 recruits different sets of coat proteins to form distinct transport vesicles that control protein transport at different intracellular organelles (Palacios, Price, Schweitzer, Collard, & D'Souza-Schorey, 2001; Spang, 2002; Stearns, Willingham, Botstein, & Kahn, 1990). For example, ARF1 recruits coatomers in the formation of COPI vesicles, which mediate protein transport from the Golgi to the ER, from the ERGIC to the Golgi, and intra-Golgi traffic, whereas ARF1-mediated recruitment of adaptor proteins and GGA, leading to the formation of the clathrin-coated vesicles on the TGN controls post-Golgi transport between the TGN, the plasma membrane and the endosomal compartment (Bonifacino, 2004). Based on the sequence homology, it is believed that ARF1 and ARF3 share the same function. In contrast, the function of ARF4 and ARF5 remains largely unknown.

We have recently determined the role of each ARF GTPase in the cell surface targeting of α_{2B}-AR (Dong et al., 2010b). Our studies demonstrated that expression of the GDP-bound, GTP-bound, and guanine nucleotide-deficient mutants of both ARF1 and ARF3 produced a profound inhibitory effect on the cell surface expression of α_{2B}-AR, whereas ARF4, ARF5, and ARF6 mutants produced only moderate or no effect. These data indicate that five human ARF GTPases differentially modulate α_{2B}-AR cell surface transport and that ARF1 and ARF3 are the primary ones regulating α_{2B}-AR export trafficking. Interestingly, we have demonstrated that ARF1 is able to physically associate with α_{2B}-AR as measured by coimmunoprecipitation and GST fusion protein pull-down assay and the interaction domain has been mapped to the C-terminus of the receptor (Dong et al., 2010b). These studies suggest that regulation of α_{2B}-AR transport by ARF1 may be mediated through their direct interaction.

It appears that ARF1 GTPase modulates the cell surface transport of α_{2B}-AR at multiple transport steps as the GDP- and GTP-bound ARF1 mutants arrested the receptors in distinct intracellular compartments (Dong et al., 2010b).

Whereas expression of the GDP-bound mutant ARF1T31N arrested α_{2B}-AR in the ER, the GTP-bound mutant ARF1Q71L induced an accumulation of the receptors in the post-ER compartments, including ERGIC, Golgi, and TGN (Dong et al., 2010b). These data indicate that expression of different ARF1 mutants blocks the export of the cargo receptors from different subcellular compartments. Such differential regulation of α_{2B}-AR export by the ARF1 GDP- and GTP mutants could be explained by their effects on the formation of transport vesicles from the different intracellular compartments. Expression of the GDP-bound mutant ARF1T31N would block the formation of COPI vesicles to disrupt the retrograde transport system, which will impair the reuse of components of transport machinery and induces defective anterograde trafficking of newly synthesized cargo. On the contrary, expression of the GTP-bound ARF1Q71L mutant would influence the release of the COPI vesicles from the ERGIC and the Golgi or the clathrin-coated vesicles from the TGN, resulting in the accumulation of α_{2B}-AR in these compartments.

Our studies have demonstrated that ARF1 may play a general role in the anterograde trafficking of the GPCR superfamily. In addition to α_{2B}-AR, we have also measured the effect of the ARF1 mutants on the cell surface transport and subcellular distribution of several other GPCRs including β_2-AR, AT1R, and C-X-C chemokine receptor type 4. Similar to their effects on α_{2B}-AR, expression of the ARF1 mutants markedly inhibited the cell surface expression of all three receptors examined and the GDP- and GTP-bound mutants arrested these receptors in different intracellular compartments (Dong et al., 2010b).

C. *Rab GTPases in the ER-Golgi-Cell Surface Transport of α_{2B}-AR*

Consisting of more than 60 members in mammals and 11 in yeast, Rab GTPases form the largest subfamily of the Ras-related GTPases and function as traffic ``cops" to coordinate almost every step of vesicle-mediated transport, particularly the targeting, tethering, and fusion of the transport vesicles. Each Rab GTPase has a distinct subcellular distribution pattern that correlates with the compartments between which it coordinates the transport (Takai et al., 2001). There are at least three Rab GTPases, Rab1, Rab2, and Rab6, which coordinate protein transport in the early secretory pathway. Rab1 is localized at the ER and the Golgi, and regulates the anterograde transport of proteins from the ER to the Golgi. Rab2 is localized to the ERGIC that works as the first station sorting cargo into anterograde or retrograde transport pathway and coordinates the early event between the ERGIC and the ER. Rab6 mainly locates in the Golgi and regulates the trafficking from the late to early Golgi cisternae and from the Golgi to the ER. In contrast, Rab8 mediates the vesicle-mediated trafficking from the Golgi/TGN to the plasma membrane.

Most studies on the roles of Rab GTPases in the intracellular trafficking of GPCRs have been focused on the events involved in the internalization (Fan, Lapierre, Goldenring, & Richmond, 2003; Murph, Scaccia, Volpicelli, & Radhakrishna, 2003; Seachrist, Anborgh, & Ferguson, 2000). In contrast, much less is known about the involvement of Rab GTPases in GPCR export to the plasma membrane. As an initial approach to investigate the anterograde transport pathways of GPCRs, we have determined the role of Rab1, Rab2, Rab6, and Rab8 GTPases in the cell surface transport of α_{2B}-AR by transiently expressing dominant-negative mutants and siRNA-mediated depletion of individual Rab GTPases. We found that these Rab GTPases differentially modulate α_{2B}-AR transport to the cell surface. Specifically, inhibition of Rab2 and Rab8 function significantly inhibited α_{2B}-AR transport to the cell surface, whereas inhibition of Rab1 and Rab6 function did not produce any effect (Dong & Wu, 2007; Dong et al., 2010a; Wu et al., 2003). These data demonstrate that the cell surface transport of α_{2B}-AR is dependent on the normal function of Rab2 and Rab8, but independent of Rab1 and Rab6, which have been well documented to function as generic regulators for protein transport between the ER and the Golgi.

As discussed above, the expression of GTP-bound mutant ARF1Q71L induced an extensive accumulation of α_{2B}-AR in the Golgi (Dong et al., 2010b), indicating that α_{2B}-AR actually passes the Golgi stacks *en route* to the cell surface. Therefore, Rab1/Rab6-independent transport of α_{2B}-AR strongly implies that α_{2B}-AR uses a nonconventional pathway to move from the ER to the Golgi. However, how this novel pathway operates remains unknown. Compared with other GPCRs, α_{2B}-AR is one of a few GPCRs that do not contain N-linked glycosylation sites in the N-termini. Glycosylation of the receptors occurs during their transport through the Golgi apparatus, resulting in the formation of mature receptors competent for subsequent transport to the cell surface. Whether posttranslational modifications such as N-linked glycoyslation function as one of the determinants for the selection of transport pathways and whether the N-linked glycosylation dictates the receptors into the Rab1/Rab6-coordinated transport need further investigation. In addition, to further study the function of other Rab GTPases in the ER-to-Golgi transport of α_{2B}-AR may provide important insights into this nonclassic transport pathway.

In contrast to α_{2B}-AR, the cell surface transport of other GPCRs including α_1-AR, β-AR, AT1R, AT2R, and hCaR was attenuated by functional inhibition of Rab1, Rab2, Rab6, and Rab8 (Dong & Wu, 2007; Dong et al., 2010a; Filipeanu et al., 2004, 2006; Li et al., 2010; Wu et al., 2003; Zhang et al., 2009; Zhuang, Adipietro, Datta, Northup, & Ray, 2010). These data demonstrate that Rab1 and Rab6 may selectively modulate the transport of distinct GPCRs. These data also suggest that distinct GPCRs that have common structural features, track to the cell surface and couple to heterotrimeric G proteins may utilize different pathways (i.e., Rab1/Rab6-dependent and Rab1/Rab6-independent) for their movement from the ER to the Golgi.

The function of Rab GTPases in regulating GPCR trafficking may be mediated through their direct interactions with the receptors. For example, Rab4, Rab5, Rab7, and Rab11 bind to AT1R to modulate its endocytic trafficking (Esseltine, Dale, & Ferguson, 2010; Seachrist et al., 2002). We recently demonstrated that both β_2-AR and α_{2B}-AR are able to associate with Rab8 as revealed by coimmunoprecipitation. Interestingly, these two adrenergic receptors use different motifs to bind Rab8. In contrast to β_2-AR using the LL motif to interact with Rab8, α_{2B}-AR uses multiple sites located in the membrane-proximal (TVFN) and distal (PW and QTGW) C-terminus to interact with Rab8 (Dong et al., 2010a). In particular, the residues N433 and P447 likely play a crucial role in mediating α_{2B}-AR interaction with Rab8 as mutation of either one almost abolished the interaction in GST fusion protein pull down assays. These data suggest that different GPCRs (i.e., α_{2B}-AR and β_2-AR) may provide distinct docking sites for Rab8 GTPase to coordinate their export from the TGN (Dong et al., 2010a).

V. CONCLUSIONS AND PERSPECTIVES

The players involved in the cell surface targeting of GPCRs in general or α_{2B}-AR in particular are just beginning to be revealed. Recent studies have demonstrated that export from the ER and the Golgi of α_{2B}-AR is dictated by specific amino acid residues or motifs scattered throughout the receptor and the transport of α_{2B}-AR from the ER through the Golgi to the cell surface along the secretory pathway is coordinated by multiple GTPases (Fig. 3). However, the mechanism underlying the regulation of α_{2B}-AR export trafficking is still largely unknown. First, although several essential sequences for ER export of α_{2B}-AR or many other GPCRs have been identified (Bermak et al., 2001; Duvernay et al., 2004; Oksche, Dehe, Schulein, Wiesner, & Rosenthal, 1998; Robert et al., 2005; Rodriguez et al., 1992; Schulein et al., 1998; Tai, Chuang, Bode, Wolfrum, & Sung, 1999), none of them have been shown to directly interact with components of COPII vesicles. The most interesting experiment probably is to continue to search for such motifs that are able to directly interact with the components of COPII transport vesicles and facilitate α_{2B}-AR recruitment onto the vesicles. Second, the experiment to use different protein–protein interaction strategies to look for proteins interacting with the well-defined export motifs as discussed above will help to elucidate the possible molecular mechanism for these motifs. Third, as it is clear that α_{2B}-AR uses a nonclassic pathway to

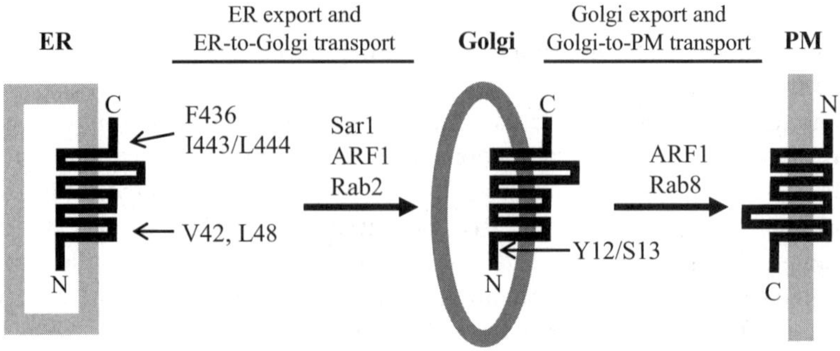

FIGURE 3 Summary of the structural basis and the roles of small GTPases in the anterograde trafficking of α_{2B}-AR. The F436, I443/L444, V42, and L48 residues regulate the exit of α_{2B}-AR from the ER and the Y12/S13 residues influence the exit from the Golgi. The small GTPase Sar1 controls α_{2B}-AR export from the ER by modulating the function of the COPII vesicles, whereas ARF1 may be involved in the export of α_{2B}-AR from multiple intracellular compartments including the ER and the Golgi. α_{2B}-AR transport from the ER to the Golgi depends on the normal function of Rab2, but independent of Rab1 and Rab6, and its transport from the Golgi to the cell surface requires Rab8.

move from the ER to the Golgi, the immediate experiments are to fully characterize this pathway.

Cell surface targeting of GPCRs is one of the important factors determining the functionality of the receptors. Indeed, dysfunction of GPCRs caused by defective cell surface trafficking is clearly associated with the development of a number of human diseases such as nephrogenic diabetes insipidus, retinitis pigmentosa, and male pseudohermaphroditism. Therefore, to thoroughly understand the mechanism underlying export trafficking of GPCRs will provide a foundation for the development of therapeutic strategies targeting on specific components of the transport pathway.

Acknowledgment

This work was supported by National Institutes of Health grant R01GM076167 (to G. Wu).

References

Angelotti, T., Daunt, D., Shcherbakova, O. G., Kobilka, B., & Hurt, C. M. (2010). Regulation of G-protein coupled receptor traffic by an evolutionary conserved hydrophobic signal. *Traffic, 11*(4), 560–578.

Bailey, S. R., Eid, A. H., Mitra, S., Flavahan, S., & Flavahan, N. A. (2004). Rho kinase mediates cold-induced constriction of cutaneous arteries: role of alpha2C-adrenoceptor translocation. *Circ Res, 94*(10), 1367–1374.

Bermak, J. C., Li, M., Bullock, C., & Zhou, Q. Y. (2001). Regulation of transport of the dopamine D1 receptor by a new membrane-associated ER protein. *Nat Cell Biol, 3*(5), 492–498.

Bonifacino, J. S. (2004). The GGA proteins: Adaptors on the move. *Nat Rev Mol Cell Biol, 5*(1), 23–32.

Boucher, R., Larkin, H., Brodeur, J., Gagnon, H., Theriault, C., & Lavoie, C. (2008). Intracellular trafficking of LRP9 is dependent on two acidic cluster/dileucine motifs. *Histochem Cell Biol, 130*(2), 315–327.

Bouvier, M. (2001). Oligomerization of G-protein-coupled transmitter receptors. *Nat Rev Neurosci, 2*(4), 274–286.

Brady, A. E., Wang, Q., Colbran, R. J., Allen, P. B., Greengard, P., & Limbird, L. E. (2003). Spinophilin stabilizes cell surface expression of alpha 2B-adrenergic receptors. *J Biol Chem, 278*(34), 32405–32412.

Carrel, D., Hamon, M., & Darmon, M. (2006). Role of the C-terminal di-leucine motif of 5-HT1A and 5-HT1B serotonin receptors in plasma membrane targeting. *J Cell Sci, 119*(Pt 20), 4276–4284.

Chen, H. J., Yuan, J., & Lobel, P. (1997). Systematic mutational analysis of the cation-independent mannose 6-phosphate/insulin-like growth factor ii receptor cytoplasmic domain. an acidic cluster containing a key aspartate is important for function in lysosomal enzyme sorting. *J. Biol. Chem., 272*(11), 7003–7012.

Daunt, D. A., Hurt, C., Hein, L., Kallio, J., Feng, F., & Kobilka, B. K. (1997). Subtype-specific intracellular trafficking of alpha2-adrenergic receptors. *Mol Pharmacol, 51*(5), 711–720.

DeGraff, J. L., Gurevich, V. V., & Benovic, J. L. (2002). The third intracellular loop of alpha 2-adrenergic receptors determines subtype specificity of arrestin interaction. *J Biol Chem, 277*(45), 43247–43252.

Dong, C., Filipeanu, C. M., Duvernay, M. T., & Wu, G. (2007). Regulation of G protein-coupled receptor export trafficking. *Biochim Biophys Acta, 1768*(4), 853–870.

Dong, C., & Wu, G. (2006). Regulation of anterograde transport of alpha2-adrenergic receptors by the N termini at multiple intracellular compartments. *J Biol Chem, 281*(50), 38543–38554.

Dong, C., & Wu, G. (2007). Regulation of anterograde transport of adrenergic and angiotensin II receptors by Rab2 and Rab6 GTPases. *Cell Signal, 19*(11), 2388–2399.

Dong, C., Yang, L., Zhang, X., Gu, H., Lam, M. L., & Claycomb, W. C., et al., (2010a). Rab8 interacts with distinct motifs in {alpha}2B- and {beta}2-adrenergic receptors and differentially modulates their transport. *J Biol Chem, 285*(26), 20369–20380.

Dong, C., Zhang, X., Zhou, F., Dou, H., Duvernay, M. T., & Zhang, P., et al., (2010b). ADP-ribosylation factors modulate the cell surface transport of G protein-coupled receptors. *J Pharmacol Exp Ther, 333*(1), 174–183.

Dong, C., Zhou, F., Fugetta, E. K., Filipeanu, C. M., & Wu, G. (2008). Endoplasmic reticulum export of adrenergic and angiotensin II receptors is differentially regulated by Sar1 GTPase. *Cell Signal, 20*(6), 1035–1043.

Duvernay, M. T., Dong, C., Zhang, X., Robitaille, M., Hebert, T. E., & Wu, G. (2009a). A single conserved leucine residue on the first intracellular loop regulates ER export of G protein-coupled receptors. *Traffic, 10*(5), 552–566.

Duvernay, M. T., Dong, C., Zhang, X., Zhou, F., Nichols, C. D., & Wu, G. (2009b). Anterograde trafficking of G protein-coupled receptors: function of the C-terminal F(X)6LL motif in export from the endoplasmic reticulum. *Mol Pharmacol, 75*(4), 751–761.

Duvernay, M. T., Zhou, F., & Wu, G. (2004). A conserved motif for the transport of G protein-coupled receptors from the endoplasmic reticulum to the cell surface. *J Biol Chem, 279*(29), 30741–30750.

Edwards, S. W., & Limbird, L. E. (1999). Role for the third intracellular loop in cell surface stabilization of the alpha2A-adrenergic receptor. *J Biol Chem, 274*(23), 16331–16336.

Esseltine, J. L., Dale, L. B., & Ferguson, S. S. (2010). Rab GTPases bind at a common site within the angiotensin II type I receptor carboxyl-terminal tail: Evidence that Rab4 regulates receptor phosphorylation, desensitization, and resensitization. *Mol Pharmacol, 79*(1), 175–184.

Fan, G. H., Lapierre, L. A., Goldenring, J. R., & Richmond, A. (2003). Differential regulation of CXCR2 trafficking by Rab GTPases. *Blood, 101*(6), 2115–2124.

Farhan, H., Reiterer, V., Korkhov, V. M., Schmid, J. A., Freissmuth, M., & Sitte, H. H. (2007). Concentrative export from the endoplasmic reticulum of the {gamma}-aminobutyric acid transporter 1 requires binding to SEC24D. *J Biol Chem, 282*(10), 7679–7689.

Filipeanu, C. M., Zhou, F., Claycomb, W. C., & Wu, G. (2004). Regulation of the cell surface expression and function of angiotensin II type 1 receptor by Rab1-mediated endoplasmic reticulum-to-Golgi transport in cardiac myocytes. *J Biol Chem, 279*(39), 41077–41084.

Filipeanu, C. M., Zhou, F., Fugetta, E. K., & Wu, G. (2006). Differential regulation of the cell-surface targeting and function of beta- and alpha1-adrenergic receptors by Rab1 GTPase in cardiac myocytes. *Mol Pharmacol, 69*(5), 1571–1578.

Gaborik, Z., Mihalik, B., Jayadev, S., Jagadeesh, G., Catt, K. J., & Hunyady, L. (1998). Requirement of membrane-proximal amino acids in the carboxyl-terminal tail for expression of the rat AT1a angiotensin receptor. *FEBS Lett, 428*(3), 147–151.

Gimelbrant, A. A., Haley, S. L., & McClintock, T. S. (2001). Olfactory receptor trafficking involves conserved regulatory steps. *J Biol Chem, 276*(10), 7285–7290.

Gurkan, C., Stagg, S. M., Lapointe, P., & Balch, W. E. (2006). The COPII cage: Unifying principles of vesicle coat assembly. *Nat Rev Mol Cell Biol, 7*(10), 727–738.

Hague, C., Chen, Z., Pupo, A. S., Schulte, N. A., Toews, M. L., & Minneman, K. P. (2004). The N terminus of the human alpha1D-adrenergic receptor prevents cell surface expression. *J Pharmacol Exp Ther, 309*(1), 388–397.

Hanyaloglu, A. C., & von Zastrow, M. (2008). Regulation of GPCRs by endocytic membrane trafficking and its potential implications. *Annu Rev Pharmacol Toxicol, 48*, 537–568.

Heymann, J. A., & Subramaniam, S. (1997). Expression, stability, and membrane integration of truncation mutants of bovine rhodopsin. *Proc Natl Acad Sci USA, 94*(10), 4966–4971.

Hirst, J., Lui, W. W. Y., Bright, N. A., Totty, N., Seaman, M. N. J., & Robinson, M. S. (2000). A family of proteins with {gamma}-adaptin and VHS domains that facilitate trafficking between the trans-Golgi network and the vacuole/lysosome. *J Cell Biol, 149*(1), 67–80.

Hou, J. C., Suzuki, N., Pessin, J. E., & Watson, R. T. (2006). A specific dileucine motif is required for the GGA-dependent entry of newly synthesized insulin-responsive aminopeptidase into the insulin-responsive compartment. *J Biol Chem, 281*(44), 33457–33466.

Jewell-Motz, E. A., Small, K. M., Theiss, C. T., & Liggett, S. B. (2000). alpha 2A/alpha 2C-adrenergic receptor third loop chimera show that agonist interaction with receptor subtype backbone establishes G protein-coupled receptor kinase phosphorylation. *J Biol Chem, 275* (37), 28989–28993.

Jeyaraj, S. C., Chotani, M. A., Mitra, S., Gregg, H. E., Flavahan, N. A., & Morrison, K. J. (2001). Cooling evokes redistribution of alpha2C-adrenoceptors from Golgi to plasma membrane in transfected human embryonic kidney 293 cells. *Mol Pharmacol, 60*(6), 1195–1200.

Johnson, K. F., & Kornfeld, S. (1992). The cytoplasmic tail of the mannose 6-phosphate/insulin-like growth factor-II receptor has two signals for lysosomal enzyme sorting in the Golgi. *J Cell Biol, 119*(2), 249–257.

Kappeler, F., Klopfenstein, D. R. C., Foguet, M., Paccaud, J. -P., & Hauri, H. -P. (1997). The recycling of ERGIC-53 in the early secretory pathway. ERGIC-53 carries a cytosolic endoplasmic reticulum-exit determinant interacting with COPII. *J Biol Chem, 272*(50), 31801–31808.

Keefer, J. R., Kennedy, M. E., & Limbird, L. E. (1994). Unique structural features important for stabilization versus polarization of the alpha 2A-adrenergic receptor on the basolateral membrane of Madin–Darby canine kidney cells. *J Biol Chem, 269*(23), 16425–16432.

Keefer, J. R., & Limbird, L. E. (1993). The alpha 2A-adrenergic receptor is targeted directly to the basolateral membrane domain of Madin–Darby canine kidney cells independent of coupling to pertussis toxin-sensitive GTP-binding proteins. *J Biol Chem, 268*(15), 11340–11347.

Li, Y., Wang, G., Lin, K., Yin, H., Zhou, C., & Liu, T., et al., (2010). Rab1 GTPase promotes expression of beta-adrenergic receptors in rat pulmonary microvascular endothelial cells. *Int J Biochem Cell Biol, 42*(7), 1201–1209.

Lori, L. T., Florencia, B. S., & Frederick, R. M. (2007). Role of an acidic cluster/dileucine motif in cation-independent mannose 6-phosphate receptor traffic. *Traffic, 8*(4), 402–413.

Ma, D., Zerangue, N., Lin, Y. F., Collins, A., Yu, M., & Jan, Y. N., et al., (2001). Role of ER export signals in controlling surface potassium channel numbers. *Science, 291*(5502), 316–319.

Marchese, A., Chen, C., Kim, Y. M., & Benovic, J. L. (2003). The ins and outs of G protein-coupled receptor trafficking. *Trends Biochem Sci, 28*(7), 369–376.

Moore, C. A., Milano, S. K., & Benovic, J. L. (2007). Regulation of receptor trafficking by GRKs and arrestins. *Annu Rev Physiol, 69*, 451–482.

Murph, M. M., Scaccia, L. A., Volpicelli, L. A., & Radhakrishna, H. (2003). Agonist-induced endocytosis of lysophosphatidic acid-coupled LPA1/EDG-2 receptors via a dynamin2- and Rab5-dependent pathway. *J Cell Sci, 116*(Pt 10), 1969–1980.

Nishimura, N., & Balch, W. E. (1997). A di-acidic signal required for selective export from the endoplasmic reticulum. *Science, 277*(5325), 556–558.

Nishimura, N., Bannykh, S., Slabough, S., Matteson, J., Altschuler, Y., & Hahn, K., et al., (1999). A di-acidic (DXE) code directs concentration of cargo during export from the endoplasmic reticulum. *J Biol Chem, 274*(22), 15937–15946.

Nishimura, N., Plutner, H., Hahn, K., & Balch, W. E. (2002). The delta subunit of AP-3 is required for efficient transport of VSV-G from the trans-Golgi network to the cell surface. *Proc Natl Acad Sci USA, 99*(10), 6755–6760.

Nufer, O., Guldbrandsen, S., Degen, M., Kappeler, F., Paccaud, J. P., & Tani, K., et al., (2002). Role of cytoplasmic C-terminal amino acids of membrane proteins in ER export. *J Cell Sci, 115*(Pt 3), 619–628.

Nufer, O., Kappeler, F., Guldbrandsen, S., & Hauri, H. P. (2003). ER export of ERGIC-53 is controlled by cooperation of targeting determinants in all three of its domains. *J Cell Sci, 116* (Pt 21), 4429–4440.

Oksche, A., Dehe, M., Schulein, R., Wiesner, B., & Rosenthal, W. (1998). Folding and cell surface expression of the vasopressin V2 receptor: Requirement of the intracellular C-terminus. *FEBS Lett, 424*(1–2), 57–62.

Palacios, F., Price, L., Schweitzer, J., Collard, J. G., & D'Souza-Schorey, C. (2001). An essential role for ARF6-regulated membrane traffic in adherens junction turnover and epithelial cell migration. *EMBO J, 20*(17), 4973–4986.

Pankevych, H., Korkhov, V., Freissmuth, M., & Nanoff, C. (2003). Truncation of the A1 adenosine receptor reveals distinct roles of the membrane-proximal carboxyl terminus in receptor folding and G protein coupling. *J Biol Chem, 278*(32), 30283–30293.

Pao, C. S., & Benovic, J. L. (2005). Structure/function analysis of alpha2A-adrenergic receptor interaction with G protein-coupled receptor kinase 2. *J Biol Chem, 280*(12), 11052–11058.

Petaja-Repo, U. E., Hogue, M., Laperriere, A., Walker, P., & Bouvier, M. (2000). Export from the endoplasmic reticulum represents the limiting step in the maturation and cell surface expression of the human delta opioid receptor. *J Biol Chem, 275*(18), 13727–13736.

Pierce, K. L., Premont, R. T., & Lefkowitz, R. J. (2002). Seven-transmembrane receptors. *Nat Rev Mol Cell Biol, 3*(9), 639–650.

Prezeau, L., Richman, J. G., Edwards, S. W., & Limbird, L. E. (1999). The zeta isoform of 14-3-3 proteins interacts with the third intracellular loop of different alpha2-adrenergic receptor subtypes. *J Biol Chem, 274*(19), 13462–13469.

Pucadyil, T. J., & Schmid, S. L. (2009). Conserved functions of membrane active GTPases in coated vesicle formation. *Science, 325*(5945), 1217–1220.

Puertollano, R., Aguilar, R. C., Gorshkova, I., Crouch, R. J., & Bonifacino, J. S. (2001a). Sorting of mannose 6-phosphate receptors mediated by the GGAs. *Science, 292*(5522), 1712–1716.

Puertollano, R., Randazzo, P. A., Presley, J. F., Hartnell, L. M., & Bonifacino, J. S. (2001b). The GGAs promote ARF-dependent recruitment of clathrin to the TGN. *Cell, 105*(1), 93–102.

Richman, J. G., Brady, A. E., Wang, Q., Hensel, J. L., Colbran, R. J., & Limbird, L. E. (2001). Agonist-regulated Interaction between alpha2-adrenergic receptors and spinophilin. *J Biol Chem, 276*(18), 15003–15008.

Robert, J., Clauser, E., Petit, P. X., & Ventura, M. A. (2005). A novel C-terminal motif is necessary for the export of the vasopressin V1b/V3 receptor to the plasma membrane. *J Biol Chem, 280*(3), 2300–2308.

Rodriguez, M. C., Xie, Y. B., Wang, H., Collison, K., & Segaloff, D. L. (1992). Effects of truncations of the cytoplasmic tail of the luteinizing hormone/chorionic gonadotropin receptor on receptor-mediated hormone internalization. *Mol Endocrinol, 6*(3), 327–336.

Rosenbaum, D. M., Rasmussen, S. G., & Kobilka, B. K. (2009). The structure and function of G-protein-coupled receptors. *Nature, 459*(7245), 356–363.

Salahpour, A., Angers, S., Mercier, J. -F., Lagace, M., Marullo, S., & Bouvier, M. (2004). Homodimerization of the {beta}2-adrenergic receptor as a prerequisite for cell surface targeting. *J Biol Chem, 279*(32), 33390–33397.

Saunders, C., Keefer, J. R., Bonner, C. A., & Limbird, L. E. (1998). Targeting of G protein-coupled receptors to the basolateral surface of polarized renal epithelial cells involves multiple, non-contiguous structural signals. *J Biol Chem, 273*(37), 24196–24206.

Saunders, C., & Limbird, L. E. (2000). Microtubule-dependent regulation of alpha(2B) adrenergic receptors in polarized MDCKII cells requires the third intracellular loop but not G protein coupling. *Mol Pharmacol, 57*(1), 44–52.

Schulein, R., Hermosilla, R., Oksche, A., Dehe, M., Wiesner, B., & Krause, G., et al., (1998). A dileucine sequence and an upstream glutamate residue in the intracellular carboxyl terminus of the vasopressin V2 receptor are essential for cell surface transport in COS.M6 cells. *Mol Pharmacol, 54*(3), 525–535.

Seachrist, J. L., Anborgh, P. H., & Ferguson, S. S. (2000). beta 2-adrenergic receptor internalization, endosomal sorting, and plasma membrane recycling are regulated by rab GTPases. *J Biol Chem, 275*(35), 27221–27228.

Seachrist, J. L., Laporte, S. A., Dale, L. B., Babwah, A. V., Caron, M. G., & Anborgh, P. H., et al., (2002). Rab5 association with the angiotensin II type 1A receptor promotes Rab5 GTP binding and vesicular fusion. *J Biol Chem, 277*(1), 679–685.

Sevier, C. S., Weisz, O. A., Davis, M., & Machamer, C. E. (2000). Efficient export of the vesicular stomatitis virus G protein from the endoplasmic reticulum requires a signal in the cytoplasmic tail that includes both tyrosine-based and di-acidic motifs. *Mol Biol Cell, 11*(1), 13–22.

Small, K. M., Brown, K. M., Forbes, S. L., & Liggett, S. B. (2001). Polymorphic deletion of three intracellular acidic residues of the alpha 2B-adrenergic receptor decreases G protein-coupled receptor kinase-mediated phosphorylation and desensitization. *J Biol Chem, 276*(7), 4917–4922.

Spang, A. (2002). ARF1 regulatory factors and COPI vesicle formation. *Curr Opin Cell Biol, 14*(4), 423–427.

Stearns, T., Willingham, M. C., Botstein, D., & Kahn, R. A. (1990). ADP-ribosylation factor is functionally and physically associated with the Golgi complex. *Proc Natl Acad Sci USA, 87*(3), 1238–1242.

Tai, A. W., Chuang, J. Z., Bode, C., Wolfrum, U., & Sung, C. H. (1999). Rhodopsin's carboxy-terminal cytoplasmic tail acts as a membrane receptor for cytoplasmic dynein by binding to the dynein light chain Tctex-1. *Cell, 97*(7), 877–887.

Takai, Y., Sasaki, T., & Matozaki, T. (2001). Small GTP-binding proteins. *Physiol Rev, 81*(1), 153–208.

Tan, C. M., Brady, A. E., Nickols, H. H., Wang, Q., & Limbird, L. E. (2004). Membrane trafficking of G protein-coupled receptors. *Annu Rev Pharmacol Toxicol, 44*, 559–609.

Tetsuka, M., Saito, Y., Imai, K., Doi, H., & Maruyama, K. (2004). The basic residues in the membrane-proximal C-terminal tail of the rat melanin-concentrating hormone receptor 1 are required for receptor function. *Endocrinology, 145*(8), 3712–3723.

Venkatesan, S., Petrovic, A., Van Ryk, D. I., Locati, M., Weissman, D., & Murphy, P. M. (2002). Reduced cell surface expression of CCR5 in CCR5Delta 32 heterozygotes is mediated by gene dosage, rather than by receptor sequestration. *J Biol Chem, 277*(3), 2287–2301.

Votsmeier, C., & Gallwitz, D. (2001). An acidic sequence of a putative yeast Golgi membrane protein binds COPII and facilitates ER export. *EMBO J, 20*(23), 6742–6750.

Wade, S. M., Lim, W. K., Lan, K. L., Chung, D. A., Nanamori, M., & Neubig, R. R. (1999). G(i) activator region of alpha(2A)-adrenergic receptors: Distinct basic residues mediate G (i) versus G(s) activation. *Mol Pharmacol, 56*(5), 1005–1013.

Wade, S. M., Scribner, M. K., Dalman, H. M., Taylor, J. M., & Neubig, R. R. (1996). Structural requirements for G(o) activation by receptor-derived peptides: Activation and modulation domains of the alpha 2-adrenergic receptor i3c region. *Mol Pharmacol, 50*(2), 351–358.

Wang, Q., & Limbird, L. E. (2002). Regulated interactions of the alpha 2A adrenergic receptor with spinophilin, 14-3-3zeta, and arrestin 3. *J Biol Chem, 277*(52), 50589–50596.

Wang, Q., Zhao, J., Brady, A. E., Feng, J., Allen, P. B., & Lefkowitz, R. J., et al., (2004a). Spinophilin blocks arrestin actions in vitro and in vivo at G protein-coupled receptors. *Science, 304*(5679), 1940–1944.

Wang, X., Matteson, J., An, Y., Moyer, B., Yoo, J. S., & Bannykh, S., et al., (2004b). COPII-dependent export of cystic fibrosis transmembrane conductance regulator from the ER uses a di-acidic exit code. *J Cell Biol, 167*(1), 65–74.

Wendeler, M. W., Paccaud, J. P., & Hauri, H. P. (2007). Role of Sec24 isoforms in selective export of membrane proteins from the endoplasmic reticulum. *EMBO Rep, 8*(3), 258–264.

Wozniak, M., & Limbird, L. E. (1996). The three alpha 2-adrenergic receptor subtypes achieve basolateral localization in Madin–Darby canine kidney II cells via different targeting mechanisms. *J Biol Chem, 271*(9), 5017–5024.

Wu, G., Benovic, J. L., Hildebrandt, J. D., & Lanier, S. M. (1998). Receptor docking sites for G-protein betagamma subunits. Implications for signal regulation. *J Biol Chem, 273*(13), 7197–7200.

Wu, G., Bogatkevich, G. S., Mukhin, Y. V., Benovic, J. L., Hildebrandt, J. D., & Lanier, S. M. (2000). Identification of Gbetagamma binding sites in the third intracellular loop of the M(3)-muscarinic receptor and their role in receptor regulation. *J Biol Chem, 275*(12), 9026–9034.

Wu, G., Krupnick, J. G., Benovic, J. L., & Lanier, S. M. (1997). Interaction of arrestins with intracellular domains of muscarinic and alpha2-adrenergic receptors. *J Biol Chem, 272*(28), 17836–17842.

Wu, G., Zhao, G., & He, Y. (2003). Distinct pathways for the trafficking of angiotensin II and adrenergic receptors from the endoplasmic reticulum to the cell surface: Rab1-independent transport of a G protein-coupled receptor. *J Biol Chem, 278*(47), 47062–47069.

Xia, Z., Gray, J. A., Compton-Toth, B. A., & Roth, B. L. (2003). A direct interaction of PSD-95 with 5-HT2A serotonin receptors regulates receptor trafficking and signal transduction. *J Biol Chem, 278*(24), 21901–21908.

Zhang, X., Wang, G., Dupre, D. J., Feng, Y., Robitaille, M., & Lazartigues, E., et al., (2009). Rab1 GTPase and dimerization in the cell surface expression of angiotensin II type 2 receptor. *J Pharmacol Exp Ther, 330*(1), 109–117.

Zhou, F., Filipeanu, C. M., Duvernay, M. T., & Wu, G. (2006). Cell-surface targeting of alpha2-adrenergic receptors – inhibition by a transport deficient mutant through dimerization. *Cell Signal, 18*(3), 318–327.

Zhu, L., Imanishi, Y., Filipek, S., Alekseev, A., Jastrzebska, B., & Sun, W., et al., (2006). Autosomal recessive retinitis pigmentosa and E150K mutation in the opsin gene. *J Biol Chem, 281*(31), 22289–22298.

Zhuang, X., Adipietro, K. A., Datta, S., Northup, J. K., & Ray, K. (2010a). Rab1 small GTP-binding protein regulates cell surface trafficking of the human calcium-sensing receptor. *Endocrinology, 151*(11), 5114–5123.

Zhuang, X., Chowdhury, S., Northup, J. K., & Ray, K. (2010b). Sar1-dependent trafficking of the human calcium receptor to the cell surface. *Biochem Biophys Res Commun, 396*(4), 874–880.

Zuzarte, M., Rinne, S., Schlichthorl, G., Schubert, A., Daut, J., & Preisig-Muller, R. (2007). A di-acidic sequence motif enhances the surface expression of the potassium channel TASK-3. *Traffic, 8*(8), 1093–1100.

CHAPTER 5

Recording Kinetics of Adrenergic Receptor Activation in Live Cells

Jean-Pierre Vilardaga

Laboratory for GPCR Biology, Department of Pharmacology and Chemical Biology, University of Pittsburgh, School of Medicine, Pittsburgh, PA, USA

I. Overview
II. Introduction
III. Principle of the Experiment
IV. Receptor Activation in Live Cells
  A. Kinetics of Receptor Activation
  B. Recording Intrinsic Efficacy of Ligands
  C. Modulation of Ligand Efficacy by Receptor Polymorphism and Receptor Heteromers
  Acknowledgment
  References

I. OVERVIEW

G protein-coupled receptors (GPCR) biosensors have been recently developed that measure intramolecular conformational changes by recording fluorescence resonance energy transfer (FRET) between fluorescent protein tags introduced at two intracellular sites in the receptor. This technique allows the spatial and temporal recording of agonist (full, partial, and inverse)-induced receptor conformational changes in live cells in real time. This review discusses the kinetics of receptor activation, the direct measurement of ligand efficacy at the level of the receptor, and how ligand efficacy can be modulated by receptor heteromers.

II. INTRODUCTION

Specialized transmembrane proteins known as G protein-coupled receptors (GPCRs) serve as cell surface switches to transmit extracellular signals into cells (Miyawaki & Tsien, 2000; Pierce, Premont, & Lefkowitz, 2002). This large

Current Topics in Membranes, Volume 67 0065-230X/10 $35.00

DOI: 10.1016/B978-0-12-384921-2.00001-X

receptor family includes receptors for small neurotransmitters (e.g., adrenaline, noradrenaline, dopamine), and larger peptide hormones (e.g., parathyroid hormone, vasopressin) involved in the function of human body systems as diverse as the endocrine, the skeletal, and the cardiovascular systems among others. As key initiators of a large palette of physiological and psychological functions, these receptors are also involved in many diseases, and are the targets of many approved drugs (e.g., β-blockers). Signal transduction through GPCRs proceed through a succession of biochemical events that take place initially at the cell membrane and involve ligand binding ($L + R \rightarrow LR$, where L and R represents a ligand and a GPCR, respectively), which switch the receptor into an active-state conformation ($LR \rightarrow LR^*$) (Wess, Han, Kim, Jacobson, & Li, 2008). The activated receptor can then interact with heterotrimeric G proteins (Gαβγ) to form a transient LR*G complex, which exhibits higher affinity for the "agonist" ligand than does the initial L-R state. This interaction further promotes a conformational change-induced exchange of GDP for GTP on the Gα subunit with concomitant release of the activated GTP-bound Gα (Gα-GTP) along with the Gβγ subunits from the LR complex. The subsequent activation by Gα-GTP of cell membrane-bound effectors such as adenylyl cyclases catalyses the synthesis of second messengers such as cAMP. Signaling responses are rapidly attenuated by receptor desensitization, typically involving receptor phosphorylation of intracellular parts of the receptor (loops and C-tail) by GRKs, which facilitates interaction of β-arrestins to the receptor (Lefkowitz, Hausdorff, & Caron, 1990; Lefkowitz & Shenoy, 2005; Lohse, Benovic, Codina, Caron, & Lefkowitz, 1990; Reiter & Lefkowitz, 2006). This interaction results in the physical uncoupling of G proteins from the receptor and thus terminates agonist-mediated signaling. Arrestin translocation to the receptor can also recruit cAMP-specific phosphodiesterase 4 (PDE4) at the plasma membrane to rapidly degrade cAMP (Hanyaloglu & von Zastrow, 2008; Perry et al., 2002), and can mediate receptor endocytosis (Hanyaloglu & von Zastrow, 2008). Receptor endocytosis has at least two outcomes. The first is directing the receptor to intracellular compartments, thus contributing to signal desensitization by reducing the number of cell surface receptors. The second is the movement of the receptor to lysosomes for degradation. These processes are thought to stop production of second messengers once receptors are internalized.

Receptors are kept in an inactive state (R) until the binding of a ligand switches them into an active state (R*). The molecular nature of conformational changes associated with β_2-adrenergic receptor (β_2-AR) activation was initially observed in purified and reconstituted receptors labeled with fluorescent molecules sensitive to environmental and/or conformational changes (Gether, Lin, & Kobilka, 1995). These changes involve a structural rearrangement of several transmembrane helices, in particular helices 3 and 6 (Sheikh et al., 1999; Sheikh, Zvyaga, Lichtarge, Sakmar, & Bourne, 1996). The importance of the relative motion between helices 3 and 6 associated with receptor activation has also been confirmed for other different GPCRs such as the PTHR (Sheikh et al., 1999; Vilardaga, Frank, Krasel, Dees, Nissenson, & Lohse, 2001). The transition between the inactive and active states is associated with a relative movement and rotation of the cytoplasmic part of helix 3 away from the cytoplasmic part of helix 6 (Farrens, Altenbach, Yang, Hubbell, & Khorana, 1996). This helical movement is triggered by the release of at least two molecular constraints. The first is the disruption of an ionic interaction between the cytoplasmic face of helix 3 and helix 6 (known as ionic lock); the second involves a rotamer toggle switch in helix 6 (modulation of the helix conformation around a conserved proline-kink) (Yao et al., 2006). These transmembrane movements presumably expose receptor epitopes at the cytosolic side, which are then recognized by heterotrimeric G proteins. Direct observations of conformational changes in GPCRs can also be observed in living cells by using an optical technique involving fluorescence resonance energy transfer (FRET) that permits spatial and temporal recordings of the activation/deactivation steps of diverse adrenergic receptors (Lohse, Nikolaev, Hein, Hoffmann, Vilardaga, & Bunemann, 2008; Vilardaga et al., 2009). This review discusses the principle of these FRET experiments and kinetics of receptor activation in response to ligand of diverse efficacies.

III. PRINCIPLE OF THE EXPERIMENT

The experimental system developed for recording GPCR activation mediated by ligand binding relies on an intramolecular FRET-based approach that involves the incorporation of two variants of the green fluorescent protein (GFP) to the intracellular part of the same receptor molecule (Fig. 1) (Hoffmann et al., 2005; Vilardaga, Bunemann, Krasel, Castro, & Lohse, 2003). The cyan (CFP) and the yellow (YFP) variants share a spectral overlap between the CFP's emission and YFP's absorption spectrums, and is a well-documented FRET pair used in live cell-based assays. FRET occurs when the excited CFP molecule transfers nonradiative energy to a YFP molecule in close proximity resulting in decreased CFP emission and increased YFP emission. The efficiency of FRET is very sensitive to the distance and orientation between CFP and YFP molecules, and falls off with the sixth power of the distance between the two fluorophores. FRET is therefore well suited to measure protein–protein interactions when the two fluorophores are in two separate proteins (intermolecular FRET), or conformational changes when the two fluorophores are inserted in two separate structural domains of the same protein (intramolecular FRET). The YFP/CFP FRET pair can be easily incorporated into proteins by using well-established DNA recombinant techniques and the engineered

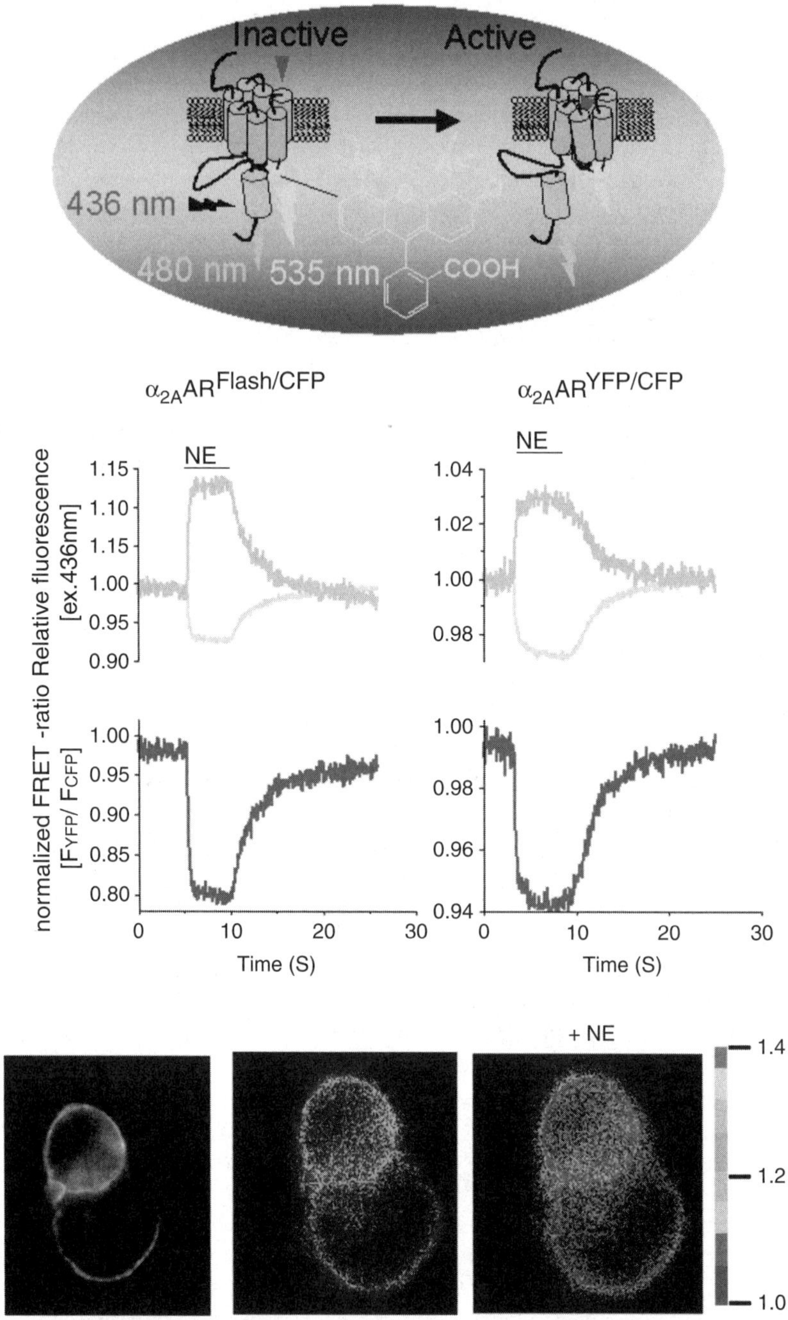

FIGURE 1 Recording and watching α_{2A}-adrenergic receptor activation. The *upper panel* represents the principle of FRET experiment with the chemical structure of Flash. The *central*

proteins can be expressed in live cells to monitor high temporal resolution protein–protein interaction, or to quantify kinetics in protein conformational changes (Miyawaki & Tsien, 2000).

FRET-based GPCR biosensors have been engineered by introducing CFP and YFP into domains of the receptor known to be conformationally sensitive. Constructs can be made with either CFP in the third intracellular loop and YFP in the C-terminus or *vice et versa* (Fig. 1). Each receptor construct (referred to as $GPCR^{CFP/YFP}$) is usually well expressed and functional, albeit G protein coupling may be reduced in some cases.

Experiments are performed with live cells placed under a fluorescence microscope, and selectively excited (light at 440 nm) to induce the CFP, and hence YFP emission fluorescence of the $GPCR^{CFP/YFP}$. Upon fast addition of an agonist, the YFP signal decreased and simultaneously the CFP signal increased, indicating a decrease in FRET, monitored as the ratio of emission intensities of YFP and CFP fluorescence. This FRET reduction reflects a conformational switch of the receptor that is compatible with receptor activation leading to G protein signaling. This strategy has been successfully applied to the α_{2A}-, β_1-, and β_2-adrenergic receptors among other GPCRs (Vilardaga et al., 2009). By exploiting this novel FRET-based approach, several studies addressed critically important questions on the mechanism of adrenergic receptor signaling. What is the kinetics of ligand-mediated receptor activation in live cells? Can receptors adopt multiple active states to induce distinct signaling pathways? How do inverse agonists act on receptors?

IV. RECEPTOR ACTIVATION IN LIVE CELLS

A. *Kinetics of Receptor Activation*

Kinetics of FRET changes of $GPCR^{CFP/YFP}$ in response to increasing concentrations of agonist revealed that the rate constant (k_{obs}) of receptor activation follows a hyperbolic dependence on ligand concentrations and reaches a

Figure 1 (Continued). panels show changes in the fluorescence of CFP and Flash (*left*) or CFP and YFP (*right*), and corresponding FRET ratio F_{YFP}/F_{CFP} in response to saturating concentration of norepinephrine (NE, 100 μM) recorded from a single HEK-293 cell expressing $\alpha_{2A}AR^{Flash/CFP}$ or $\alpha_{2A}AR^{YFP/CFP}$. Initial values of relative fluorescence (dark grey traces for CFP, and light grey traces for YFP or Flash) and the FRET ratio were set to one. Horizontal bars represent the duration of ligand application. The *lower panels* show FRET imaging of receptor activation in HEK293 cells transiently transfected with $\alpha_{2A}AR^{Flash/CFP}$. The *left* panel shows the epifluorescence image and the next two panels present the pseudocolored FRET ratio before and after stimulation by NE.

maximal value at saturating concentrations of agonist. This is indicative of a two-step process where a fast binding step is followed by a slower conformational change. At low agonist concentrations, the rate constants increase in proportion to agonist concentrations, indicating that agonist binding to the receptor is the rate-limiting step. At saturating concentrations of the agonist, the rate constant reached a limit, indicating that a step other than the binding between agonist-receptor is rate limiting. This limit is compatible with the ability of the agonist to mediate a conformational change of the receptor. The time constant (τ) of α_{2A}-, β_1-, and β_2-adrenergic receptor activation is $\approx$50 ms at a saturating concentration of a full agonist (Fig. 2) (Hoffmann et al., 2005; Reiner, Ambrosio, Hoffmann, & Lohse, 2010; Rochais, Vilardaga, Nikolaev, Bunemann, Lohse, & Engelhardt, 2007; Vilardaga et al., 2003; Vilardaga, Steinmeyer, Harms, & Lohse, 2005). This speed is considerably faster than that recorded for peptide hormone receptors such as the parathyroid hormone receptor ($\tau \approx$1 s). This difference in the activation time course might depend on intrinsic properties of receptors themselves (family A versus family B GPCRs), and also on the type of ligand and its mode of binding to the receptor (small molecules versus peptides) (Ferrandon et al., 2009).

B. Recording Intrinsic Efficacy of Ligands

Ligands can either stimulate fully (full agonists) or partially (partial agonists), or reduce (inverse agonists) the basal receptor activity. The term intrinsic efficacy was introduced as a fundamental parameter to differentiate and to

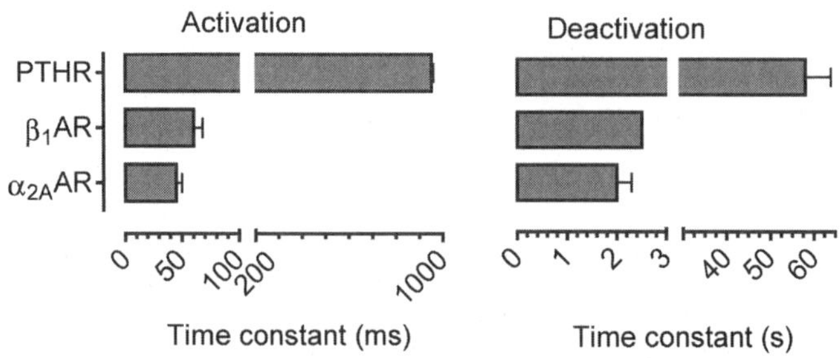

FIGURE 2 Kinetics of receptor activation/deactivation measured by FRET. Time constants of PTHR, α_{2A}AR, and β_1AR activation (measured in HEK293 cells expressing their corresponding FRET-based biosensors as shown in Fig. 1) in response to a saturating concentration of PTH (for PTHR), or norepinephrine (for adrenergic receptors). Deactivation time constants were measured after ligand washout.

classify the varying action of ligands when they occupy the same fraction of a single receptor to produce a functional response (Kenakin, 2002a, 2002b). The determination of intrinsic efficacies currently relies on indirect measurements of downstream physiological or biochemical responses (e.g., measurements of second messengers production such as cAMP among others, protein phosphorylation, level of gene expression, or smooth muscle cell relaxation) resulting from receptor activation that are dependent on receptor, G protein, and transducer expression levels. Using FRET-based GPCR biosensors can circumvent these difficulties by directly measuring the ligand-induced change in the receptor conformation itself, and which is independent from variation in either receptor number or cell conditions.

Studies with the α_{2A}-adrenergic receptor biosensor (α_{2A}-AR$^{CFP/FlAsH}$) expressed in human embryonic cells (HEK293) or neuron-like PC12 cells revealed that full, partial, and inverse agonists of different chemical structures and efficacies produced full, partial, and reverse FRET signals that correspond exactly to their predicted properties (Nikolaev, Hoffmann, Bunemann, Lohse, & Vilardaga, 2006; Vilardaga, Nikolaev, Lorenz, Ferrandon, Zhuang, & Lohse, 2008; Vilardaga et al., 2005). These studies not only revealed that ligands of different efficacies induce receptor's conformational changes of a different nature, magnitude, and kinetics, but also showed a direct correlation between the intrinsic efficacy of ligands and the kinetics of the conformational change in receptors (Figs. 3 and 4): fast conformational changes for full agonists (time constant $t \approx 50$ ms), progressively slower and smaller changes for partial agonists ($t_{1/2} < 1$ s), and even slower changes ($t \approx 1$ s) in the opposite direction with inverse agonists. The direct correlation between the relative agonist efficacy (measured by the extent of G protein activation with a FRET-based Gi biosensor; Bunemann, Frank, & Lohse, 2003) and the corresponding half-time of receptor activation and Gi activation is interpreted as evidence for the stabilization of distinct active states of the receptors, which in turn set the kinetics of Gi activation, as a mechanism to modulate the threshold of the cellular response. Given that an intramolecular FRET assay with GPCR biosensors is not dependent on transducer (G-proteins), effector proteins or receptor expression levels, the amplitude and the kinetics of changes in FRET can directly monitor intrinsic efficacy of compounds acting at a given GPCR.

C. Modulation of Ligand Efficacy by Receptor Polymorphism and Receptor Heteromers

The direct recording of ligand efficacy at the level of the receptor has permitted to associate intrinsic efficacies of β-blockers with the effects of a frequent β_1-adrenergic receptor (β_1-AR) polymorphism found in heart failure patients

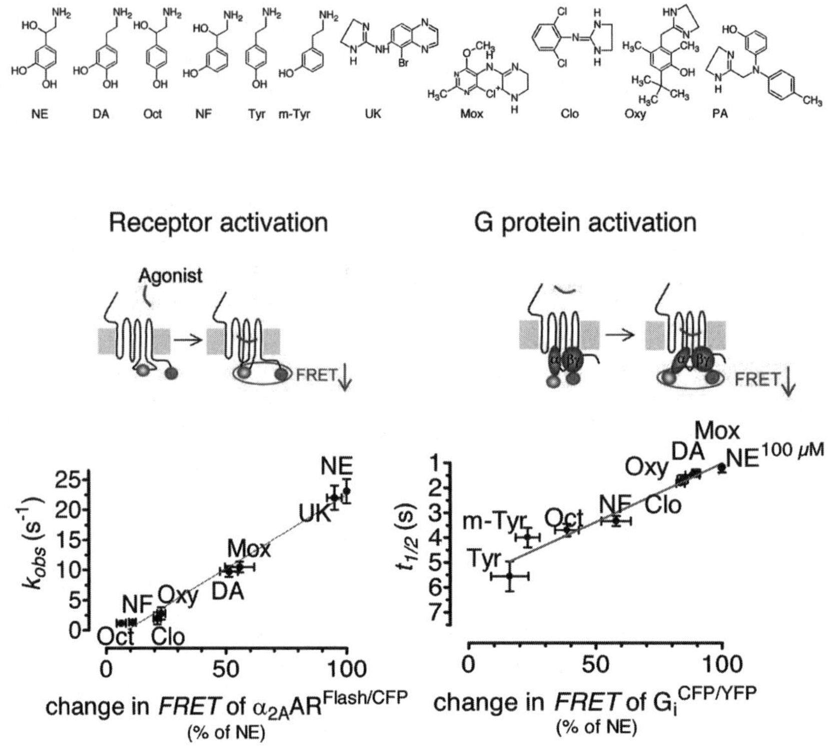

FIGURE 3 Action of ligand of different intrinsic efficacy on a FRET-based α_{2A}AR biosensor. FRET signals mediated by a saturating concentration of the agonist norepinephrine (NE), the partial agonist clonidine (Clo), and the inverse agonist rauwolscine (Rau). (Adapted from Vilardaga et al., 2005).

(Rochais et al., 2007). Patients with the Arg389 in the carboxy terminus tail of the β_1AR showed a better survival rate upon treatment with the β-blocker bucindolol than those with the Gly389 variant (Bristow, 2000). Other β-blockers currently in clinical use such as metoprolol, bisoprolol, and carvedilol act as inverse agonists of the human β_1AR, and have demonstrated significant improvement in survival in patients with severe chronic heart failure. However, the association between the genetic heterogeneity of the β_1AR and the effect of these β-blockers on receptor signaling has been unclear until a FRET study done with the $\beta_1AR^{CFP/YFP}$ biosensor revealed that carvedilol differentiated itself from both metoprolol and bisoprolol by a selective and strong inverse agonist effect at the more frequent Arg389-variant (Rochais et al., 2007). This FRET study thus suggests that carvedilol could be

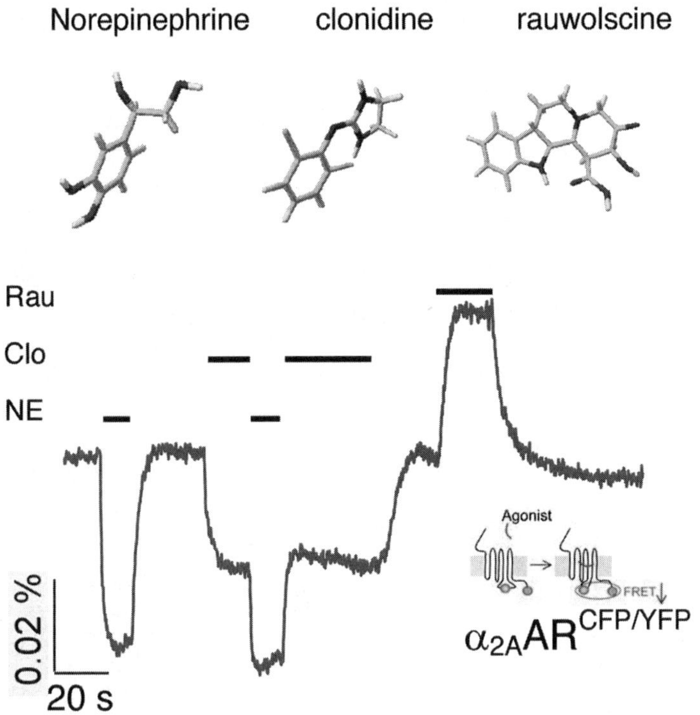

FIGURE 4 Multiple receptor conformations determine distinct kinetics of G-protein activation. *Top*: chemical structures of α_{2A}AR agonists: norepinephrine (NE), dopamine (DA), octopamine (Oct), norphenylephrine (NF), tyramine(Tyr), m-tyramine (m-Tyr), UK-14,304 (UK) moxonidine (Mox), clonidine (Clo), oxymetazoline (Oxy), and α_{2A}AR-antagonist phentolamine (PA). *Bottom left*: correlation between the rate constant of receptor activation (mean $\pm$ SE) and respective extent of FRET amplitude seen with different partial agonists (mean $\pm$ S.E.). *Bottom right*: correlation between half-time of G_i protein activation and agonist efficacy (where efficacy is defined as the ability of an agonist to produce a FRET response compared with the maximal response achieved by NE) analyzed by linear regression (red line, $r^2 = 0.95$, $p < 0.0001$). (Adapted from Nikolaev et al., 2006). (For interpretation of the references to colour in this figure legend, the reader is referred to the web version of this book.)

a better treatment for patients carrying the Arg389-variant β_1-AR. The FRET-based GPCR biosensors might thus be a valuable tool to fine-tune a therapy targeted at a receptor genotype.

Ligand efficacy can be modulated not only by point mutations encountered in receptor polymorphism but also by cross-conformational change that propagates from one receptor to another. This fascinating process has been experimentally demonstrated for the α_{2A}-adrenergic receptor (α_{2A}AR), which can form heterodimer (also named heteromer) complexes with the μ-opioid receptor

(α_{2A}AR/μOR) in neurons and cultured cells (Vilardaga et al., 2008). These receptors act either singly or as heteromer complexes that stimulate common signaling pathways through the inhibitory Gi protein. When α_{2A}AR/μOR heteromers are formed, morphine binding to the μ-opioid receptor decreases NE-mediated activation of both Gi-protein signaling and MAP kinases (ERK1/2) phosphorylation. The kinetic of this change is rapid ($t_{1/2} < 400$ ms) and faster than that of G protein activation ($t_{1/2} = 500$ ms). Thus, the conformational cross-talk between α_{2A}AR and μOR serves as a likely means to prevent over-stimulation of signaling pathways by rapidly adjusting both the intrinsic efficacy (the efficacy of NE is decreased in the morphine bound α_{2A}AR/μOR heteromer) and the extent of G protein activation to the two different ligands acting on the receptor heteromer.

Acknowledgment

This work was supported by the National Institutes of Health (NIH) grant DK087688.

References

Bristow, M. R. (2000). beta-adrenergic receptor blockade in chronic heart failure. *Circulation, 101*(5), 558–569.

Bunemann, M., Frank, M., & Lohse, M. J. (2003). Gi protein activation in intact cells involves subunit rearrangement rather than dissociation. *Proc Natl Acad Sci USA, 100*(26), 16077–16082.

Farrens, D. L., Altenbach, C., Yang, K., Hubbell, W. L., & Khorana, H. G. (1996). Requirement of rigid-body motion of transmembrane helices for light activation of rhodopsin. *Science, 274*(5288), 768–770.

Ferrandon, S., Feinstein, T. N., Castro, M., Wang, B., Bouley, R., & Potts, J. T., et al., (2009). Sustained cyclic AMP production by parathyroid hormone receptor endocytosis. *Nat Chem Biol, 5*(10), 734–742.

Gether, U., Lin, S., & Kobilka, B. K. (1995). Fluorescent labeling of purified beta 2 adrenergic receptor. Evidence for ligand-specific conformational changes. *J Biol Chem, 270*(47), 28268–28275.

Hanyaloglu, A. C., & von Zastrow, M. (2008). Regulation of GPCRs by endocytic membrane trafficking and its potential implications. *Annu Rev Pharmacol Toxicol, 48*, 537–568.

Hoffmann, C., Gaietta, G., Bunemann, M., Adams, S. R., Oberdorff-Maass, S., & Behr, B., et al., (2005). A FlAsH-based FRET approach to determine G protein-coupled receptor activation in living cells. *Nat Methods, 2*(3), 171–176.

Kenakin, T. (2002a). Drug efficacy at G protein-coupled receptors. *Annu Rev Pharmacol Toxicol, 42*, 349–379.

Kenakin, T. (2002b). Efficacy at G-protein-coupled receptors. *Nat Rev Drug Discov, 1*(2), 103–110.

Lefkowitz, R. J., Hausdorff, W. P., & Caron, M. G. (1990). Role of phosphorylation in desensitization of the beta-adrenoceptor. *Trends Pharmacol Sci, 11*(5), 190–194.

Lefkowitz, R. J., & Shenoy, S. K. (2005). Transduction of receptor signals by beta-arrestins. *Science, 308*(5721), 512–517.

Lohse, M. J., Benovic, J. L., Codina, J., Caron, M. G., & Lefkowitz, R. J. (1990). beta-Arrestin: a protein that regulates beta-adrenergic receptor function. *Science, 248*(4962), 1547–1550.

Lohse, M. J., Nikolaev, V. O., Hein, P., Hoffmann, C., Vilardaga, J. P., & Bunemann, M. (2008). Optical techniques to analyze real-time activation and signaling of G-protein-coupled receptors. *Trends Pharmacol Sci, 29*(3), 159–165.

Miyawaki, A., & Tsien, R. Y. (2000). Monitoring protein conformations and interactions by fluorescence resonance energy transfer between mutants of green fluorescent protein. *Methods Enzymol, 327*, 472–500.

Nikolaev, V. O., Hoffmann, C., Bunemann, M., Lohse, M. J., & Vilardaga, J. P. (2006). Molecular basis of partial agonism at the neurotransmitter alpha2A-adrenergic receptor and Gi-protein heterotrimer. *J Biol Chem, 281*(34), 24506–24511.

Perry, S. J., Baillie, G. S., Kohout, T. A., McPhee, I., Magiera, M. M., & Ang, K. L., et al., (2002). Targeting of cyclic AMP degradation to beta 2-adrenergic receptors by beta-arrestins. *Science, 298*(5594), 834–836.

Pierce, K. L., Premont, R. T., & Lefkowitz, R. J. (2002). Seven-transmembrane receptors. *Nat Rev Mol Cell Biol, 3*(9), 639–650.

Reiner, S., Ambrosio, M., Hoffmann, C., & Lohse, M. J. (2010). Differential signaling of the endogenous agonists at the {beta}2-adrenergic receptor. *J Biol Chem, 285*(46), 36188–36198.

Reiter, E., & Lefkowitz, R. J. (2006). GRKs and beta-arrestins: Roles in receptor silencing, trafficking and signaling. *Trends Endocrinol Metab, 17*(4), 159–165.

Rochais, F., Vilardaga, J. P., Nikolaev, V. O., Bunemann, M., Lohse, M. J., & Engelhardt, S. (2007). Real-time optical recording of beta1-adrenergic receptor activation reveals supersensitivity of the Arg389 variant to carvedilol. *J Clin Invest, 117*(1), 229–235.

Sheikh, S. P., Vilardarga, J. P., Baranski, T. J., Lichtarge, O., Iiri, T., & Meng, E. C., et al., (1999). Similar structures and shared switch mechanisms of the beta2-adrenoceptor and the parathyroid hormone receptor. Zn(II) bridges between helices III and VI block activation. *J Biol Chem, 274* (24), 17033–17041.

Sheikh, S. P., Zvyaga, T. A., Lichtarge, O., Sakmar, T. P., & Bourne, H. R. (1996). Rhodopsin activation blocked by metal-ion-binding sites linking transmembrane helices C and F. *Nature, 383*(6598), 347–350.

Vilardaga, J. P., Bunemann, M., Feinstein, T. N., Lambert, N., Nikolaev, V. O., & Engelhardt, S., et al., (2009). GPCR and G proteins: Drug efficacy and activation in live cells. *Mol Endocrinol, 23* (5), 590–599.

Vilardaga, J. P., Bunemann, M., Krasel, C., Castro, M., & Lohse, M. J. (2003). Measurement of the millisecond activation switch of G protein-coupled receptors in living cells. *Nat Biotechnol, 21* (7), 807–812.

Vilardaga, J. P., Frank, M., Krasel, C., Dees, C., Nissenson, R. A., & Lohse, M. J. (2001). Differential conformational requirements for activation of G proteins and the regulatory proteins arrestin and G protein-coupled receptor kinase in the G protein-coupled receptor for parathyroid hormone (PTH)/PTH-related protein. *J Biol Chem, 276*(36), 33435–33443.

Vilardaga, J. P., Nikolaev, V. O., Lorenz, K., Ferrandon, S., Zhuang, Z., & Lohse, M. J. (2008). Conformational cross-talk between alpha2A-adrenergic and mu-opioid receptors controls cell signaling. *Nat Chem Biol, 4*(2), 126–131.

Vilardaga, J. P., Steinmeyer, R., Harms, G. S., & Lohse, M. J. (2005). Molecular basis of inverse agonism in a G protein-coupled receptor. *Nat Chem Biol, 1*(1), 25–28.

Wess, J., Han, S. J., Kim, S. K., Jacobson, K. A., & Li, J. H. (2008). Conformational changes involved in G-protein-coupled-receptor activation. *Trends Pharmacol Sci, 29*(12), 616–625.

Yao, X., Parnot, C., Deupi, X., Ratnala, V. R., Swaminath, G., & Farrens, D., et al., (2006). Coupling ligand structure to specific conformational switches in the beta2-adrenoceptor. *Nat Chem Biol, 2* (8), 417–422.

CHAPTER 6

Modulation of Immune Cell Function by α_1-Adrenergic Receptor Activation

Laurel A. Grisanti[1], Dianne M. Perez[2] and James E. Porter[1]

[1]Department of Pharmacology, Physiology, and Therapeutics, School of Medicine and Health Sciences, University of North Dakota, Grand Forks, North Dakota, USA
[2]Department of Molecular Cardiology, Lerner Research Institute, The Cleveland Clinic Foundation, Cleveland, Ohio, USA

I. Overview
II. Introduction
III. α_1-Adrenergic Receptor Expression in the Immune System
IV. The Role of α_1-Adrenergic Receptors in the Innate Immune System
  A. Monocytes
  B. Macrophages
  C. Dendritic Cells
  D. Neutrophils
  E. Mast Cells
  F. Natural Killer Cells
V. α_1-Adrenergic Receptor Influences on the Adaptive Immune System
  A. Peripheral Blood Mononuclear Cells
  B. T lymphocytes
  C. B lymphocytes
VI. α_1-Adrenergic Receptors in Immune Tissues
  A. Spleen
  B. Thymus
  C. Blood/Circulating Cytokines
  D. Nonimmune Tissue
VII. α_1-Adrenergic Receptors in Disease States
VIII. Conclusions
  References

I. OVERVIEW

The sympathetic nervous system regulates human immune system functions through epinephrine (Epi) and norepinephrine (NE) activation of adrenergic receptors (ARs) expressed on immunocompetent cell populations. The

Current Topics in Membranes, Volume 67 0065-230X/10 $35.00

DOI: 10.1016/B978-0-12-384921-2.00006-9

anti-inflammatory effects that are most often attributed to increased sympathetic activity have been shown to occur through β_2- and α_2-AR stimulation. However, dichotomous AR effects on immune system responses are becoming increasingly apparent. Reports of α_1-AR expression on immune cell populations have been conflicting due to a lack of specific antibodies or subtype-selective receptor ligands. This has made α_1-AR identification difficult and limits further characterization of α_1-AR subtype expression. Nevertheless, there is some evidence suggesting an induction of α_1-AR expression on immunocompetent cells under certain physiological conditions and disease states. Also, the function of α_1-AR activation to modulate immune responses is just beginning to emerge in the literature. Changes in the secretion of inflammatory mediators as well as increased cell migration and differentiation have been described following α_1-AR stimulation on immunocompetent cells. These observations demonstrate the significance of α_1-AR activity in immune cell biology and emphasize the importance for understanding α_1-AR effects on the immune system.

II. INTRODUCTION

The endogenous catecholamines Epi and NE are critical for initiating the "fight or flight" response of the sympathetic nervous system. Epi and NE are released from peripheral neurons and the adrenal medulla in response to physical as well as psychological stress to regulate a number of physiological functions including energy metabolism, cardiovascular homeostasis, and thermal adaptation. There are extensive interactions of the central nervous system with the immune system and all immune organs are innervated by postganglionic sympathetic fibers. Furthermore, sympathetic nerve terminals are located in the vicinity of immune cells that comprise both the innate and adaptive immune system. Moreover, macrophages have recently been shown to synthesize and release catecholamines *in vivo* (Flierl et al., 2007). Consequently, the close propinquity of catecholamines released onto the cells of the immune system introduces an opportunity for these endogenous AR agonists to regulate immune cell functions.

AR-mediated sympathetic responses to stress are a result of receptor agonist stimulation caused by the increased release of Epi and NE. The AR family is classified according to type (α_1-, α_2-, and β-AR), which can be further characterized into nine distinct receptor subtypes (α_{1A}-, α_{1B}-, α_{1D}; α_{2A}-, α_{2B}-, α_{2C}-; β_1-, β_2-, and β_3-AR; see review by Guimarães & Moura, 2001). All three AR types are expressed in the immune system and like glucocorticoid receptors are considered immunosuppressive when activated by Epi or NE. However, there is a growing body of evidence to suggest that AR activation influences the immune response in a less monochromatic way.

AR activation serves many functions in the immune system including modifying the number or proportion of cells participating in an immune response as well as altering individual immune cell responsiveness (Calcagni & Elenkov, 2006; Bao et al., 2007; Pešić et al., 2009). In addition, a variety of immune cell activities are modulated by AR stimulation including cell proliferation, cytokine production, lytic activity, migration, and antibody production (Maestroni, 2000; Seiffert et al., 2002; Pešić et al., 2009; Grisanti et al., 2010). Studies examining the β-AR family are the most extensive, with the "anti-inflammatory" β_2-AR subtypes thought to be the predominant AR expressed in the immune system (Elenkov et al., 2000). However, there is growing evidence to suggest a "pro-inflammatory" function of β-AR activation, which is mediated through the β_1-AR subtype (Grisanti et al., 2010). The α_2-AR family has also been extensively investigated and again is regarded as having anti-inflammatory effects when activated (Elenkov et al., 2000). The α_1-AR family is the least characterized AR in the immune system, which is likely due to conflicting reports of their expression as well as function on immune cells (Ricci et al., 1999; Elenkov et al., 2000; Tayebati et al., 2000).

III. α_1-ADRENERGIC RECEPTOR EXPRESSION IN THE IMMUNE SYSTEM

The three characterized α_1-AR subtypes (α_{1A}-, α_{1B}-, and α_{1D}-) are differentially expressed in many organs and cells of the immune system. Investigation of α_1-AR expression in immune tissues has relied heavily on RT-PCR analysis. Little information is available about α_1-AR subtype localization at the protein level in the immune system since commercially available antibodies have been shown to be nonselective in wild-type and transgenic animal models (Jensen et al., 2009). Therefore, most studies have been performed utilizing PCR techniques, which are prone to contamination, or radioligand binding studies that used nonselective ligands. α_1-AR expression is found in murine hematopoietic stem cell progenitor cells during all stages of development from bone marrow to monocytes/macrophages (Muthu et al., 2007). Murine bone marrow expresses α_{1A}- and α_{1B}-AR mRNA while human bone marrow transcriptionally expresses the α_{1B}- and α_{1D}-AR subtypes (Maestroni et al., 1992; Kavelaars, 2002). High levels of α_{1A}- and α_{1B}-AR mRNA are present in the human spleen (Price et al., 1993; Faure et al., 1995), while others have reported the expression for all three α_1-AR subtypes (Kavelaars, 2002).

The majority of studies examining immune cell α_1-AR expression have been performed on peripheral blood mononuclear cell (PMBC) preparations, which include several blood cell types including T cells, B cells, NK cells, monocytes, and macrophages. Reported α_1-AR expression on PMBC preparations as well as many of the individual cell populations is contradictory. There are numerous reports documenting an absence of α_1-AR expression using PMBC preparations (Casale & Kaliner, 1984; Kavelaars, 2002). Conversely, others have shown no genomic α_1-AR expression on PMBC preparations under normal culturing conditions, but expression could be induced for all three α_1-AR subtypes following phytohemagglutinin (PHA) or lipopolysaccharide (LPS) stimulation (Rouppe van der Voort et al., 2000). *In situ* hybridization techniques have been used to show that the majority of cells in a PMBC preparation are positive for α_{1B}- and α_{1A}-AR expression with α_{1D}-AR subtypes found to a lesser extent (Tayebati et al., 2000). Immunocytochemistry analysis in this same study confirmed that a majority of PMBCs express the mature α_{1B}-AR protein with fewer cells expressing the α_{1A}- and α_{1D}-AR subtypes. In other immune cell types, genomic expression of the α_{1A}-AR subtype has been shown in RNA isolated from rat microglia, the resident macrophage of the brain (Mori et al., 2002). Radioligand binding analysis has described the representative expression of all three AR families on human NK cells (Jetschmann et al., 1997). Immature dendritic cells (DCs) have been shown to express α_{1B}-AR mRNA, which is lost upon maturation (Maestroni, 2000).

Studies that characterize α_1-AR expression on monocytes have been controversial. The human monocytic cell line, THP-1, has been shown to endogenously express α_{1B}- and α_{1D}-AR mRNA, while genomic α_{1A}-AR expression could be induced following treatment with tissue necrosis factor (TNF)-α or interleukin (IL)-1β (Heijnen et al., 2002). Furthermore, evidence for functional α_1-AR expression on primary monocytes isolated from human blood has been described (Takahashi et al., 2005). Conversely, other reports have documented no detectible α_1-AR mRNA from human monocytes (Rouppe van der Voort et al., 1999). However, this same comprehensive study demonstrated that monocytes cultured in the presence of a glucocorticoid, dexamethasone, or the β_2-AR agonist, terbutaline, resulted in the induced expression of α_{1B}- and α_{1D}-AR mRNA. In addition, upregulation of cAMP-dependent protein kinases using dibutyryl cAMP specifically increased α_{1B}-AR mRNA expression. These authors also utilized immunoblot techniques and radioligand binding analysis to confirm α_1-AR expression changes at the protein level.

The thymus, an important organ of the adaptive immune system where T cell maturation and differentiation occurs, has been reported to have low mRNA expression for all three α_1-AR subtypes (Kavelaars, 2002). Further translational analysis using immunohistochemistry techniques confirmed α_1-AR thymus expression, predominantly in the subcapsulary/subtrabecular cortex and corticomedullary junctions, but rarely in the thymic medulla (Pešić et al., 2009). α_1-AR expression was found in this study primarily on thymic epithelial cells, but also was shown on clusters of differentiated (CD)68$^+$ cells, a monocyte/macrophage marker. Only a small portion of T cell precursors or thymocytes, showed mature α_1-AR protein expression and of this population, the majority were CD3$^-$ cells with lower expression found on CD3^{low} and CD3^{high} thymocytes. Others have demonstrated that matured thymocytes or lymphocytes, isolated from the blood of healthy human patients, express mRNA for all three α_1-AR subtypes, with α_{1B}-AR expression being the highest (Ricci et al., 1999). Radioligand binding analysis in this investigation confirmed mature translational expression on lymphocytes with a calculated [^{3}H]-prazosin affinity (K_d) value of 0.65 ± 0.05 nmol/L and total receptor density (B_{max}) of 175 ± 20 fmol/10^6 cells for these α_1-ARs.

IV. THE ROLE OF α_1-ADRENERGIC RECEPTORS IN THE INNATE IMMUNE SYSTEM

The human innate immune system is a nonspecific means of defense against pathogenic challenges. This generic means of defense is thought to be a more evolutionary primitive design compared to the adaptive immune system. Protective mechanisms initiated by innate immune responses include recruitment of immunocompetent cells at sites of infection, production of chemical mediators such as cytokines, activation of the complement cascade to identify and clear invading pathogens, the removal of foreign substances by white blood cells, and activation of the adaptive immune system through antigen presentation. Physical barriers such as epithelial cells as well as homeostatic mechanisms such as peristaltic movement, tears, and mucus production help to prevent colonization and expedite removal of invading pathogens. Furthermore, innate inflammatory responses create a biological barrier through the release of chemical factors from injured cells, which establishes an additional obstruction against the spread of infection while promoting healing by increasing pathogen clearance. Cells of the innate immune system provide the first line of defense against invading pathogens through recognition of pathogen-associated molecular patterns (PAMPs) by pattern recognition receptors (PRRs), which initiates cellular and humoral responses. The complement system is a protease C3-convertase cascade, which when activated leads to the recruitment of inflammatory cells while at the same time opsonizing infected cells for destruction through disruption of the plasma membrane resulting in cytolysis.

A. Monocytes

Monocytes are a type of white blood cell that respond rapidly to inflammatory signals by moving into the affected tissue and differentiating into macrophages and DCs. Monocytes are involved in phagocytosis, antigen presentation, and are major producers of proinflammatory cytokines. Reported α_1-AR expression on monocytes has been variable. Using RT-PCR to examine transcriptional expression levels in the human monocytic cell line, THP-1, it was shown that these cells express mRNA for α_{1B}- and α_{1D}-AR subtypes only (Heijnen et al., 2002). Conversely, other reports using primary human monocytes demonstrated no detectable levels for any α_1-AR mRNA (Rouppe van der Voort et al., 1999). There is some evidence suggesting that monocyte α_1-AR expression levels change during certain culturing conditions, possibly explaining the variable reports found in the literature. For example, culturing primary monocytes in the presence of dexamethasone or terbutaline induces α_{1B}- and α_{1D}-AR mRNA expression (Rouppe van der Voort et al., 1999). In other reports, addition of the proinflammatory cytokines TNF-α or IL-1β into the media induced genomic α_{1A}-AR expression, while at the same time decreasing α_{1D}-AR subtype expression (Heijnen et al., 2002).

Little evidence is found in the literature as to the function of α_1-AR activation on monocytes. There has been some suggestion that inhibition of α_1-AR signaling on monocytes regulates migration. For example, migration of THP-1 or PMBCs in response to monocyte chemotactic protein-1 (MCP-1) is dose-dependently attenuated by administration of the α_1-AR antagonists doxazosin or phenoxybenzamine (Kintscher et al., 2001). However, this inhibitory effect on the monocyte migratory response was suggested to occur independent of α_1-AR blockade, possibly through enhanced expression of tissue inhibitor of metalloproteinases 1 (TIMP-1). There is also compelling evidence suggesting that α_1-AR stimulation by Epi and NE enhance compliment synthesis. Studies have shown that enhanced complement component 2 (C2) was synthesized from PBMCs treated with increasing concentrations of the selective α_1-AR agonist, phenylephrine (PE), but not when the selective β-AR agonist isoproterenol was used (Lappin & Whaley, 1982). Furthermore, use of the selective α-AR antagonist, phentolamine as well as the selective α_1-AR antagonist, prazosin but not the selective α_2- or β-AR antagonists, yohimbine and propranolol respectively, abrogated the increased C2 synthesis observed in monocytes treated with Epi, NE, and PE. Moreover, other complement cascade components including C4, C3, C5, factor B, properdin, C3bINA, and β1H were also observed to be increased following PE treatment. Inhibition of monocyte α_1-AR activation also influences the expression of signaling components related to T cell activation. Preparations of PBMCs treated with the quinazoline-based α_1-AR antagonists doxazosin, prazosin, and terazosin, induced the expression of intercellular adhesion molecule-1 (ICAM-1) and CD40 in a concentration-dependent manner (Takahashi et al., 2005). This study also demonstrated decreased production of the proinflammatory cytokine IL-18 from PBMCs using these same receptor antagonists. Alternatively the presence of selective α_2-, β_1-, or β_2-AR antagonists did not change levels of ICAM-1, CD40, or IL-18. Recent studies have shown that murine hematopoietic progenitor cells, ER-MP20$^+$, which are monocyte committed cells, express α_1-ARs on their cell surface that function to increase LPS-mediated TNF-α secretion (Muthu et al., 2007).

Our laboratory has also examined the influence of α_1-AR activation to modulate proinflammatory cytokine production from pathogenically challenged human monocytes. Using LPS, which is a component of Gram negative bacterial cell walls and a potent endotoxin, to model inflammation, we investigated changes in the cytokine profile generated from monocytes concurrently treated with PE. An antibody array containing 40 specific antibodies for known mediators of inflammation immobilized on a membrane support (Table I; RayBiotech, Norcross, GA) was used to screen for modulated cytokines/chemokines in response to individual or combined 3 h treatments with 10 μM PE and 25 ng/mL LPS as described previously (Fig. 1; Grisanti et al., 2010). Expression levels of inflammatory mediators from quiescent THP-1 cells were relatively low with the strongest basal expression observed for IL-8 and macrophage inflammatory protein (MIP)-1β (Fig. 1A). Although a majority of inflammatory mediators did not change for monocytes treated with PE when compared to control, minor qualitative changes in the levels of MIP-1α, MIP-1β, and IL-8 were observed (Fig. 1B). As expected, treatment with LPS qualitatively increased monocyte secretion for many inflammatory mediators including IL-1β, TNF-α, IL-6, the IL-6 receptor, IL-8, IL-10, ICAM-1, TIMP-2, and RANTES when compared to basal levels (Fig. 1C). Finally, there were characteristic changes in the inflammatory protein expression pattern secreted from monocytes treated concomitantly with PE and LPS (Fig. 1D). There was a significant qualitative increase in the levels of IL-1β released from PE plus LPS treated monocytes when compared to LPS alone. Conversely, there was a qualitative loss of LPS-mediated TNF-α, IL-8, and MIP-1β levels in the presence of PE plus LPS when compared to monocytes treated with LPS alone. These results initially characterize the functional regulation of α_1-AR activation for multiple inflammatory factors from LPS-challenged monocytes and demonstrate the utility of antibody arrays to analyze several mediators of the innate immune response simultaneously.

B. Macrophages

Macrophages are phagocytes residing in tissues functioning as antigen presentation cells to stimulate responses of the adaptive immune system. They also have an important regulatory role in the development of innate immune responses by producing chemical substances including complement proteins, cytokines, chemokines, and proteolytic enzymes. There are limited

TABLE I

Map of cytokine/chemokine antibody array

POS	POS	NEG	NEG	EOTAXIN	EOTAXIN-2	GCSF	GM-CSF	ICAM-1	IFN-γ	I-309	IL-1α
POS	POS	NEG	NEG	EOTAXIN	EOTAXIN-2	GCSF	GM-CSF	ICAM-1	IFN-γ	I-309	IL-1α
IL-1β	IL-2	IL-3	IL-4	IL-6	IL-6Sr	IL-7	IL-8	IL-10	IL-11	IL-12 p40	IL-12 p70
IL-1β	IL-2	IL-3	IL-4	IL-6	IL-6Sr	IL-7	IL-8	IL-10	IL-11	IL-12 p40	IL-12 p70
IL-13	IL-15	IL-16	IL-17	IP-10	MCP-1	MCP-2	M-CSF	MIG	MIP-1α	MIP-1β	MIP-1δ
IL-13	IL-15	IL-16	IL-17	IP-10	MCP-1	MCP-2	M-CSF	MIG	MIP-1α	MIP-1β	MIP-1δ
RANTES	TGF-β1	TNF-α	TNF-β	s TNF RI	s TNF RII	PDGR-BB	TIMP-2	BLANK	BLANK	NEG	POS
RANTES	TGF-β1	TNF-α	TNF-β	s TNF RI	s TNF RII	PDGR-BB	TIMP-2	BLANK	BLANK	NEG	POS

Positions on membrane support of antibodies that recognize specific mediators of inflammation. Abbreviations used are: POS, positive biotinylated protein control; NEG, negative BSA control; GCSF, granulocyte-colony stimulating factor; GM-CSF, granulocyte-macrophage colony stimulating factor; ICAM, intercellular adhesion molecule; IFN, interferon; IL, interleukin; IL-6Sr, interleukin-6 soluble receptor; IL-12 p40, interleukin-12 p40 subunit; IL-12 p70, interleukin-12 p70 subunit; IP-10, IFN receptor inducible protein 10; MCP, monocyte chemoattractant protein; M-CSF, macrophage-colony stimulating factor; MIG, monokine induced by γ-interferon; MIP, macrophage inflammatory protein; TGF, tissue growth factor; TNF, tissue necrosis factor; s TNF R, soluble tumor necrosis factor receptor; PDGR, platelet-derived growth factor; TIMP, tissue inhibitor of metalloproteinases.

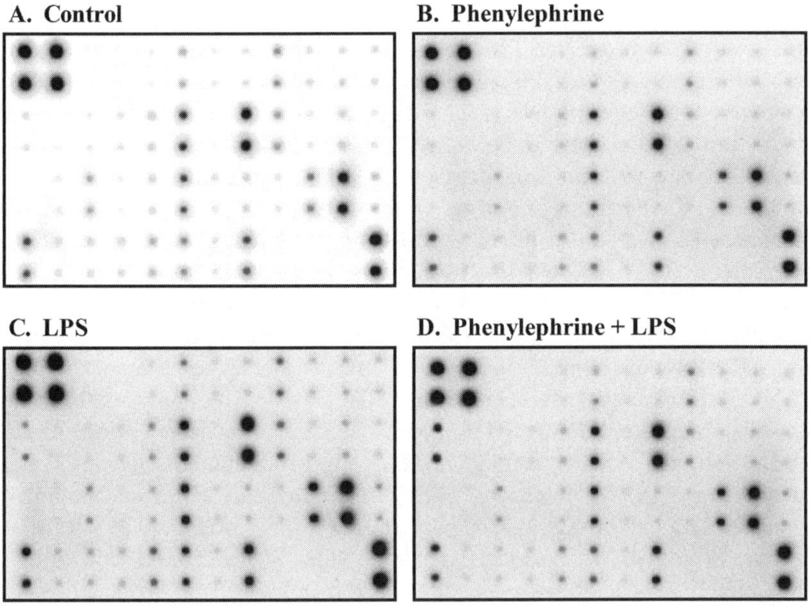

FIGURE 1 Representative antibody array membrane for specific inflammatory cytokines (see Table I for layout) incubated with conditioned media taken from THP-1 cells treated with (A) culture media only, (B) 10 μM PE, (C) 25 ng/mL LPS, or (D) 10 μM PE plus 25 ng/mL LPS.

investigations characterizing the expression profile or function of α_1-ARs on macrophages. The previously described ER-MP20$^+$ monocyte committed progenitor cells in mice, which only differentiate into monocytes and macrophages, express functional α_1-ARs that lead to increases in TNF-α secretion through a cooperative mechanism with Toll-like receptor (TLR)4 (Muthu et al., 2007). In the rat thymus, immunohistochemistry techniques co-localized α_1-AR expression with the monocyte/macrophage marker CD68 (Pešić et al., 2009). Functional α_1-ARs were identified on murine RAW264 macrophages when PE and other protein kinase C (PKC) activating agents were used to initiate cell spreading (Petty, 1989). In other studies, PE was also shown to increase primary peritoneal macrophage phagocytosis (Javierre et al., 1975). Kupffer cells are resident macrophages of the liver, which are important for cell–cell communication and are essential for the physiological immune response of the liver. Kupffer cell preparations from control and tumor-bearing rats increased production of prostaglandin (PG)E_2 when treated with PE, which could be blocked by the administration of prazosin (Seelaender et al., 1999). Resident macrophages of the brain (microglia) in the rat have been shown to transcriptionally express the α_{1A}-AR subtype (Mori et al., 2002). These studies also demonstrated that culturing microglia with LPS and L-serine then treating with PE significantly decreased TNF-α and IL-6 at the transcriptional and translational levels as well as inhibited the production of nitric oxide (NO). PE treatments have also been shown to have an inhibitory effect on LPS-induced NO production in the murine N9 microglia cell line (Chang & Liu, 2000). Osteoclasts are resident bone macrophages that function to resorb old bone formations. Using the receptor activator of nuclear factor *kappa*-light-chain-enhancer of activated B cells (NF-κB) ligand (RANKL) to differentiate murine macrophage RAW264 cells into osteoclasts, investigations have characterized α_{1A}-AR mRNA expression in macrophages with higher levels observed in osteoclasts (Suga et al., 2010). Conversely, transcriptional expression of the α_{1B}-AR subtype was only observed in RAW264 cells and was not found in RANKL differentiated osteoclasts (Suga et al., 2010). This study also examined the postulated α_1-AR-mediated neuro-osteogenic network using primary murine superior cervical ganglia and RANKL differentiated osteoclasts cocultures. Neurite activation was evoked by treatment with scorpion venom (SV), which subsequently resulted in osteoclast activation as measured by Ca^{2+} mobilization. Treatment of osteoclasts alone with SV exhibited no response. Pretreatment of cocultures with prazosin did not affect SV-mediated neurite activation but did inhibit the osteoclast Ca^{2+} response. Subsequent treatment of RANKL differentiated osteoclasts with PE increased the synthesis of IL-6 validating functional α_1-AR expression on these cells.

C. Dendritic Cells

Dendritic cells (DCs) are important components of the innate immune system, which process and present antigen material to activate cells of the adaptive immune system. Immature murine DCs express mRNA for the α_{1B}-AR subtype, which is lost upon maturation in the lymph nodes (Maestroni, 2000). Correspondingly, migration of immature Langerhans cells, skin DCs, to the lymph nodes in response to NE was inhibited by prazosin, but not by propranolol. Interestingly, pretreatment with yohimbine had the opposite effect by increasing NE-mediated Langerhans cell migration. Conversely, other investigations have identified α_{1A}-AR mRNA expression in murine Langerhans cells as well as in a DC line (Seiffert et al., 2002).

D. Neutrophils

Neutrophils are the most abundant white blood cell in mammals and are vital to the innate immune response. Under normal conditions, neutrophils reside in the blood and upon initiation of inflammation migrate towards the site of injury. An important function of neutrophils at the site of injury is to release cytokines, which amplify the inflammatory response. Additionally, neutrophils are phagocytes that ingest and destroy microorganisms or cell debris. Like many cells of the innate immune system, literature reports of neutrophil α_1-AR expression are mixed. Early radioligand binding investigations of polymorphonuclear leukocyte (PMN) preparations isolated from human blood, which is a mixed population of cells comprised of eosinophils, basophils, and neutrophils, demonstrated no specific binding for [^{3}H]-prazosin (Casale & Kaliner, 1984). While there is little evidence to suggest functional α_1-ARs on neutrophils, α_1-AR activation has a positive effect on neutrophilia. For example, LPS dramatically increases the number of circulating neutrophils in rats 3 h following injection, while pretreating these animals with reserpine, which depletes catecholamine levels by blocking the vesicular monoamine transporter, significantly decreased blood neutrophilia as a result of LPS (Altenburg et al., 1997). Similarly, pretreatment with phentolamine or prazosin inhibited the LPS-induced increase in blood neutrophil counts, while use of PE alone caused neutrophilia in the absence of LPS. The effects of LPS on neutrophilia in rats were not affected by pretreating animals with yohimbine or propranolol.

E. Mast Cells

Mast cells are resident immune cells that contain numerous secretory granules and are best known for their role in allergic and anaphylactic responses. However, mast cells also play an important role in wound healing and host defense mechanisms against pathogens. Mast cell activation by cross-linking immunoglobulin (Ig)E receptors or complement proteins, causes cell degranulation releasing inflammatory mediators into the interstitium. Mature α_1-AR expression has been shown on mast cells from cultured neonatal rat hearts using immunocytochemistry techniques (Schulze & Fu, 1996). In a murine mast cell line, increasing concentrations of PE or NE enhanced [^{14}C]-histamine release, which was blocked in the presence of phentolamine (Moroni et al., 1977). Other investigations have shown a correlation between decreased amounts of the degranulation marker, mast cell peroxidase (MPO), and the protective effects of NE pretreatment on a rat model of heart ischemia-reperfusion injury (Parikh & Singh, 1999). Following NE pretreatment, prazosin administration during the ischemia manipulations reversed the MPO decrease observed in ischemic-NE preconditioned animals, suggesting an α_1-AR mediated response.

F. Natural Killer Cells

Natural killer (NK) cells are a large, granular type of cytotoxic lymphocyte that play a major role in the rejection of tumors and viral infected cells through release of cytotoxic secretory granules (Paust et al., 2010). They were named NK cells because they do not require any preceding pathogenic stimulation to initiate cell killing. Activation of NK cells by cytokines, the Fc portion of antibodies that bind to Fc receptors, or other activating/inhibiting receptors leads to the release of secretory granules containing perforin and granzyme causing the target cell to die by apoptosis. Radioligand binding has been performed on $CD16^+$ cells isolated from human blood, which identified α_1-, α_2-, and β-AR expression on NK cells (Jetschmann et al., 1997). Expression of α-ARs on NK cells appears to vary depending on the external stimuli. For example, infusion of Epi decreased expression of β_2- and α_1- but not α_2-ARs while NE had no effect on AR expression (Jetschmann et al., 1997). Morphine exposure suppresses splenic NK activity, which occurs, in part through α_1-ARs. In lymphocyte populations isolated from mouse spleen, phentolamine or prazosin administration reverses the effects of morphine on NK cell activity, while yohimbine showed no change from morphine administration alone (Carr et al., 1993).

To circumvent problems with nonspecific antibodies and nonselective ligands, our laboratory used α_{1A}-AR-enhanced green fluorescent protein (EGFP) tagged transgenic mice to assess lymphocyte populations (Papay et al., 2006). These mice are under the control of the endogenous promoter and therefore express the α_{1A}-AR subtype in all naturally occurring cell types throughout the body (Rorabaugh et al., 2005). We focused our investigations on the liver not only because this organ harbors a large population of innate immune cells, but also because of the bright green cells present in the sinusoids, which play a key role in systemic innate immunity regulated by the liver.

DX5 antibody recognizes the CD49b antigen (Arase et al., 2001) that is expressed on the vast majority of mouse NK cells as well as on 5% of $CD8^+$ cytotoxic T cells (Kambayashi et al., 2001). $DX5^+$ NK cells also display an increased cytotoxicity and indicate that functional subsets exist among NK cell population (Arase et al., 2001). B (bone marrow-derived) lymphocytes (cells) not only play a pivotal role in humoral immunity through the production of antibodies, but are also involved in antigen presentation and regulation of T-cell function (LeBien & Tedder, 2008). CD19 is a specific B cell marker and is present on the earliest B lineage cells during development. CD3 is a general marker for most T cells as it is part of the T cell receptor (TCR) complex present on adult T cells. Both T and B cells also possess the ability to remember encountered antigens in the form of memory cells, which mediates adaptive immunity (i.e., subsequent immune responses to an encounter, which is different from the first).

We found that the α_{1A}-AR subtype were not expressed in $CD3^+$ T cells (Fig. 2A), but were instead expressed in $CD19^+$ B cells (Fig. 2B) and in $DX5^+$ NK cells (Fig. 2C). As expected, the α_{1A}-AR subtype was also highly expressed in liver vasculature (Fig. 2D). Hepatic NK cells are located in the sinusoids as indicated in Figure 2C and high levels are present in the liver, more than any other organ (Nemeth et al., 2009). Since the liver is a target organ for the metastasis of many cancers and for innate immunity, high levels of NK cells may have an effective antitumor effect (Subleski et al., 2006). Localization of the α_{1A}-AR in liver vasculature and immune cells may also account for the high expression levels of this subtype found in liver membrane preparations analyzed by ligand binding (Rorabaugh et al., 2005) even though the α_{1B}-AR subtype is dominant in the rodent liver (Yang et al., 1998). B cells are of low abundance in the liver comprising less than 10% of the lymphocyte population (Nemeth

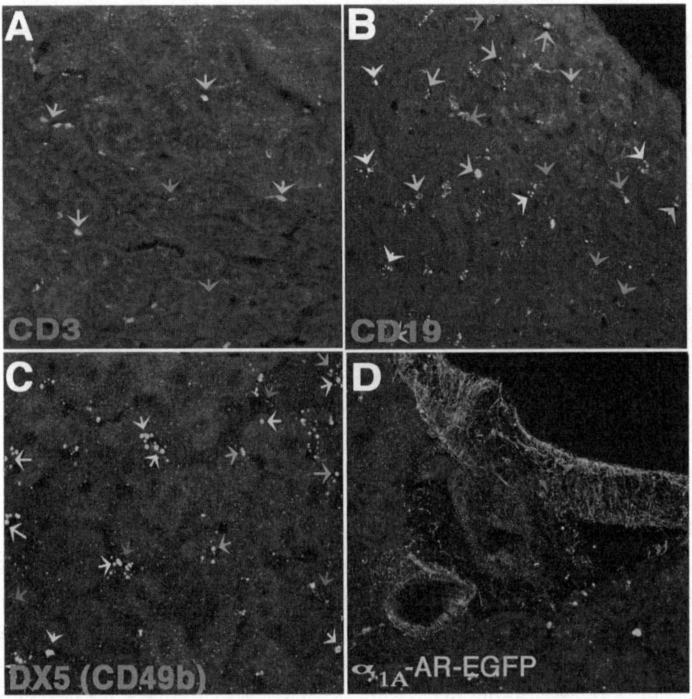

FIGURE 2 Immunohistochemistry of α_{1A}-AR-EGFP liver tissue sections reacted with (A) CD3 antibody (T cells), (B) CD19 antibody (B cells), or (C) DX5 antibody (NK cells). (D) Designates α_{1A}-AR-EGFP expression in the liver vasculature. Red immunofluorescence indicates the antibody and cell type as indicated. Yellow indicates that the α_{1A}-AR co-localized with this cell type. Green indicates cells expressing only the α_{1A}-AR. (For interpretation of the references to colour in this figure legend, the reader is referred to the web version of this book.)

et al., 2009). However, recent evidence suggests that B cell lymphopoiesis occurs in the liver sinusoids by endothelial cells and we therefore speculate this may account for the high degree of colocalization in smaller cells expressing the α_{1A}-AR (Wittig et al., 2010).

V. α_1-ADRENERGIC RECEPTOR INFLUENCES ON THE ADAPTIVE IMMUNE SYSTEM

The highly specialized adaptive immune system allows the host to recognize and remember specific pathogens so that a strong attack can be mounted every time the pathogen is reencountered. The system is highly pliant allowing a small number of host genes to generate huge numbers of diverse antigen receptors uniquely expressed on individual lymphocytes. The adaptive immune system functions to recognize specific nonself antigens to generate maximal effective responses tailored to eliminate specific pathogens or pathogen-infected cells. Through this initial response the adaptive immune system also develops immunological memory by forming unique antibodies so that memory cells can be called upon to quickly eliminate the pathogen upon subsequent infections. Lymphocytes are the effector cells of the adaptive immune system of which there are two main types, B cells and T cells. Mature lymphocytes that have left the bone marrow or thymus and entered into the lymphatic system are naïve and have yet to encounter their cognate antigen. Upon activation by the B or T cell's cognate antigen, they become effector cells, which are actively involved in eliminating the invading pathogen. Memory cells are long lived lymphocyte survivors of past infections that can recognize specific pathogenic antigens.

A. Peripheral Blood Mononuclear Cells

As stated previously, the majority of studies examining immune cell α_1-AR expression have been performed on PMBC preparations, which include a variety of blood cell types. In studies that induced PMBC α_1-AR expression using PHA or LPS, subsequent addition of NE increased extracellular signal-regulated kinase (ERK) activation, which could be blocked by pretreating with the irreversible α_1-AR antagonist benextramine, but not yohimbine (Rouppe van der Voort et al., 2000). The functional outcome of NE-initiated ERK activation was not further characterized in this investigation. PE treatment of PMBCs isolated from patients with juvenile rheumatoid arthritis induced the mature expression of IL-6, which is commonly secreted by T cells, when compared to preparations from healthy individuals that showed little change in cytokine levels (Heijnen et al., 1996).

B. T lymphocytes

T (thymus-derived) lymphocytes or T cells are a type of white blood cell that plays an important role in the adaptive immune response. As stated previously, expression of the TCR complex is a marker that distinguished T cells from other lymphocytes. There are several types of mature T cells including T helper cells (T_H cells), cytotoxic T cells (T_C cells), memory T cells, regulatory T cells (T_{reg} cells), natural killer T cells (NKT cells), and $\delta\gamma$ T cells (Chaplin, 2010). T_H cells are $CD4^+$ cells that aid other immunocompetent cells in their function such as maturation of B cells as well as activation of T_C cells and macrophages. Activation of macrophage and T_C cells occurs through the presentation of peptide antigens by major histocompatibility complex (MHC) class II molecules on the surface of T_H and other antigen presenting cells. Activation results in rapid cell division as well as cytokine secretion which facilitate a variety of immune responses. T_C cells are $CD8^+$ cells that destroy tumor and viral infected cells through TCRs that recognize specific antigenic peptides bound to MHC class I and CD8 glycoproteins. Memory T cells are antigen-specific $CD4^+$ or $CD8^+$ cells that remain following termination of the infection, which can quickly expand into effector T cells (T_H or T_C) following re-exposure to their cognate antigen. T_{reg} cells are crucial for immunological tolerance by functioning to end T-cell-mediated immunity following an immune reaction as well as suppressing auto-reactive T cells that may escape the negative selection process in the thymus. NKT cells are unique in that they bridge the innate and adaptive immune responses. Unlike most T cells, which recognize peptide antigens presented by MHC molecules, NKT cells recognize glycolipid antigens presented by CD1d. Following activation, NKT cells have similar functions as T_H and T_C cells in that they contribute to both cytokine production and release of cytolytic molecules. $\delta\gamma$ T cells are a small T cell subset primarily found in the gut mucosa having a distinct TCR made from one δ- and one γ-chain glycolipid.

While α_1-AR expression is not commonly found on T cells under normal conditions, some investigations suggest that α_1-AR expression may be regulated in certain lymphoid compartments or under certain pathologic conditions. For example, lymphocytes from rat mesenteric lymph nodes have been shown to transcriptionally express the α_1-AR (Bao et al., 2007). Activation of these lymphocytes with a T cell mitogen, concanavalin A (Con A), increased the quantity of α_1-AR mRNA over resting lymphocytes. However, this study was unable to determine a potential function for these α_1-AR transcripts in that PE had no effect on Con A-induced proliferation or interferon (IFN)-γ and IL-4 production. In other investigations, transcriptional α_{1A}- and α_{1D}-AR subtype expression was also detected in rat lymphocyte populations from the thymus, spleen, and peripheral blood (Schauenstein et al., 2000). However, this study documented a decrease in α_{1A}- and α_{1D}-AR mRNA following peripheral blood lymphocyte (PBL) treatment with Con A. These investigators also showed no differences in the mRNA expression of α_{1A}- and α_{1D}-AR subtypes between $CD4^+$ and $CD8^+$ T cells. Radioligand binding studies found a correlation of [^{3}H]-prazosin binding site densities between PBL preparations isolated from spontaneously hypertensive rats (SHR) and humans diagnosed with essential hypertension (Veglio et al., 2001). In this investigation, PBL isolated from both human hypertensive and SHR showed significant decreases in the [^{3}H]-prazosin B_{max} when compared with Wistar-Kyoto (WKY) rat controls or normotensive individuals. Using subtype-selective α_1-AR antagonists to characterize these specific [^{3}H]-prazosin binding sites in humans, the authors described the mature expression for all three α_1-AR subtypes in both hypertensive and normotensive subjects. The α_{1B}-AR subtype was the highest expressed receptor in normal patients, while PBLs isolated from hypertensive individuals showed no change in the α_{1A}-AR subtype density when compared to control. However, there was decreased α_{1B}-AR subtype expression with a concomitant increase in the α_{1D}-AR population from hypertensive patients when compared with normotensive individuals.

Early investigations have also demonstrated the role for α_1-AR activation to inhibit proliferative T cell responses (Heilig et al., 1993). In this study, [^{3}H]-thymidine incorporation of primary murine lymphocytes isolated from immunized animals decreased with increasing concentrations of PE, which could be blocked with phentolamine. A more recent comprehensive study using flow cytometry has shown that 11.3% of isolated rat thymus cells express the α_1-AR (Pešić et al., 2009). This α_1-AR expressing cell population primarily consisted of the least mature $CD3^-$ (51.2%) and $CD3^{low}$ (33.2%) to the most mature $CD3^{high}$ (14.2%) differentiating and proliferating thymocytes prior to leaving the thymus. Chronic treatment with the selective α_1-AR antagonist, urapidil, increased the absolute and relative thymic weight in these animals. This result correlated with both absolute and relative thymocyte numbers resulting from a decrease in cell apoptosis. There was also a greater frequency of the nuclear cell proliferation-associated antigen, Ki-67, on thymocytes after urapidil treatment, which is another indication for a negative regulatory function of α_1-AR activation in T cells.

The overall number of progenitor $TCR\alpha\beta^-$ thymocytes, which give rise to the distinct CD4 or CD8 functional T cell subsets, was likewise increased in urapidil-treated rats (Pešić et al., 2009). In this group, the number of $CD4^-CD8^+$ single-positive (SP) and $CD4^-CD8^-$ double-negative (DN) cells remained unaltered. However, a rise in $CD4^+CD8^+$ double-positive (DP) and $CD4^+CD8^-$SP cell subsets was the reason for an increased number of urapidil-treated $TCR\alpha\beta^-$ thymocytes. Urapidil treatment also increased the number of immature $TCR\alpha\beta^{low}$ thymocytes undergoing the selection process. In this group, absolute numbers of $CD4^+CD8^+$DP and $CD4^+CD8^-$SP thymocyte subsets were increased while the cellularity of other subsets remained unaltered. Similarly, urapidil treatment increased the overall frequency of the most mature, postselected TCR$\alpha\beta^{high}$ cells, which is reflected as large increases in the $CD4^+CD8^-$SP subset, a decrease in $CD4^-CD8^-$DN thymocytes, while the $CD4^-CD8^+$SP and $CD4^+CD8^+$DP populations remained the same.

The Ig Thy-1 (CD90) has been shown to modulate TCR$\alpha\beta$ signaling and selection thresholds (Hueber et al., 1997). In urapidil-treated thymocytes, CD90 expression was increased in all TCR$\alpha\beta$ groups examined over control, again supporting a role for the inhibition of T cell proliferation by α_1-AR signaling (Pešić et al., 2009). The impact of urapidil treatment on T_{reg} maturation in the thymus was similarly examined using this animal model. The unique "self-antigen" RT6.1 previously identified on peripheral T lymphocytes was used to distinguish maturing $CD4^+CD25^+$ T_{reg} thymocytes from cells re-entering the thymus (Agus et al., 1991). In urapidil treated animals, the relative and absolute numbers of $CD4^+CD25^+RT6.1^-$ thymocytes was greater when compared to control. Finally, chronic urapidil treatment also increased both the relative and absolute numbers of maturing $CD161^+TCR\alpha\beta^+$ NKT cells relative to control animals. These results together point to an α_1-AR mechanism that negatively regulates maturation of T cells in the thymus.

C. B lymphocytes

B cells are the major producers of antibodies that circulate in the blood plasma and lymphatic system. Upon activation, B cells generate antibodies that recognize unique antigens which neutralize specific pathogens. Each B cell expresses a unique B cell receptor (BCR) that recognizes and binds a particular antigen. Upon antigen recognition, B cells differentiate into effector plasma cells. Plasma cells secrete these specific antibodies, which bind the unique pathogenic antigens on cells to initiate the complement cascade as well as targeting these antigenic cells for phagocytes. While there are varied reports of α_1-AR expression on isolated PMBCs, a mixed cell preparation which contains B cells, there are no reports to suggest α_1-AR expression specifically on B cells (Casale & Kaliner, 1984; Rouppe van der Voort et al., 2000; Tayebati et al., 2000).

VI. α_1-ADRENERGIC RECEPTORS IN IMMUNE TISSUES

A. Spleen

The spleen is an important immune system organ responsible for removing old red blood cells, maintaining a blood reserve, recycling elemental iron, synthesizing antibodies as well as retaining half the body's monocytes, which allows them to move into injured tissues for differentiation into DCs and macrophages. The spleen is richly innervated by the sympathetic nerves, which affects its physiology (Felten et al., 1987). Reports of high transcriptional expression for all three α_1-AR subtypes in the spleen have been published (Alonso-Llamazares et al., 1995; Kavelaars, 2002). However, the spleen was one of the first tissues used to demonstrate translational α_1-AR homogeneity (Han et al., 1987). Subsequent radioligand binding analysis in bovine and guinea pig spleen also demonstrated a homogenous α_{1B}-AR subtype population (Büscher et al., 1996). Conversely, no specific [^{3}H]-prazosin binding could be observed in a murine (strain HLG) broken cell spleen preparation (Yang et al., 1998).

There is also evidence to suggest functional α_1-AR expression in the spleen. Electrical stimulation (ES) of isolated murine spleen slices inhibits basal IL-6 secretion, which is attenuated by phentolamine (Straub et al., 1997). Application of the α_1-AR agonist methoxamine mimicked the inhibitory response of ES on basal IL-6 levels. In other studies, NE treatment in the presence of propranolol enhanced the murine IgM antibody response in primary spleen cells immunized with sheep erythrocytes *in vitro* (Sanders & Munson, 1984). In a subsequent investigation, methoxamine was used to demonstrate that early IgM increases from immunized murine spleen cells were mediated through α_1-AR activation, while late IgM changes observed in the presence of clonidine were facilitated by α_2-AR stimulation (Sanders & Munson, 1985). Additionally, there is evidence linking changes in spleen α_1-AR activation with chronic inflammatory disease states (Straub et al., 2008). In this study, ES of the splenic nerve in an early type II collagen-induced arthritis (CIA) mouse model showed a decrease in IFN-γ secretion compared to control animals, which was partially reversed in the presence of the α_1-AR antagonist benoxathian.

B. Thymus

The thymus is an organ important for T cell maturation and differentiation as well as contribution to the production and secretion of additional factors that influence immune system function. The thymus comprises a central medulla and peripheral cortex, which is entirely surrounded by an outer capsule. The peripheral cortex is where thymocyte development begins and TCR gene rearrangement and positive T cell selection occurs as well. Conversely, the medulla is the location of late T cell development where a majority of negative selection happens. There are two main thymus cell types; thymic stromal cells, which include cortical epithelial cells and thymic medullary epithelial cells; and cells of hematopoietic origin such as DCs, thymocytes or T cell precursors. The thymus is innervated by postganglionic sympathetic nerve fibers that use NE as the primary neurotransmitter (Felten et al., 1987).

In addition to the previously described expression of α_1-ARs on thymocytes, which are the hematopoietic progenitor T cell precursors (Pešić et al., 2009), mRNA for all three α_1-AR subtypes is reportedly expressed in the human thymus (Kavelaars, 2002). Further analysis using immunohistochemistry delineated the location of α_1-ARs predominantly in the subcapsulary/subtrabeculary cortex and corticomedullary junction, with rare expression in the thymic medulla (Pešić et al., 2009). Cell-specific α_1-AR expression occurs primarily on thymic epithelial cells, but can also be found on CD68$^+$ cells, a monocyte/macrophage marker, located in the outer cortex and corticomedullary junction (Pešić et al., 2009).

α_1-AR expression in the subcapsulary/subtrabeculary cortex and corticomedullary junction suggests a role for these receptors in early T cell development and proliferation. For example, previous studies have demonstrated increased lymphopoiesis and greater mitogen reactivity over control from cultured fetal thymus explants incubated in the presence of PE (Singh, 1979). Additionally, α_1-AR blockade using urapidil has been shown to decrease the proportional thymus weight in immature rat pups, which was the result of reduced total thymocyte number (Plećaš-Solarović et al., 2005). This decrease in thymocyte cell number was localized to the cortex, but not observed in the medullary compartment. Chronic urapidil treatment also resulted in a decreased population of CD4$^+$CD8$^-$SP cells with a concomitant increase in CD4$^-$CD8$^+$SP cells, supporting the idea that α_1-ARs influence thymocyte proliferation.

In contrast to immature pups, chronic urapidil treatment increased absolute and relative thymic weight in adult rats (Plećaš-Solarović et al., 2005; Pešić et al., 2009). Both absolute and relative thymocyte numbers were also increased in urapidil-treated rats (Pešić et al., 2009). Using annexin V as an indicator of cell death, this study showed a decreased frequency of annexin V$^+$ thymocytes in urapidil-treated animals when compared with control. Using a marker of nuclear cell proliferation, Ki-67 immunostaining was significantly greater in the subcapsular/subtrabecular cortex of rats treated with urapidil as compared to control. As described previously, this investigation also demonstrated an overall increase in thymocyte numbers, which resulted in proportional changes of CD4CD8 T cell populations.

C. Blood/Circulating Cytokines

α_1-AR investigations using individual cell populations and isolated tissues provide a detailed account of their expression and potential function. However, the immune system is a complex network of interacting cells and it is therefore important to know how α_1-ARs are affecting immune responses under physiological conditions *in vivo*. For example, prazosin treatment alone has no effect on murine basal plasma IL-1β levels, but prazosin pretreatment could block increases in plasma IL-1β as a result of intraperitoneal LPS injection (Dong et al., 2002). Similarly, another investigation demonstrated decreased TNF-α levels in mice pretreated with prazosin prior to LPS injection when compared to LPS treatment alone (Sugino et al., 2009). Conversely, both studies showed that prazosin pretreatment further increased levels of the anti-inflammatory cytokine IL-10 in LPS treated animals when compared to LPS treatment only.

D. Nonimmune Tissue

α_1-ARs in tissues not considered part of the immune system have been shown to influence immune processes in a number of ways. In transfected α_{1A}-AR rat fibroblasts, oligonucleotide microarray technology demonstrated that following Epi treatment, several mediators of inflammation, cell motility, and adhesion were temporally altered, which was confirmed using RT-PCR and immunoblot analysis (Shi et al., 2006). Some notable observations included increased levels of IL-6 following 1 h of Epi treatment, followed by a decreased level that was still significantly above levels detected from nonstimulated control cells after 18 h. Likewise, levels of the neutrophil chemoattractant CXC chemokine, Gro (CXCL1) were increased after 1 h of Epi treatment, but alternatively dropped to levels significantly below basal after 18 h. No change was noted after 1 h for the T cell proliferation cytokine IL-15 and the inflammatory transcriptional regulator, high mobility group box 2 (Hmgb2) protein when compared to the control, but transcriptional levels for these targets were significantly decreased and increased, respectively after 18 h. Similarly, hyaluronan synthase 2 that synthesizes a component of the extracellular matrix hyaluronan and the hyaluronan receptor (CD44), which mediates lymphocyte binding showed no change from basal levels after 1 h, but both were significantly increased after 18 h of Epi treatment. These changes in hyaluronan targets were also observed in a treated rat thoracic aorta smooth muscle cell line (A-10) as well as in DDT1-MF2 hamster smooth muscle cells transfected with the human α_{1A}-AR subtype.

Increased IL-6 levels following α_1-AR activation were also shown in primary murine neonatal cardiomyocytes (Perez et al., 2009). This study further elucidated the molecular mechanisms responsible for the observed increased cytokine levels, which included stabilization of the IL-6 transcript. In addition, unique α_1-AR-mediated signaling pathways were found to be necessary for the increased cardiomyocyte IL-6 translation observed in this study. Specifically, pharmacological inhibition of p38 mitogen-activated protein kinase (MAPK) and NF-κB significantly decreased IL-6 levels when compared to Epi treatment. Transgenic mice overexpressing a constitutively active (CAM) α_{1A}-AR construct under control of the endogenous mouse promoter also showed significantly increased IL-6 serum levels when compared with CAM α_{1B}-AR transgenic or nontransgenic animals (Perez et al., 2009). Remarkably, these CAM α_{1A}-AR mice showed no signs of inflammation and were protected against myocardial ischemia suggesting that increased IL-6 levels may be a beneficial and adaptive α_1-AR cardioprotective mechanism (Rorabaugh et al., 2005).

VII. α_1-ADRENERGIC RECEPTORS IN DISEASE STATES

Given the summary of information to date, an immunomodulatory role under pathophysiological conditions can be brought forward for α_1-ARs expressed in the immune system. For example, α_1-AR activation has been proposed to be important in the development of experimental autoimmune encephalomyelitis (EAE), an inflammatory demyelinating disease of the central nervous system that is often used as a model for multiple sclerosis (Brosnan et al., 1985). Development of EAE results from T lymphocyte sensitization against the myelin basic protein (MBP) and is a typical inflammatory delayed-type hypersensitivity response. Rats sensitized to MBP were analyzed histologically for assessment of inflammatory cell infiltration as well as clinically (e.g., muscle weakness, ataxia, impaired respiration) using a scaled index to grade EAE development. In MBP-sensitized male rats, maximal signs of the disease were observed at 13 days post-inoculation (dpi) with a peak clinical index score of 3.9. Conversely, in sensitized prazosin treated animals, there was a dose-dependent increase when peak clinical signs were observed (15 dpi) as well as a decreased peak clinical index (2.2). These observations were specific for α_1-AR antagonism because treatment of MBP sensitized rats with yohimbine or propranolol had the opposite effect by increasing and extending the signs and duration of the disease. This study also described a dose-dependent decrease of brain and spinal cord immune cell infiltration from prazosin treated animals when compared with control. Other investigations demonstrated changes in the blood–brain barrier permeability following induction of EAE (Goldmuntz et al., 1986). In this study, immune cell infiltration occurred less rapidly in prazosin treated animals, with no differences from control in cell infiltration at the disease peak. As a result, it is unclear if prazosin affects EAE disease progression through changes in the vasculature or alterations in immune cell function.

There are several studies correlating α_1-AR expression with disease onset or severity. For instance, PMBC preparations are most often reported to have little to no α_1-AR expression. However, PMBCs isolated from patients with juvenile rheumatoid arthritis have increased IL-6 production following PE treatment, which was abolished in the presence of doxazosin (Heijnen et al., 1996). Conversely, PE treatment of PMBCs isolated from normal patients demonstrated a decrease in the generation of IL-6. Using a CIA animal model of autoimmune disease, other investigations have focused around effects of the sympathetic nervous system on splenic function in relation to INF-γ secretion (Straub et al., 2008). In this study, ES was used to release NE from splenic sympathetic nerve terminals in an isolated perfused slice preparation. ES significantly decreased basal IFN-γ levels in CIA mice, which was partially reversed in the presence of benoxathian, indicating that NE acts on α_1-ARs to inhibit INF-γ secretion. In parallel experiments, T cell depletion with anti-CD3 antibodies completely eliminated the basal response from CIA spleens, indicating INF-γ secretion is a T-cell-dependent process.

One of the most common causes of intensive care unit patient death is shock due to sepsis, in which cytokine overproduction by the immune system results in systemic vasodilation and circulatory failure due to decreased vasoconstrictor reactivity (Russell, 2006). Patients diagnosed with sepsis require increasing doses of NE in order to maintain blood pressure via α_1-AR activation on vascular smooth muscle cells. In a rat model of LPS-induced endotoxemia, increased levels of TNF-α and IL-1β were associated with decreased mRNA levels for all three α_1-AR subtypes (Bucher et al., 2003). Addition of TNF-α and IL-1β to rat renal cells decreased levels of α_{1B}-AR subtype expression as assessed by [^{3}H]-prazosin binding. In other studies, blocking vasculature α_1-ARs using prazosin had the same cardiovascular effects on control mice as observed in cecal ligation and puncture (CLP)-induced septic mice (Schmidt et al., 2009). Dexamethasone treatment or RNA interference technology to decrease levels of cytokine expression caused an attenuated cardiovascular effect and α_1-AR downregulation in CLP mice when compared to the control. Small interfering RNA treatment specific for NF-κB also prevented downregulation of α_1-ARs and inhibited cardiovascular dysfunction of CLP mice in this study.

VIII. CONCLUSIONS

α_1-AR expression on various immunocompetent cell populations has been reported and shown to be regulated during pathophysiological processes. However, α_1-AR function to modulate immune cell responses is just beginning to be understood. α_1-AR activation appears to alter production of inflammatory mediators from certain cell types including monocytes, macrophages, and myocytes. Additionally, α_1-AR signaling plays a role in DC migration, lymphopoiesis, and mast cell degranulation. A better understanding of how α_1-ARs regulate immune system function could uncover potential therapeutic strategies for modulating pathophysiological responses in human diseases where chronic, inappropriate, or hyperactive inflammation is an underlying etiology.

References

Agus, D. B., Surh, C. D., & Sprent, J. (1991). Reentry of T cells to the adult thymus is restricted to activated T cells. *J Exp Med, 173*, 1039–1046.

Alonso-Llamazares, A., Zamanillo, D., Casanova, E., Ovalle, S., Calvo, P., & Chinchetru, M. A. (1995). Molecular cloning of α_{1d}-adrenergic receptor and tissue distribution of three α_1-adrenergic receptor subtypes in mouse. *J Neurochem, 65*, 2387–2392.

Altenburg, S. P., Martins, M. A., Silva, A. R., Cordeiro, R. S., & Castro-Faria-Neto, H. C. (1997). LPS-induced blood neutrophilia is inhibited by α_1-adrenoceptor antagonists: A role for catecholamines. *J Leukoc Biol, 61*, 689–694.

Arase, H., Saito, T., Phillips, J. H., & Lanier, L. L. (2001). Cutting edge: The mouse NK cell-associated antigen recognized by DX5 monoclonal antibody is CD49b (α integrin, very late antigen-2). *J Immunol, 167*, 1141–1144.

Bao, J. Y., Huang, Y., Wang, F., Peng, Y. P., & Qiu, Y. H. (2007). Expression of α-AR subtypes in T lymphocytes and role of the α-ARs in mediating modulation of T cell function. *Neuroimmunomodulation, 14*, 344–353.

Brosnan, C. F., Goldmuntz, E. A., Cammer, W., Factor, S. M., Bloom, B. R., & Norton, W. T. (1985). Prazosin, an α_1-adrenergic receptor antagonist, suppresses experimental autoimmune encephalomyelitis in the Lewis rat. *Proc Natl Acad Sci USA, 82*, 5915–5919.

Bucher, M., Kees, F., Taeger, K., & Kurtz, A. (2003). Cytokines down-regulate α_1-adrenergic receptor expression during endotoxemia. *Crit Care Med, 31*, 566–571.

Büscher, R., Heeks, C., Taguchi, K., & Michel, M. C. (1996). Comparison of guinea-pig, bovine and rat α_1-adrenoceptor subtypes. *Br J Pharmacol, 117*, 703–711.

Calcagni, E., & Elenkov, I. (2006). Stress system activity, innate and T helper cytokines, and susceptibility to immune-related diseases. *Ann N Y Acad Sci, 1069*, 62–76.

Carr, D. J., Gebhardt, B. M., & Paul, D. (1993). *Alpha* adrenergic and *mu*-2 opioid receptors are involved in morphine-induced suppression of splenocyte natural killer activity. *J Pharmacol Exp Ther, 264*, 1179–1186.

Casale, T. B., & Kaliner, M. (1984). Demonstration that circulating human blood cells have no detectable α_1-adrenergic receptors by radioligand binding analysis. *J Allergy Clin Immunol, 74*, 812–818.

Chang, J. Y., & Liu, L. Z. (2000). Catecholamines inhibit microglial nitric oxide production. *Brain Res Bull, 52*, 525–530.

Chaplin, D. D. (2010). Overview of the immune response. *J Allergy Clin Immunol, 125*, S3–S23.

Dong, J., Mrabet, O., Moze, E., Li, K., & Neveu, P. J. (2002). Lateralization and catecholaminergic neuroimmunomodulation: Prazosin, an α_1/α_2-adrenergic receptor antagonist, suppresses interleukin-1 and increases interleukin-10 production induced by lipopolysaccharides. *Neuroimmunomodulation, 10*, 163–168.

Elenkov, I. J., Wilder, R. L., Chrousos, G. P., & Vizi, E. S. (2000). The sympathetic nerve – an integrative interface between two supersystems: The brain and the immune system. *Pharmacol Rev, 52*, 595–638.

Faure, C., Gouhier, C., Langer, S. Z., & Graham, D. (1995). Quantification of α_1-adrenoceptor subtypes in human tissues by competitive RT-PCR analysis. *Biochem Biophys Res Commun, 213*, 935–943.

Felten, D. L., Felten, S. Y., Bellinger, D. L., Carlson, S. L., Ackerman, K. D., & Madden, K. S., et al., (1987). Noradrenergic sympathetic neural interactions with the immune system: Structure and function. *Immunol Rev, 100*, 225–260.

Flierl, M. A., Rittirsch, D., Nadeau, B. A., Chen, A. J., Sarma, J. V., & Zetoune, F. S., et al., (2007). Phagocyte-derived catecholamines enhance acute inflammatory injury. *Nature, 449*, 721–725.

Goldmuntz, E. A., Brosnan, C. F., & Norton, W. T. (1986). Prazosin treatment suppresses increased vascular permeability in both acute and passively transferred experimental autoimmune encephalomyelitis in the Lewis rat. *J Immunol, 137*, 3444–3450.

Grisanti, L. A., Evanson, J., Marchus, E., Jorissen, H., Woster, A. P., & DeKrey, W., et al., (2010). Pro-inflammatory responses in human monocytes are β_1-adrenergic receptor subtype dependent. *Mol Immunol, 47*, 1244–1254.

Guimarães, S., & Moura, D. (2001). Vascular adrenoceptors: An update. *Pharmacol Rev, 53*, 319–356.

Han, C., Abel, P. W., & Minneman, K. P. (1987). Heterogeneity of α_1-adrenergic receptors revealed by chlorethylclonidine. *Mol Pharmacol, 32*, 505–510.

Heijnen, C. J., Rouppe van der Voort, C., van de Pol, M., & Kavelaars, A. (2002). Cytokines regulate α_1-adrenergic receptor mRNA expression in human monocytic cells and endothelial cells. *J Neuroimmunol, 125*, 66–72.

Heijnen, C. J., Rouppe van der Voort, C., Wulffraat, N., van der Net, J., Kuis, W., & Kavelaars, A. (1996). Functional α_1-adrenergic receptors on leukocytes of patients with polyarticular juvenile rheumatoid arthritis. *J Neuroimmunol, 71*, 223–226.

Heilig, M., Irwin, M., Grewal, I., & Sercarz, E. (1993). Sympathetic regulation of T-helper cell function. *Brain Behav Immun, 7*, 154–163.

Hueber, A. O., Bernard, A. M., Battari, C. L., Marguet, D., Massol, P., & Foa, C., et al., (1997). Thymocytes in *Thy-1*$^{-/-}$ mice show augmented TCR signaling and impaired differentiation. *Curr Biol, 7*, 705–708.

Javierre, M. Q., Pinto, L. V., Lima, A. O., & Sassine, W. A. (1975). Immunologic phagocytosis by macrophages: Effect by stimulation of alpha adrenergic receptors. *Rev Bras Pesqui Med Biol, 8*, 271–274.

Jensen, B. C., Swigart, P. M., & Simpson, P. C. (2009). Ten commercial antibodies for alpha-1-adrenergic receptor subtypes are nonspecific. *Naunyn Schmiedebergs Arch Pharmacol, 379*, 409–412.

Jetschmann, J. U., Benschop, R. J., Jacobs, R., Kemper, A., Oberbeck, R., & Schmidt, R. E., et al., (1997). Expression and *in-vivo* modulation of α- and β-adrenoceptors on human natural killer ($CD16^+$) cells. *J Neuroimmunol, 74*, 159–164.

Kambayashi, T., Assarsson, E., Chambers, B. J., & Ljunggren, H. G. (2001). Expression of the DX5 antigen on $CD8^+$ T cells is associated with activation and subsequent cell death or memory during influenza virus infection. *Eur J Immunol, 31*, 1523–1530.

Kavelaars, A. (2002). Regulated expression of α-1 adrenergic receptors in the immune system. *Brain Behav Immun, 16*, 799–807.

Kintscher, U., Kon, D., Wakino, S., Goetze, S., Graf, K., & Fleck, E., et al., (2001). Doxazosin inhibits monocyte chemotactic protein 1-directed migration of human monocytes. *J Cardiovasc Pharmacol, 37*, 532–539.

Lappin, D., & Whaley, K. (1982). Adrenergic receptors on monocytes modulate complement component synthesis. *Clin Exp Immunol, 47*, 606–612.

LeBien, T. W., & Tedder, T. F. (2008). B lymphocytes: How they develop and function. *Blood, 112*, 1570–1580.

Maestroni, G. J. (2000). Dendritic cell migration controlled by α_{1b}-adrenergic receptors. *J Immunol, 165*, 6743–6747.

Maestroni, G. J., Conti, A., & Pedrinis, E. (1992). Effect of adrenergic agents on hematopoiesis after syngeneic bone marrow transplantation in mice. *Blood, 80*, 1178–1182.

Mori, K., Ozaki, E., Zhang, B., Yang, L., Yokoyama, A., & Takeda, I., et al., (2002). Effects of norepinephrine on rat cultured microglial cells that express α_1, α_2, β1 and β2 adrenergic receptors. *Neuropharmacology, 43*, 1026–1034.

Moroni, F., Fantozzi, R., Masini, E., & Mannaioni, P. F. (1977). The modulation of histamine release by *alpha*-adrenoceptors: Evidences in murine neoplastic mast cells. *Agents Actions, 7*, 57–61.

Muthu, K., Iyer, S., He, L. K., Szilagyi, A., Gamelli, R. L., & Shankar, R., et al., (2007). Murine hematopoietic stem cells and progenitors express adrenergic receptors. *J Neuroimmunol, 186*, 27–36.

Nemeth, E., Baird, A. W., & O'Farrelly, C. (2009). Microanatomy of the liver immune system. *Semin Immunopathol, 31*, 333–343.

Papay, R., Gaivin, R., Jha, A., McCune, D. F., McGrath, J. C., & Rodrigo, M. C., et al., (2006). Localization of the mouse α_{1A}-adrenergic receptor (AR) in the brain: α_{1A}AR is expressed in neurons, GABAergic interneurons, and NG2 oligodendrocyte progenitors. *J Comp Neurol, 497*, 209–222.

Parikh, V., & Singh, M. (1999). Possible role of adrenergic component and cardiac mast cell degranulation in preconditioning-induced cardioprotection. *Pharmacol Res, 40*, 129–137.

Paust, S., Senman, B., & von Andrian, U. H. (2010). Adaptive immune responses mediated by natural killer cells. *Immunol Rev, 235*, 286–296.

Perez, D. M., Papay, R. S., & Shi, T. (2009). α_1-Adrenergic receptor stimulates interleukin-6 expression and secretion through both mRNA stability and transcriptional regulation: Involvement of p38 mitogen-activated protein kinase and nuclear factor-κB. *Mol Pharmacol, 76*, 144–152.

Pešić, V., Kosec, D., Radojević, K., Pilipović, I., Perišić, M., & Vidić-Danković, B., et al., (2009). Expression of α_1-adrenoceptors on thymic cells and their role in fine tuning of thymopoiesis. *J Neuroimmunol, 214*, 55–66.

Petty, H. R. (1989). Regulation of RAW264 macrophage morphology and spreading: Studies with protein kinase C activators, inhibitors and a cyclic AMP analog. *Biochim Biophys Acta, 1012*, 284–290.

Plećaš-Solarović, B., Hristić-Živković, I., Radojević, K., Kosec, D., & Leposavić, G. (2005). Chronic α_1-adrenoreceptor blockade produces age-dependent changes in rat thymus structure and thymocyte differentiation. *Histol Histopathol, 20*, 833–841.

Price, D. T., Lefkowitz, R. J., Caron, M. G., Berkowitz, D., & Schwinn, D. A. (1993). Localization of mRNA for three distinct α_1-adrenergic receptor subtypes in human tissues: Implications for human α-adrenergic physiology. *Mol Pharmacol, 45*, 171–175.

Ricci, A., Bronzetti, E., Conterno, A., Greco, S., Mulatero, P., & Schena, M., et al., (1999). α_1-Adrenergic receptor subtypes in human peripheral blood lymphocytes. *Hypertension, 33*, 708–712.

Rorabaugh, B. R., Ross, S. A., Gaivin, R. J., Papay, R. S., McCune, D. F., & Simpson, P. C., et al., (2005). α_{1A}- but not α_{1B}-adrenergic receptors precondition the ischemic heart by a staurosporine-sensitive, chelerythrine-insensitive mechanism. *Cardiovasc Res, 65*, 436–445.

Rouppe van der Voort, C., Kavelaars, A., van de Pol, M., & Heijnen, C. J. (1999). Neuroendocrine mediators up-regulate α_{1b}- and α_{1d}-adrenergic receptor subtypes in human monocytes. *J Neuroimmunol, 95*, 165–173.

Rouppe van der Voort, C., Kavelaars, A., van de Pol, M., & Heijnen, C. J. (2000). Noradrenaline induces phosphorylation of ERK-2 in human peripheral blood mononuclear cells after induction of α_1-adrenergic receptors. *J Neuroimmunol, 108*, 82–91.

Russell, J. A. (2006). Management of sepsis. *N Engl J Med, 355*, 1699–1713.

Sanders, V. M., & Munson, A. E. (1984). *Beta* adrenoceptor mediation of the enhancing effect of norepinephrine on the murine primary antibody response *in vitro*. *J Pharmacol Exp Ther, 230*, 183–192.

Sanders, V. M., & Munson, A. E. (1985). Role of *alpha* adrenoceptor activation in modulating the murine primary antibody response *in vitro*. *J Pharmacol Exp Ther, 232*, 395–400.

Schauenstein, K., Felsner, P., Rinner, I., Liebmann, P. M., Stevenson, J. R., & Westermann, J., et al., (2000). *In vivo* immunomodulation by peripheral adrenergic and cholinergic agonists/antagonists in rat and mouse models. *Ann N Y Acad Sci, 917*, 618–627.

Schmidt, C., Kurt, B., Hocherl, K., & Bucher, M. (2009). Inhibition of NF-κB activity prevents downregulation of α_1-adrenergic receptors and circulatory failure during CLP-induced sepsis. *Shock, 32*, 239–246.

Schulze, W., & Fu, M. L. (1996). Localization of α_1-adrenoceptors in rat and human hearts by immunocytochemistry. *Mol Cell Biochem, 163–164*, 159–165.

Seelaender, M. C., Kazantzis, M., & Costa Rosa, L. F. (1999). The effect of adrenaline and Walker-256 tumour-induced cachexia upon Kupffer cell metabolism. *Cell Biochem Funct, 17*, 151–156.

Seiffert, K., Hosoi, J., Torii, H., Ozawa, H., Ding, W., & Campton, K., et al., (2002). Catecholamines inhibit the antigen-presenting capability of epidermal langerhans cells. *J Immunol, 168*, 6128–6135.

Shi, T., Duan, Z. H., Papay, R., Pluskota, E., Gaivin, R. J., & de la Motte, C. A., et al., (2006). Novel α_1-adrenergic receptor signaling pathways: Secreted factors and interactions with the extracellular matrix. *Mol Pharmacol, 70*, 129–142.

Singh, U. (1979). Effect of catecholamines on lymphopoiesis in fetal mouse thymic explants. *J Anat, 129*, 279–292.

Straub, R. H., Herrmann, M., Berkmiller, G., Frauenholz, T., Lang, B., & Schölmerich, J., et al., (1997). Neuronal regulation of interleukin 6 secretion in murine spleen: Adrenergic and opioidergic control. *J Neurochem, 68*, 1633–1639.

Straub, R. H., Rauch, L., Fassold, A., Lowin, T., & Pongratz, G. (2008). Neuronally released sympathetic neurotransmitters stimulate splenic interferon-γ secretion from T cells in early type II collagen-induced arthritis. *Arthritis Rheum, 58*, 3450–3460.

Subleski, J. J., Hall, V. L., Back, T. C., Ortaldo, J. R., & Wiltrout, R. H. (2006). Enhanced antitumor response by divergent modulation of natural killer and natural killer T cells in the liver. *Cancer Res, 66*, 11005–11012.

Suga, S., Goto, S., & Togari, A. (2010). Demonstration of direct neurite-osteoclastic cell communication in vitro via the adrenergic receptor. *J Pharmacol Sci, 112*, 184–191.

Sugino, H., Futamura, T., Mitsumoto, Y., Maeda, K., & Marunaka, Y. (2009). Atypical antipsychotics suppress production of proinflammatory cytokines and up-regulate interleukin-10 in lipopolysaccharide-treated mice. *Prog Neuropsychopharmacol Biol Psychiatry, 33*, 303–307.

Takahashi, H. K., Iwagaki, H., Tamura, R., Katsuno, G., Xue, D., & Sugita, S., et al., (2005). α_1-Adrenergic receptor antagonists induce production of IL-18 and expression of ICAM-1 and CD40 in human monocytes. *J Immunother, 28*, 40–43.

Tayebati, S. K., Bronzetti, E., Morra Di Cella, S., Mulatero, P., Ricci, A., & Rossodivita, I., et al., (2000). *In situ* hybridization and immunocytochemistry of α_1-adrenoceptors in human peripheral blood lymphocytes. *J Auton Pharmacol, 20*, 305–312.

Veglio, F., Tayebati, S. K., Schiavone, D., Ricci, A., Mulatero, P., & Bronzetti, E., et al., (2001). α_1-Adrenergic receptor subtypes in peripheral blood lymphocytes of essential hypertensives. *J Hypertens, 19*, 1847–1854.

Wittig, O., Paez-Cortez, J., & Cardier, J. E. (2010). Liver sinusoidal endothelial cells promote B lymphopoiesis from primitive hematopoietic cells. *Stem Cells Dev, 19*, 341–350.

Yang, M., Reese, J., Cotecchia, S., & Michel, M. C. (1998). Murine $alpha_1$-adrenoceptor subtypes. I. Radioligand binding studies. *J Pharmacol Exp Ther, 286*, 841–847.

CHAPTER 7

Pre- versus Postsynaptic Signaling by α_2-Adrenoceptors

Ralf Gilsbach[1], Julián Albarrán-Juárez[1] and Lutz Hein[1,2]

[1]Institute of Experimental and Clinical Pharmacology, University of Freiburg, Freiburg, Germany
[2]BIOSS Centre for Biological Signaling Studies, University of Freiburg, Freiburg, Germany

I. Overview
II. Introduction
III. Functions of α_2-Adrenoceptor Subtypes
    A. α_{2A}-Adrenoceptor Functions
    B. α_{2B}-Adrenoceptor Functions
    C. α_{2C}-Adrenoceptor Functions
IV. Transgenic Dissection of Pre- Versus Postsynaptic α_2-Adrenoceptor Functions
    A. α_2-Adrenoceptor-Mediated Sedation and Hypnosis
    B. α_2-Adrenoceptor-Mediated Antinociception
    C. Hypothermia
    D. Cognitive Functions
    E. Behavior and Depression
    F. Cardiovascular Effects of α_2-Agonists
V. Conclusions
    Acknowledgment
    References

I. OVERVIEW

The adrenergic system is an important modulator of synaptic transmission in the central and peripheral nervous system. The endogenous catecholamines epinephrine and norepinephrine activate multiple G protein-coupled receptors (GPCR) to transmit their signal within the neurons. Previous studies highlighted the important role of the three subtypes of α_2-adrenoceptors, α_{2A}, α_{2B}, α_{2C}, for neuronal function. However, the detailed knowledge about the brain regions, types of neurons and cells, subcellular localization, and intracellular signaling events has only been acquired recently. The present review summarizes recent insight derived from transgenic mouse models to distinguish α_2-adrenoceptor functions mediated by the classical presynaptic α_2-feedback autoreceptors in

Current Topics in Membranes, Volume 67 0065-230X/10 $35.00

DOI: 10.1016/B978-0-12-384921-2.00007-0

noradrenergic neurons versus α_2-functions as heteroreceptor in nonadrenergic neurons and cells. Interestingly, only a few functions previously ascribed to α_2-adrenoceptors were mediated by autoreceptors on adrenergic neurons, including feedback inhibition of norepinephrine release from sympathetic nerves and spontaneous locomotor activity. Most of the pharmacological effects of α_2-agonists, including analgesia, hypothermia, sedation, anesthetic-sparing, modulation of working memory were mediated by α_2-receptors in nonadrenergic cells. These findings extend the current view of the α_2-adrenergic receptor synaptic localization and function in the nervous system and provide important new avenues for future drug development.

II. INTRODUCTION

G protein-coupled receptors (GPCR) are important mediators of neurotransmitter function in the central and peripheral nervous system. In addition to their action as postsynaptic receptors, they also modulate neurotransmitter release as presynaptic receptors (Fig. 1). Presynaptic receptors that inhibit neurotransmitter release were first discovered for acetylcholine, GABA, and norepinephrine (for reviews see Starke, Endo, & Taube, 1975; Langer, 1997; Starke, 2001). However, presynaptic modulation is not only limited to inhibition, for example, by α_2-adrenoceptors, but also includes facilitation of transmitter exocytosis, for example, by β-adrenoceptors. Presynaptic receptors may be grouped into auto- and heteroreceptors (Fig. 1) (Gilsbach & Hein, 2008). "Autoreceptors" are those presynaptic receptors that control release of their own neurotransmitter. In contrast, "heteroreceptors" modulate exocytosis of other neurotransmitters by direct presynaptic action (Fig. 1). Every neuron terminal or postsynaptic membrane area may contain a large variety of pre- and postsynaptic receptor subtypes to facilitate complex interactions between adjacent neurons. Two sets of recent experimental data highlight this complexity of pre- and postsynaptic receptors. In a quantitative proteomic analysis 13 GPCRs were identified in the rat brain which were contained in nanodomains together with Cav2 Ca^{2+}-channels that supply the Ca^{2+} influx required for transmitter exocytosis (Muller et al., 2010). In mouse sympathetic ganglia, expression of 57 GPCRs was detected by microarray analysis (Gilsbach, Schneider, Lother, Schickinger, Leemhuis, & Hein, 2010). Thus, pre- and postsynaptic membranes contain a multitude of GPCRs and other receptors and channels to control and fine-tune synaptic communication.

The aim of the present review is to give a comprehensive overview of the pre- versus postsynaptic and cellular localization of α_2-adrenoceptors involved in the main physiological and pharmacological effects of α_2-agonist drugs. α_2-receptor functions outside of the nervous system and the physiological significance of α_2-adrenoceptor subtypes in transgenic mouse models have been summarized in recent reviews (MacDonald, Kobilka, & Scheinin, 1997; Kable, Murrin, & Bylund, 2000; Hein, 2001; Philipp & Hein, 2004; Gilsbach & Hein, 2008).

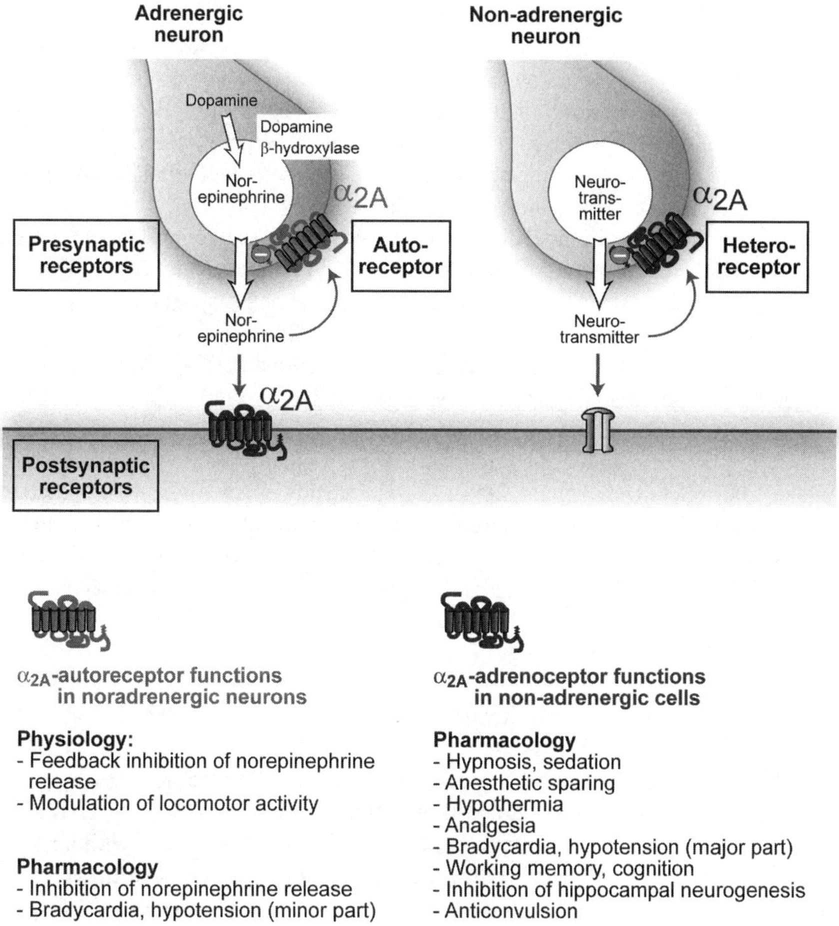

FIGURE 1 Pre- versus postsynaptic α_{2A}-adrenoceptor functions. Presynaptic α_{2A}-adrenoceptors may inhibit exocytosis of norepinephrine from adrenergic neurons as "autoreceptors" or they can reduce release of other neurotransmitters as "heteroreceptors" in nonadrenergic neurons. In addition to their presynaptic action, α_2-receptors have been identified in postsynaptic membranes. A summary of α_{2A}-adrenoceptor functions in adrenergic versus nonadrenergic cells is given in the *lower panel* (for references see text).

III. FUNCTIONS OF α_2-ADRENOCEPTOR SUBTYPES

The family of adrenoceptors comprises nine GPCR members that can be categorized into three different subgroups containing three α_1-subtypes (α_{1A}, α_{1B}, and α_{1D}), three α_2-subtypes (α_{2A}, α_{2B}, and α_{2C}), and three β-receptors (β_1, β_2, and β_3) (Bylund et al., 1994). The three-dimensional structures of β_1- and β_2-adrenoceptors have been reported, including active and inactive receptors (Cherezov et al., 2007; Rasmussen et al., 2007, 2011; Rosenbaum et al., 2007, 2011; Warne et al., 2008). However, the precise spatial conformation of α_1- and α_2-adrenoceptors remains to be solved. The physiological and pharmacological functions of these individual receptor subtypes have mostly been unraveled by generating mouse models with targeted deletions or transgenic overexpression of the respective genes *in vivo* (MacDonald et al., 1997; Kable et al., 2000; Hein, 2001; Philipp & Hein, 2004).

However, the first drugs eliciting their pharmacological action via adrenoceptors were developed and introduced into medical therapy several decades before these molecular discoveries. For the α_2-adrenoceptor group, the agonist clonidine, which was developed and characterized in the early 1960 s, was the first clinically used selective α_2-agonist (Hoefke & Kobinger, 1966). Clonidine elicits a number of pharmacological actions, most of which are still used in current pharmacological treatment, including sedation and hypnosis, sympathetic inhibition, hypotension, analgesia, reduction of intraocular pressure (Kamibayashi & Maze, 2000; Scholz & Tonner, 2000; Maze, Scarfini, & Cavaliere, 2001; Sanders & Maze, 2007). However, the large spectrum of clinical effects of clonidine has limited its more widespread use in clinical medicine. In particular cardiovascular side effects, notably hypotension and bradycardia, as well as sedation have been restrictions to application of clonidine in other medical disciplines.

A. α_{2A}-Adrenoceptor Functions

With the cloning of nine adrenoceptor subtypes, the idea arose to optimize drug therapy by generating drugs that activate or inhibit only one or limited numbers of these adrenoceptor subtypes. To this end, mouse lines with targeted deletions in the α_2-adrenoceptor genes have facilitated the identification of specific physiological and pharmacological roles of these α_2-adrenoceptors (MacDonald et al., 1997; Kable et al., 2000; Hein, 2001; Philipp & Hein, 2004). Surprisingly, α_{2A}-adrenoceptors gathered most of the pharmacological functions of nonselective α_2-agonist drugs, including bradycardia, hypotension (MacMillan, Hein, Smith, Piascik, & Limbird, 1996), sedation, antinociception (Lakhlani et al., 1997), and consolidation of working memory (Wang et al., 2007). Some of the cardiovascular actions of α_2-agonists could be ascribed to activation of α_{2B}-receptors.

B. α_{2B}-Adrenoceptor Functions

Stimulation of α_{2B}-receptors counteracted the hypotension elicited by α_{2A}-adrenoceptors (Link, Desai, Hein, Stevens, Chruscinski, & Bernstein,

1996). In addition, α_{2B}-receptors were essential mediators of the analgesic effect of nitrous oxide (Guo, Davies, Kingery, Patterson, Limbird, & Maze, 1999; Sawamura et al., 2000; Kingery et al., 2002) and they were required for the development of the placenta vasculature (Philipp, Brede, & Hein, 2002; Muthig et al., 2007) and the lung (Haubold, Gilsbach, & Hein, 2010). Newborn mice lacking α_{2B}-receptors suffered from early postnatal respiratory failure (Haubold et al., 2010). Within the first hours after birth α_{2B}-mutant mice developed cyanosis that could be traced back to a retardation in lung development with significantly reduced alveolar volume and thickened interalveolar septi. Ablation of α_{2B}-expression led to upregulation of the morphogen sonic hedgehog and its receptor patched, resulting in mesenchymal proliferation (Haubold et al., 2010).

C. α_{2C}-Adrenoceptor Functions

Several peripheral and central nervous system functions were discovered to be mediated by α_{2C}-receptors. *In vitro*, α_{2C}-receptors cooperated with α_{2A}- and α_{2B}-subtypes to inhibit neuronal norepinephrine exocytosis (Hein, Altman, & Kobilka, 1999; Trendelenburg, Philipp, Meyer, Klebroff, Hein, & Starke, 2003). However, ablation of α_{2C}-expression *in vivo* did not lead to an increase in circulating norepinephrine levels but rather caused elevated epinephrine plasma concentrations (Brede, Wiesmann, Jahns, Hadamek, Arnolt, & Neubauer, 2002; Brede, Nagy, Philipp, Sorensen, Lohse, & Hein, 2003). Following this observation, α_{2C}-adrenoceptors were identified in chromaffin cells to inhibit adrenal catecholamine release (Brede et al., 2003). Mice with partial or complete loss of α_{2C}-receptor function showed acceleration of heart failure development after chronic pressure overload (Brede et al., 2002; Gilsbach et al., 2007; Lymperopoulos, Rengo, Funakoshi, Eckhart, & Koch, 2007a). Recent studies have shown that high levels of catecholamines may desensitize adrenal α_{2C}-receptor signaling by increasing protein levels of G protein-coupled receptor kinase 2 (Grk2) thus accelerating the progression of cardiac hypertrophy and failure (Lymperopoulos et al., 2007a; Lymperopoulos, Rengo, & Koch, 2007b; Lymperopoulos, Rengo, Zincarelli, Soltys, & Koch, 2008; Lymperopoulos, Rengo, Gao, Ebert, Dorn, & Koch, 2010; Rengo et al., 2010). In addition, internalization of α_2-adrenoceptors may contribute to reduced feedback inhibition of catecholamine release from sympathetic nerves in chronic heart failure (Gilsbach et al., 2010).

Taken together, considerable knowledge about subtype-specific functions of α_2-adrenoceptor subtypes has been obtained based on studies in gene-targeted mouse models. However despite this progress, the cellular localization of α_2-receptors involved in these biological functions was largely unknown until recently.

IV. TRANSGENIC DISSECTION OF PRE- VERSUS POSTSYNAPTIC α_2-ADRENOCEPTOR FUNCTIONS

Initially, all three α_2-adrenoceptor genes were targeted for deletion by homologous recombination to yield mouse lines with ablation of expression of specific α_2-receptor subtypes (Link, Stevens, Kulatunga, Scheinin, Barsh, & Kobilka, 1995; Link et al., 1996; Altman et al., 1999). Mice with homozygous deletions developed apparently normally, although mice lacking α_{2B}-adrenergic receptors were not born at the expected Mendelian ratios indicating that this receptor may play a role during embryonic development (Link et al., 1996; Cussac, Schaak, Denis, Flordellis, Calise, & Paris, 2001). Deletion of all three α_2-receptor subtypes in mice by cross-breeding caused embryonic lethality due to severe defects of placental angiogenesis (Philipp, Brede, Hadamek, Gessler, Lohse, & Hein, 2002). The α_{2B}-receptor was found to be essential for mouse placenta and lung development (Philipp et al., 2002; Muthig et al., 2007; Haubold et al., 2010). These findings indicate that the α_{2B}-adrenoceptor acts primarily as a "postsynaptic receptor" outside of the central nervous system.

Homologous recombination via "hit and run" gene targeting in mouse embryonic stem cells was also used to generate a mutated α_{2A}-adrenoceptor allele carrying a D79N mutation in the genome (MacMillan et al., 1996). Mutation of the highly conserved aspartate residue (D79) in the second transmembrane domain of the α_{2A}-adrenoceptor has been shown to eliminate receptor regulation by monovalent cations and to interfere with receptor–G protein–effector coupling in cell lines *in vitro* (Surprenant, Horstman, Akbarali, & Limbird, 1992). α_{2A}-D79N mice showed two alterations that are important for the interpretation of the phenotypic results. On the one hand, electrophysiological experiments in α_{2A}-D79N mice indicated that mutant receptors did not activate K^+ currents in locus coeruleus (LC) neurons after agonist stimulation (Lakhlani et al., 1997). In addition, agonist-induced inhibition of voltage-gated Ca^{2+} channels was greatly blunted in α_{2A}-D79N mice (Lakhlani et al., 1997). However, in dissociated superior cervical ganglion neurons the α_2-agonist effect that remained in the α_{2A}-D79N mice could be blocked by the α-antagonist prazosin, suggesting that some neurons may also express α_{2B}- and/or α_{2C}-subtypes (Lakhlani et al., 1997). The presence of all three α_2-adrenoceptor subtypes in postganglionic sympathetic ganglia was later confirmed by studies in mice carrying targeted deletions in their α_2-adrenoceptor genes (Trendelenburg et al., 2003). A second, more unexpected finding in α_{2A}-D79N mice was a dramatic reduction in receptor expression levels. In brain membranes from α_{2A}-D79N mice, the total α_{2A}-receptor density was reduced to $<20\%$ of the level in wild-type mice (MacMillan et al., 1996). Thus, the α_{2A}-D79N mutation resulted in a hypomorphic allele with disrupted receptor–effector coupling and greatly reduced receptor expression. It should be noted, however, that young α_{2A}-D79N mice had normal presynaptic function in a vas deferens contraction assay (Altman et al., 1999).

In order to dissect, which of the physiological and pharmacological functions of α_2-adrenoceptors are mediated by presynaptic or postsynaptic functions, a transgenic mouse model has recently been generated and extensively studied (Gilsbach et al., 2009, 2010). To this end, transgenic mice that expressed α_{2A}-receptors under control of the dopamine β-hydroxylase promoter were crossed with mice with a constitutive deletion of their α_{2A}- and α_{2C}-adrenoceptor genes (Gilsbach et al., 2009). Through this approach, we could generate mice that exclusively expressed α_{2A}-adrenoceptors in adrenergic cells (by means of the dopamine β-hydroxylase promoter) but were deficient in α_{2A}- and α_{2C}-receptors in all other cell types of the body. These Dbh-α_{2A} transgenic mice were compared with wild-type mice and with mice deficient in α_{2A}- and α_{2C}-subtypes (Gilsbach et al., 2009, 2010).

The Dbh-α_{2A} transgene rescued the loss of α_{2A} mRNA expression in α_{2A}-deficient mice in the LC and in sympathetic ganglia but not in other tissues, including spinal cord, cerebellum, or hypothalamus (Gilsbach et al., 2009). α_{2A}-receptor protein could be detected in the presynaptic plasma-membrane of adrenergic nerve termini in the hippocampus by immunohistochemistry (Gilsbach et al., 2009). Furthermore, the Dbh-α_{2A} transgene rescued the loss of feedback inhibition of norepinephrine exocytosis and of Ca^{2+}-current inhibition in sympathetic neurons of α_{2AC}-deficient mice (Gilsbach et al., 2009). Thus, the Dbh-α_{2A} transgene fulfilled several criteria required for presynaptic α_{2A}-autoreceptors. Surprisingly, the majority of α_2-agonist effects that were lost after constitutive deletion of the α_{2A}- and α_{2C}-adrenoceptor genes were not rescued by the Dbh-α_{2A} transgene but seemed to be regulated by α_2-heteroreceptors (Gilsbach et al., 2009, 2010).

A. α_2-Adrenoceptor-Mediated Sedation and Hypnosis

In humans, α_2-agonists are used in the postoperative phase or in intensive care as sedative, hypnotic, and analgesic agents (Kamibayashi & Maze, 2000; Scholz & Tonner, 2000; Maze et al., 2001; Sanders & Maze, 2007). In mice, the sedative effect of α_2-agonists is mediated by the α_{2A}-subtype (Hunter et al., 1997; Lakhlani et al., 1997). Also in α_{2A}-D79N mutant mice the sedative effect of dexmedetomidine as determined in the rotarod assay was completely lost (Lakhlani et al., 1997). Similarly, the anesthetic-sparing effect of dexmedetomidine was absent in α_{2A}-D79N mice (Lakhlani et al., 1997). Interestingly, heterozygous α_{2A}-deficient mice responded with hypotension to injection of full α_2-agonists but they did not show signs of sedation, indicating that a greater number of α_{2A}-adrenoceptors must be activated to evoke sedation than to lower blood pressure (Tan, Wilson, MacMillan, Kobilka, & Limbird, 2002). Similarly, partial agonists were effective in causing hypotension without sedation in the same study (Tan et al., 2002).

Surprisingly, expression of α_{2A}-receptors in adrenergic cells by the Dbh-promoter (Dbh-α_{2A}) did not restore α_2-agonist-induced sedation in α_{2AC}-knockout mice (Gilsbach et al., 2009). Even at the highest dose of 1000 μg/kg medetomidine, Dbh-α_{2A} mice did not loose their right reflex similar to α_{2AC}-deficient mice (Gilsbach et al., 2009). In addition, the anesthetic-sparing effect of medetomidine, which was lost in α_{2A}- and in α_{2AC}-mutant mice, could not be restored by the Dbh-α_{2A} transgene (Gilsbach et al., 2009). These findings were somewhat unexpected as previous data had suggested that α_2-agonists cause sedation by reducing activity and norepinephrine release from LC neurons *in vivo* (Mizobe, Maghsoudi, Sitwala, Tianzhi, Ou, & Maze, 1996). Thus, mice deficient in dopamine β-hydroxylase (Dbh) which are unable to synthesize norepinephrine (Thomas, Matsumoto, & Palmiter, 1995) were tested in the same paradigm. Both, heterozygous and homozygous Dbh-mutant mice lost their righting reflex after medetomidine injection similar to wild-type animals (Gilsbach et al., 2009). These results indicated that the sedative and anesthetic-sparing effects of α_2-agonists were mediated by α_2-adrenoceptors in nonadrenergic cells. These findings are consistent with earlier observations showing that bilateral injection of 6-hydroxydopamine or application of the neurotoxin DSP-4 to destroy the LC and deplete norepinephrine in rats did not eliminate the sedative effect of clonidine (Spyraki & Fibiger, 1982; Nassif, Kempf, Cardo, & Velley, 1983).

Extensive evidence has been accumulated to show that α_2-agonists interfere with the endogenous nonrapid eye movement (NREM) sleep-promoting pathway (Fig. 2) (Nelson, Guo, Lu, Saper, Franks, & Maze, 2002, 2003). According to this model, activation of α_2-adrenoceptors reduces the activity and neurotransmitter release from adrenergic LC neurons leading to disinhibition of GABAergic neurons in the ventrolateral preoptic (VLPO) area of the anterior hypothalamus. The VLPO, when activated, releases GABA onto neurons of the tuberomammillary nucleus (TMN). As a consequence TMN neurons may release less arousal-promoting histamine onto cortical neurons to induce loss of consciousness (Nelson et al., 2002, 2003). In addition to the LC neurons, which contain α_{2A}-autoreceptors, other nuclei of the NREM pathway may also be modulated by α_2-adrenoceptors. Nonadrenergic neurons or nerve endings projecting to the VLPO area can be tuned in their activity by norepinephrine and α_2-agonists (Matsuo, Jang, Nabekura, & Akaike, 2003; Liu, Li, & Ye, 2010). In addition, histamine release in the brain cortex can be directly inhibited by α_2-agonist activation (Hill & Straw, 1988; Gulat-Marnay, Lafitte, Arrang, & Schwartz, 1989).

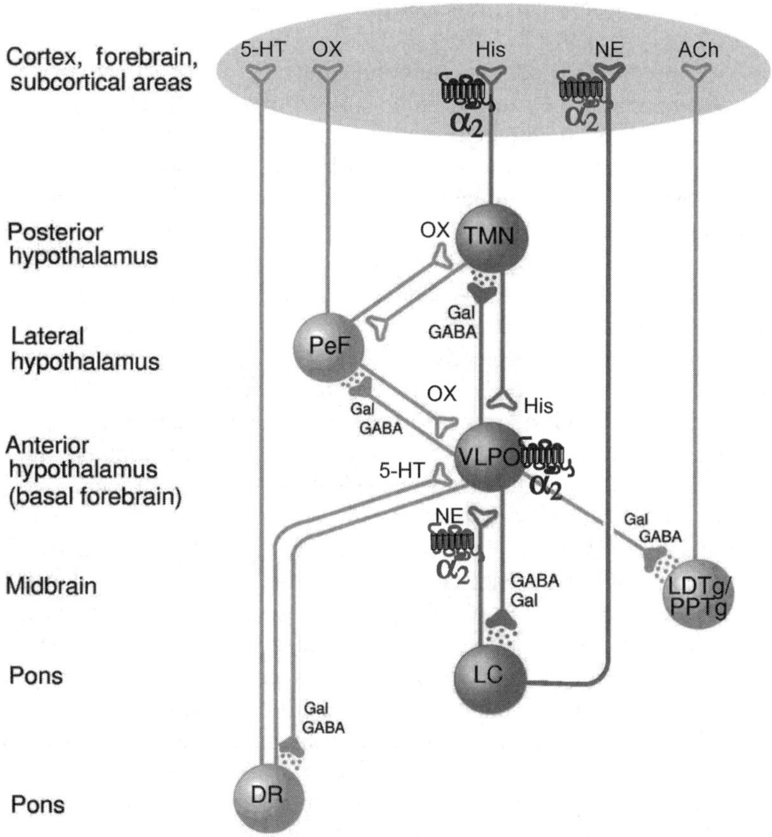

FIGURE 2 α_2-adrenoceptors involved in the nonrapid eye movement (NREM) sleep-promoting pathway (modified from Nelson et al., 2002). α_2-autoreceptors inhibiting norepinephrine release from locus coeruleus (LC) neurons are depicted in red. α_2-heteroreceptors in nonadrenergic neurons of the ventrolateral preoptic area (VLPO) and tuberomammillary neurons (TMN) are drawn in blue color. (For interpretation of the references to colour in this figure legend, the reader is referred to the web version of this book.)

Ablation of the G protein α-subunit i2 ($G\alpha_{i2}$) but not of $G\alpha_{i1}$ or $G\alpha_{i3}$ eliminated the anesthetic-sparing effect of medetomidine (Albarran-Juarez et al., 2009). Thus, either coupling of α_{2A}-receptors or other GPCRs in the sleep-promoting pathway to intracellular signaling systems via $G\alpha_{i2}$ are essential for anesthetic-sparing. In contrast, mice lacking the α_{2A}-receptor interacting protein spinophilin showed enhanced sedative/hypnotic effects of α_2-agonist stimulation (Lu et al., 2010). However, further studies are required to identify the neurons involved in α_2-mediated sedation and hypnosis in the central nervous system.

B. α_2-Adrenoceptor-Mediated Antinociception

α_2-adrenoceptors play an important role in the regulation of pain perception (for recent review, see Fairbanks, Stone, & Wilcox, 2009). The α_{2A}-receptor subtype is involved in antinociception induced by α_2-agonists (Hunter et al., 1997; Stone, MacMillan, Kitto, Limbird, & Wilcox, 1997; Fairbanks & Wilcox, 1999). Surprisingly, the antinociceptive effect of isoflurane was also altered in α_{2A}-knockout mice (Kingery et al., 2002). Some α_2-agonists, including the imidazoline/α_2-receptor ligand moxonidine, may involve additional receptor subtypes, for example, the α_{2C}-receptor, to induce antinociception (Fairbanks, Stone, Kitto, Nguyen, Posthumus, & Wilcox, 2002). α_{2B}-receptors play an essential role in the antinociceptive pathway of nitrous oxide that is used as a potent inhalative analgesic during anesthesia (Sawamura et al., 2000). In the periaqueductal gray, nitrous oxide activates endogenous endorphin release to stimulate a descending noradrenergic pathway that releases norepinephrine onto α_{2B}-receptors in the dorsal horn of the spinal cord. In α_{2B}-deficient mice, the antinociceptive effect of nitrous oxide was completely abolished (Sawamura et al., 2000). In the spinal cord of mice, the highest density of α_2-adrenoceptors could be identified in the superficial layers of the dorsal horn (Philipp et al., 2002). The mechanism of analgesic action of α_2-adrenoceptors may be preferentially presynaptic. Recent studies in substantia gelatinosa neurons of rat spinal cord slices *in vitro* indicate that norepinephrine via presynaptic α_2-receptors can inhibit glutamatergic transmission in primary afferent Aδ and C fibers (Kawasaki, Kumamoto, Furue, & Yoshimura, 2003). These presynaptic receptors are likely to be located on nonadrenergic cells, as no analgesic effect of the α_2-agonist medetomidine was observed in Dbh-α_{2A} transgenes (Gilsbach et al., 2009).

C. Hypothermia

α_2-agonists lower body core temperature in human and in animal studies (Hunter et al., 1997; Sallinen et al., 1997; Lahdesmaki, Sallinen, MacDonald, Sirvio, & Scheinin, 2003; Bexis & Docherty, 2005), affect nonshivering thermogenesis in mice (Hocker, Paris, Scholz, Tonner, Nielsen, Bein, 2008), and may be used clinically to treat postoperative shivering (Morrison, Nakamura, & Madden, 2008). In mice, both α_{2A}- and α_{2C}-adrenoceptors have been implicated in the hypothermic action of clonidine (Hunter et al., 1997; Sallinen et al., 1997; Lahdesmaki et al., 2003; Bexis & Docherty, 2005). When medetomidine was used as the α_2-agonist, α_{2A}-receptors predominated over α_{2C}-receptors to lower core body temperature (Gilsbach et al., 2009). Mice lacking both, α_{2A}- and α_{2C}-receptors showed no hypothermic response after medetomidine injection. Also α_{2AC}-deficient mice with transgenic expression of Dbh-α_{2A}-receptors did not respond to α_2-agonist injection with a change in body temperature (Gilsbach et al., 2009), indicating that α_2-receptors in nonadrenergic cells are essential for the hypothermia. This result is supported by earlier findings in cats after injection of α-methyl-*p*-tyrosine as a competitive tyrosine hydroxylase inhibitor to eliminate synthesis of norepinephrine (Myers, Beleslin, & Rezvani, 1987). Inhibition of catecholamine synthesis did not affect hypothermia after microinjection of clonidine into the anterior hypothalamus/preoptic area (Myers et al., 1987).

Recent experimental studies have greatly enhanced our understanding of the neuronal pathways involved in the regulation of body temperature especially in response to infections (for reviews, see Blatteis, 2006; Morrison et al., 2008; Morrison & Nakamura, 2011). The ventromedial preoptic-anterior hypothalamus (POA) represents the main temperature-regulating center in the CNS. Prostaglandin E2 (PGE2) may reach POA neurons either directly via the blood stream or it may indirectly activate this center via the ventral noradrenergic bundle to induce the fever response (Morrison et al., 2008; Morrison & Nakamura, 2011). Norepinephrine interacts closely with PGE2 in the POA region to modulate the thermic response via α_1- and α_2-adrenoceptors (Feleder, Perlik, & Blatteis, 2004, 2007) (Fig. 3). Experimentally, lipopolysaccharide activates the central adrenergic system to induce fever. Thus both α_1- and α_2-receptors may raise body temperature, however, with different time courses and via distinct signaling pathways (Feleder et al., 2004, 2007). Central injection of the α_2-agonist clonidine into the POA region of experimental animals causes an early fall in core body temperature which may hours later be followed by a rise in temperature (Feleder et al., 2004, 2007). According to this model, both presynaptic α_2-autoreceptors on noradrenergic nerve terminals as well as α_2-receptors on warm-sensitive POA neurons and on astrocytes are involved the thermal response to α_2-agonists.

D. Cognitive Functions

Cognitive functions relayed in the prefrontal cortex are under the control of catecholamine neurotransmitters, including norepinephrine and dopamine (for recent review, see Ramos & Arnsten, 2007). In the prefrontal cortex, environmental information is maintained to modulate "working memory" and thus affects behavior, thought, and affect (Ramos & Arnsten, 2007). Neuronal networks in the prefrontal cortex are controlled by GABA and catecholamines and many other neurotransmitters. Noradrenergic fibers originating in the LC project to the prefrontal cortex and information is relayed back to the LC. Depletion of norepinephrine in the prefrontal cortex affects working memory, causes

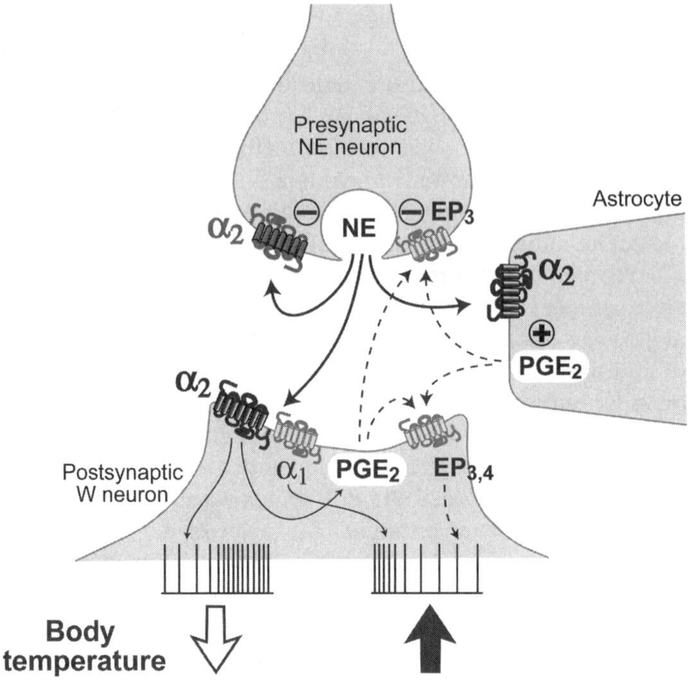

FIGURE 3 α_2-adrenoceptors in the thermoregulatory center of the preoptic-anterior hypothalamus (POA) (modified from Feleder et al., 2004). Presynaptic α_2-autoreceptors inhibiting norepinephrine release are depicted in red. Nonadrenergic cell α_2-receptors in astrocytes or warm-sensitive POA neurons ("postsynaptic W neuron") are indicated in blue color. (For interpretation of the references to colour in this figure legend, the reader is referred to the web version of this book.)

distractibility and attentional deficits in humans as well as in animal models (Carli et al., 1983; Cole & Robbins, 1992). However, working memory tasks can be improved by activating α_2-adrenoceptors, even in the absence of endogenous catecholamine release (Arnsten & Goldman-Rakic, 1985; Cai et al., 1993). The fact that α_2-agonists may restore prefrontal cortex function even in catecholamine-depleted animals shows that these receptors are postsynaptic receptors (Ramos & Arnsten, 2007). Indeed, systemic treatment with clonidine or guanfacine led to significant improvement of working memory function in monkeys with global or local catecholamine depletion (Arnsten & Goldman-Rakic, 1985). Experimental evidence suggests that α_{2A}-adrenoceptors are involved in the modulation of working memory based on the fact that the α_{2A}-preferring agonist guanfacine was active and that this beneficial effect of guanfacine was lost in α_{2A}-mutant mice carrying a point mutation (α_{2A}-D79N) (Franowicz & Arnsten, 1998; Franowicz, Kessler, Borja, Kobilka, Limbird, & Arnsten, 2002). The cellular mechanism and intracellular signaling of guanfacine to facilitate working memory has recently been uncovered (Wang et al., 2007). According to the model derived from these experiments (see Fig. 4), α_{2A}-receptors are colocalized with cAMP-responsive hyperpolarization-activated (HCN1) channels on distal dendritic spines of prefrontal pyramidal neurons. Dendritic spines receive excitatory input via glutamate released from neighboring pyramidal neurons. Under conditions of elevated intracellular cAMP levels, HCN channels are open, shunting synaptic inputs and thus reducing prefrontal network activity. Upon activation of α_{2A}-adrenoceptors, cAMP levels are decreased, HCN channels are closed and thus excitatory input is no longer suppressed so that electrical activity may proceed within the prefrontal networks (Wang et al., 2007).

These findings are of important clinical relevance since guanfacine has been documented to strengthen working memory and prefrontal cortical function in

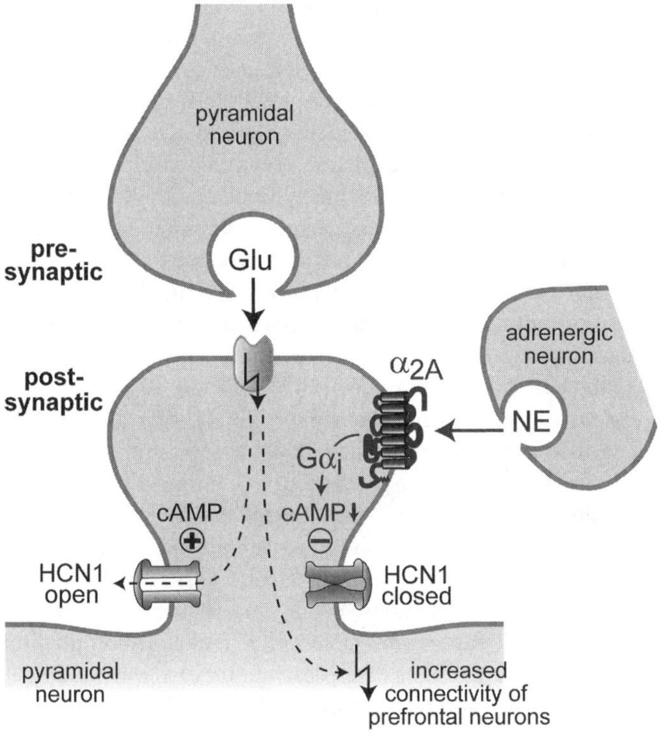

FIGURE 4 α_{2A}-adrenoceptor pathway regulating working memory in the prefrontal cortex (modified from Wang et al., 2007). The scheme depicts two dendritic spines from adjacent pyramidal prefrontal neurons and a noradrenergic nerve terminal. Postsynaptic α_A-receptors in a pyramidal dendritic spine are indicated in blue color. HCN1, hyperpolarization-activated, cyclic nucleotide-gated channel type 1. (For interpretation of the references to colour in this figure legend, the reader is referred to the web version of this book.)

patients with schizophrenia (McClure et al., 2007) or attention-deficient hyperactivity disorder (ADHD) (Scahill et al., 2001; Taylor & Russo, 2001). At least part of the beneficial effects of guanfacine may result from inhibition of cAMP levels and closure of HCN channels by postsynaptic α_{2A}-adrenoceptors in prefrontal pyramidal neurons (Wang et al., 2007).

E. Behavior and Depression

α_2-adrenoceptors modulate a number of behavioral functions in the central nervous system. In particular, the α_{2C}-receptor subtype has been shown to inhibit processing of sensory information in the central nervous system of the mouse (Scheinin, Sallinen, & Haapalinna, 2001). Activation of α_{2C}-receptors disrupts execution of spatial and nonspatial search patterns. Thus, drugs acting via the α_{2C}-receptor may gain therapeutic relevance in disorders associated with enhanced startle responses and sensorimotor gating deficits, such as schizophrenia, post-traumatic stress disorder, and drug withdrawal. α_{2A}-deficient mice were more sensitive to the behavioral and neurochemical effects of amphetamine than control mice, resulting in faster depletion of norepinephrine stores and increased startle responses (Lahdesmaki, Sallinen, MacDonald, & Scheinin, 2004). In contrast, the α_2-agonist, dexmedetomidine inhibited amphetamine-induced hyperlocomotion to a greater extent in wild-type mice than in α_{2C}-deficient mice (Sallinen, Haapalinna, Viitamaa, Kobilka, & Scheinin, 1998).

Activation of central α_{2A}-receptors elicits a powerful antiepileptogenic effect in mice (Janumpalli, Butler, MacMillan, Limbird, & McNamara, 1998) but other α_2-subtypes may also decrease the seizure threshold (Szot, Lester, Laughlin, Palmiter, Liles, & Weinshenker, 2004). The anticonvulsant effects of an α_2-agonist were not abolished in Dbh-deficient mice, indicating that postsynaptic α_2-receptors in nonadrenergic neurons regulate this action. In contrast, the proconvulsant α_2-agonist effect was ablated in Dbh-deficient mice, suggesting that activation of presynaptic α_2-autoreceptors on noradrenergic neurons decreases the seizure threshold (Szot et al., 2004).

α_2-adrenoceptors are also involved in the modulation of behavioral paradigms associated with depression. Deletion of the α_{2C}-subtype diminished the response to stress in classic tests of depression, for example the forced swim test (Sallinen et al., 1999). In a recent study in mice, α_2-agonists inhibited proliferation of hippocampal progenitor cells (Hocker et al., 2008). This effect persisted in norepinephrine-deficient mice, indicating that α_2-receptors on nonadrenergic cells are essential for this action. Isolated adult hippocampal progenitor cells expressed all three α_2-adrenoceptor subtypes and *in vitro* incubation with α_2-agonists inhibited neurosphere frequency. These results suggest that α_2-adrenoceptors on neuronal progenitor cells are directly modulating neurogenesis by inhibiting proliferation of these cells (Hocker et al., 2008). Blockade of these α_2-adrenoceptors accelerated the neurogenic and behavioral effects of the antidepressant imipramine (Hocker et al., 2008). In contrast, norepinephrine transport inhibitors enhanced progenitor proliferation via β_3-adrenoceptors (Jhaveri et al., 2010). Thus, neuronal precursor cell α_2- and β_3-adrenoceptors may represent a novel target to speed up classical antidepressant action.

F. Cardiovascular Effects of α_2-Agonists

The sympathetic system plays an essential role in the control of the cardiovascular system. Genetic variation of the G protein regulator phosducin in humans or gene ablation in mice causes sympathetic overactivity, stress-dependent hypertension, and end-organ damage (Beetz et al., 2009; Beetz & Hein, 2010). Clonidine and other α_2-agonists are potent inhibitors of sympathetic tone and they are clinically used to reduce sympathetic activation during withdrawal of illicit drugs and to treat hypertensive crisis. Thus, the question arises whether presynaptic α_2-receptors expressed in sympathetic neurons are essential for these pharmacological effects. Expression of the Dbh-α_{2A} transgene could readily be detected in sympathetic nerves innvervating the aorta and the vas deferens (Gilsbach et al., 2010). However, only a minor part of the agonist-induced reduction in heart rate and arterial blood pressure was mediated by α_2-receptors in adrenergic cells. The major part of these actions was mediated by receptors in nonadrenergic cells and could be inhibited by blockade of muscarinic acetylcholine receptors (Gilsbach et al., 2010). These results indicate that the clonidine's inhibitory effect on blood pressure and heart rate is only in part mediated by α_2-adrenoceptors in sympathetic neurons. Quantitatively, activation of the parasympathetic system by α_2-adrenoceptors predominated (Gilsbach et al., 2010). These findings are entirely consistent with clinical observations demonstrating that clonidine-induced bradycardia and hypotension can be diminished by atropine (Tank, Diedrich, Szczech, Luft, & Jordan, 2004; Tank, Jordan, Diedrich, Obst, Plehm, & Luft, 2004). Also, chemical sympathectomy by 6-hydroxydopamine only slightly reduced the cardiovascular effects of clonidine injected into the brain stem (Hamilton & Longman, 1982). Thus, α_2-agonists act as central inhibitors to lower sympathetic tone, blood pressure, and heart rate primarily via α_{2A}-adrenoceptors on nonadrenergic cells (for review, see Guyenet, 2006).

The major role of presynaptic α_2-adrenoceptors seems to be related to inhibition of basal norepinephrine release from sympathetic neurons. General ablation of α_{2AC}-receptor expression led to increased circulation catecholamine levels at rest which could be normalized by the Dbh-α_{2A} transgene (Gilsbach et al., 2010). In parallel, the increase in systolic blood pressure due to loss of α_2-adrenoceptor expression in α_{2AC}-/- mice and the resulting cardiac hypertrophy was rescued by the Dbh-α_{2A} transgene. However, during chronic sympathetic activation as induced by left ventricular pressure overload, sympathetic α_{2A}-adrenoceptors in the Dbh-α_{2A} transgene did no longer prevent increased plasma catecholamine levels or cardiac remodeling (Gilsbach et al., 2010). Several lines of *in vitro* evidence could be obtained to suggest that presynaptic sympathetic α_2-receptors are desensitized during chronic sympathetic activation. *In vitro* agonist exposure led to internalization of α_{2A}-receptors in sympathetic neurons isolated from Dbh-α_{2A} transgenice mice (Gilsbach et al., 2010). In sympathetic neurons from mice carrying a hemagglutinin-epitope tagged α_{2A}-receptor, clonidine but not guanfacine stimulation resulted in receptor endocytosis and desensitization (Lu et al., 2009). At present, it remains unknown whether receptor internalization or additional pathways are required for desensitization of sympathetic α_2-adrenoceptors *in vivo*. Taken together presynaptic sympathetic α_{2A}-adrenoceptors operate to reduce norepinephrine exocytosis at rest but are desensitized upon chronic agonist exposure. α_2-receptor desensitization may thus contribute to increased circulating catecholamine levels during the development and progression of chronic heart failure.

V. CONCLUSIONS

Previous studies have highlighted the importance of individual α_2-adrenoceptor subtypes for the physiology and pharmacology of the adrenergic system. However, for most functions knowledge about the brain regions, types of neurons and cells, subcellular localization, and intracellular signaling events has been lacking until recently. Sophisticated neurobiological experiments and novel transgenic models have started to provide better insight into the α_2-adrenoceptor functions mediated by the classical presynaptic α_2-feedback autoreceptors in noradrenergic neurons versus α_2-functions as heteroreceptor in nonadrenergic cells. Further mouse models with more advanced spatial and neuron-specific receptor expression will be required to provide a high-resolution map of adrenergic modulation of synaptic transmission in the nervous system. Data summarized in this review emphasize the pharmacological importance of α_2-adrenoceptors in nonadrenergic cells and neurons (Fig. 1) which may lead to the development of innovative future drugs targeting cognition, depression, sedation, analgesia, as well as central cardiovascular regulation.

Acknowledgment

This study was supported in part by grants from the Deutsche Forschungsgemeinschaft (to Ralf Gilsbach and Lutz Hein), by the Excellence Initiative of the German Federal and State Governments (EXC 294), and the DAAD Deutscher Akademischer Austauschdienst (to Julián Albarrán-Juárez).

References

Albarran-Juarez, J., Gilsbach, R., Piekorz, R. P., Pexa, K., Beetz, N., & Schneider, J., et al., (2009). Modulation of alpha2-adrenoceptor functions by heterotrimeric Galphai protein isoforms. *J Pharmacol Exp Ther, 331*, 35–44.

Altman, J. D., Trendelenburg, A. U., MacMillan, L., Bernstein, D., Limbird, L., & Starke, K., et al., (1999). Abnormal regulation of the sympathetic nervous system in alpha2A-adrenergic receptor knockout mice. *Mol Pharmacol, 56*, 154–161.

Arnsten, A. F., & Goldman-Rakic, P. S. (1985). Alpha2-adrenergic mechanisms in prefrontal cortex associated with cognitive decline in aged nonhuman primates. *Science, 230*, 1273–1276.

Beetz, N., Harrison, M. D., Brede, M., Zong, X., Urbanski, M. J., & Sietmann, A., et al., (2009). Phosducin influences sympathetic activity and prevents stress-induced hypertension in humans and mice. *J Clin Investig, 119*, 3597–3612.

Beetz, N., & Hein, L. (2010). The physiological roles of phosducin: From retinal function to stress-dependent hypertension. *Cell Mol Life Sci,*. doi: 10.1007/s00018-010-0550-0.

Bexis, S., & Docherty, J. R. (2005). Role of alpha2A-adrenoceptors in the effects of MDMA on body temperature in the mouse. *Br J Pharmacol, 146*, 1–6.

Blatteis, C. M. (2006). Endotoxic fever: New concepts of its regulation suggest new approaches to its management. *Pharmacol Ther, 111*, 194–223.

Brede, M., Nagy, G., Philipp, M., Sorensen, J. B., Lohse, M. J., & Hein, L. (2003). Differential control of adrenal and sympathetic catecholamine release by alpha 2-adrenoceptor subtypes. *Mol Endocrinol, 17*, 1640–1646.

Brede, M., Wiesmann, F., Jahns, R., Hadamek, K., Arnolt, C., & Neubauer, S., et al., (2002). Feedback inhibition of catecholamine release by two different alpha2-adrenoceptor subtypes prevents progression of heart failure. *Circulation, 106*, 2491–2496.

Bylund, D. B., Eikenberg, D. C., Hieble, J. P., Langer, S. Z., Lefkowitz, R. J., & Minneman, K. P et al., (1994). International Union of Pharmacology nomenclature of adrenoceptors. *Pharmacol Rev, 46*, 121–136.

Carli, M., Robbins, T. W., Evenden, J. L., & Everitt BJ. (1983). Effects of lesions to ascending noradrenergic neurones on performance of a 5-choice serial reaction task in rats; implications for theories of dorsal noradrenergic bundle function based on selective attention and arousal. *Behav Brain Res, 9*, 361–380.

Cole, B. J., & Robbins, T. W. (1992). Forebrain norepinephrine: role in controlled information processing in the rat. *Neuropsychopharmacology, 7*, 129–142.

Cherezov, V., Rosenbaum, D. M., Hanson, M. A., Rasmussen, S. G., Thian, F. S., & Kobilka, T. S et al., (2007). High-resolution crystal structure of an engineered human beta2-adrenergic G protein-coupled receptor. *Science, 318*, 1258–1265.

Cussac, D., Schaak, S., Denis, C., Flordellis, C., Calise, D., & Paris, H. (2001). High level of alpha2-adrenoceptor in rat foetal liver and placenta is due to alpha2B-subtype expression in haemato-poietic cells of the erythrocyte lineage. *Br J Pharmacol, 133*, 1387–1395.

Fairbanks, C. A., Stone, L. S., Kitto, K. F., Nguyen, H. O., Posthumus, I. J., & Wilcox, G. L. (2002). alpha2C-Adrenergic receptors mediate spinal analgesia and adrenergic-opioid synergy. *J Pharmacol Exp Ther, 300*, 282–290.

Fairbanks, C. A., Stone, L. S., & Wilcox, G. L. (2009). Pharmacological profiles of alpha2 adrenergic receptor agonists identified using genetically altered mice and isobolographic analysis. *Pharmacol Ther, 123*, 224–238.

Fairbanks, C. A., & Wilcox, G. L. (1999). Moxonidine, a selective alpha2-adrenergic and imidazoline receptor agonist, produces spinal antinociception in mice. *J Pharmacol Exp Ther, 290*, 403–412.

Feleder, C., Perlik, V., & Blatteis, C. M. (2004). Preoptic alpha1- and alpha2-noradrenergic agonists induce, respectively, PGE2-independent and PGE2-dependent hyperthermic responses in guinea pigs. *Am J Physiol Regul Integr Comp Physiol, 286*, R1156–R1166.

Feleder, C., Perlik, V., & Blatteis, C. M. (2007). Preoptic norepinephrine mediates the febrile response of guinea pigs to lipopolysaccharide. *Am J Physiol Regul Integr Comp Physiol, 293*, R1135–R1143.

Franowicz, J. S., & Arnsten, A. F. (1998). The alpha-2a noradrenergic agonist, guanfacine, improves delayed response performance in young adult rhesus monkeys. *Psychopharmacology, 136*, 8–14.

Franowicz, J. S., Kessler, L. E., Borja, C. M., Kobilka, B. K., Limbird, L. E., & Arnsten, A. F. (2002). Mutation of the alpha2A-adrenoceptor impairs working memory performance and annuls cognitive enhancement by guanfacine. *J Neurosci, 22*, 8771–8777.

Gilsbach, R., Brede, M., Beetz, N., Moura, E., Muthig, V., & Gerstner, C., et al., (2007). Heterozygous alpha 2C-adrenoceptor-deficient mice develop heart failure after transverse aortic constriction. *Cardiovasc Re, 75*, 728–737.

Gilsbach, R., & Hein, L. (2008). Presynaptic metabotropic receptors for acetylcholine and adrenaline/noradrenaline. *In* "Handbook of experimental pharmacology: Vol. 184. Pharmacology of neurotransmitter release" (T. C. Südhoff, and K. Starke, eds.), pp. 261–288. Springer, Heidelberg.

Gilsbach, R., Roser, C., Beetz, N., Brede, M., Hadamek, K., & Haubold, M., et al., (2009). Genetic dissection of alpha2-adrenoceptor functions in adrenergic versus nonadrenergic cells. *Mol Pharmacol, 75*, 1160–1170.

Gilsbach, R., Schneider, J., Lother, A., Schickinger, S., Leemhuis, J., & Hein, L. (2010). Sympathetic alpha2-adrenoceptors prevent cardiac hypertrophy and fibrosis in mice at baseline but not after chronic pressure overload. *Cardiovasc Res, 86*, 432–442.

Gulat-Marnay, C., Lafitte, A., Arrang, J. M., & Schwartz, J. C. (1989). Modulation of histamine release and synthesis in the brain mediated by alpha2-adrenoceptors. *J Neurochem, 53*, 513–518.

Guo, T. Z., Davies, M. F., Kingery, W. S., Patterson, A. J., Limbird, L. E., & Maze, M. (1999). Nitrous oxide produces antinociceptive response via alpha2B and/or alpha2C adrenoceptor subtypes in mice. *Anesthesiology, 90*, 470–476.

Guyenet, P. G. (2006). The sympathetic control of blood pressure. *Nat Rev Neurosci, 7*, 335–346.

Hamilton, T. C., & Longman, S. D. (1982). A comparison of the cardiovascular and sedative actions of the alpha-adrenoceptor agonists, FLA-136 and clonidine, in the rat. *Br J Pharmacol, 75*, 13–21.

Haubold, M., Gilsbach, R., & Hein, L. (2010). Alpha2B-adrenoceptor deficiency leads to postnatal respiratory failure in mice. *J Biol Chem, 285*, 34213–34219.

Hein, L. (2001). Transgenic models of alpha2-adrenergic receptor subtype function. *Rev Physiol Biochem Pharmacol, 142*, 161–185.

Hein, L., Altman, J. D., & Kobilka, B. K. (1999). Two functionally distinct alpha2-adrenergic receptors regulate sympathetic neurotransmission. *Nature, 402*, 181–184.

Hill, S. J., & Straw, R. M. (1988). Alpha 2-adrenoceptor-mediated inhibition of histamine release from rat cerebral cortical slices. *Br J Pharmacol, 95*, 1213–1219.

Hocker, J., Paris, A., Scholz, J., Tonner, P. H., Nielsen, M., & Bein, B. (2008). Differential effects of alpha 2-adrenoceptors in the modulation of the thermoregulatory response in mice induced by meperidine. *Anesthesiology, 109*, 95–100.

Hoefke, W., & Kobinger, W. (1966). Pharmacological effects of 2-(2,6-dichlorophenylamino)-2-imidazoline hydrochloride, a new, antihypertensive substance. *Arzneimittelforschung, 16*, 1038–1050.

Hunter, J. C., Fontana, D. J., Hedley, L. R., Jasper, J. R., Lewis, R., & Link, R. E., et al., (1997). Assessment of the role of alpha2-adrenoceptor subtypes in the antinociceptive, sedative and hypothermic action of dexmedetomidine in transgenic mice. *Br J Pharmacol, 122*, 1339–1344.

Janumpalli, S., Butler, L. S., MacMillan, L. B., Limbird, L. E., & McNamara, J. O. (1998). A point mutation (D79N) of the alpha2A adrenergic receptor abolishes the antiepileptogenic action of endogenous norepinephrine. *J Neurosci, 18*, 2004–2008.

Jhaveri, D. J., Mackay, E. W., Hamlin, A. S., Marathe, S. V., Nandam, L. S., & Vaidya, V. A., et al., (2010). Norepinephrine directly activates adult hippocampal precursors via beta3-adrenergic receptors. *J Neurosci, 30*, 2795–2806.

Kable, J. W., Murrin, L. C., & Bylund, D. B. (2000). *In vivo* gene modification elucidates subtype-specific functions of alpha2-adrenergic receptors. *J Pharmacol Exp Ther, 293*, 1–7.

Kamibayashi, T., & Maze, M. (2000). Clinical uses of alpha2 -adrenergic agonists. *Anesthesiology, 93*, 1345–1349.

Kawasaki, Y., Kumamoto, E., Furue, H., & Yoshimura, M. (2003). Alpha 2 adrenoceptor-mediated presynaptic inhibition of primary afferent glutamatergic transmission in rat substantia gelatinosa neurons. *Anesthesiology, 98*, 682–689.

Kingery, W. S., Agashe, G. S., Guo, T. Z., Sawamura, S., Frances Davies, M., & David Clark, J et al., (2002). Isoflurane and nociception: Spinal alpha2A adrenoceptors mediate antinociception while supraspinal alpha1 adrenoceptors mediate pronociception. *Anesthesiology, 96*, 367–374.

Lahdesmaki, J., Sallinen, J., MacDonald, E., & Scheinin, M. (2004). Alpha2A-adrenoceptors are important modulators of the effects of D-amphetamine on startle reactivity and brain monoamines. *Neuropsychopharmacology, 29*, 1282–1293.

Lahdesmaki, J., Sallinen, J., MacDonald, E., Sirvio, J., & Scheinin, M. (2003). Alpha2-adrenergic drug effects on brain monoamines, locomotion, and body temperature are largely abolished in mice lacking the alpha2A-adrenoceptor subtype. *Neuropharmacology, 44*, 882–892.

Lakhlani, P. P., MacMillan, L. B., Guo, T. Z., McCool, B. A., Lovinger, D. M., & Maze, M., et al., (1997). Substitution of a mutant a_{2a}-adrenergic receptor via "hit and run" gene targeting reveals the role of this subtype in sedative, analgesic, and anesthetic-sparing responses *in vivo*. *Proc Natl Acad Sci USA, 94*, 9950–9955.

Langer, S. Z. (1997). 25 years since the discovery of presynaptic receptors: Present knowledge and future perspectives. *Trends Pharmacol Sci, 18*, 95–99.

Link, R. E., Desai, K., Hein, L., Stevens, M. E., Chruscinski, A., & Bernstein, D., et al., (1996). Cardiovascular regulation in mice lacking alpha2-adrenergic receptor subtypes b and c. *Science, 273*, 803–805.

Link, R. E., Stevens, M. S., Kulatunga, M., Scheinin, M., Barsh, G. S., & Kobilka, B. K. (1995). Targeted inactivation of the gene encoding the mouse alpha 2c-adrenoceptor homolog. *Mol Pharmacol, 48*, 48–55.

Liu, Y. W., Li, J., & Ye, J. H. (2010). Histamine regulates activities of neurons in the ventrolateral preoptic nucleus. *J Physiol, 588*, 4103–4116.

Lu, R., Chen, Y., Cottingham, C., Peng, N., Jiao, K., & Limbird, L. E., et al., (2010). Enhanced hypotensive, bradycardic, and hypnotic responses to alpha2-adrenergic agonists in spinophilin-null mice are accompanied by increased G protein coupling to the alpha2A-adrenergic receptor. *Mol Pharmacol, 78*, 279–286.

Lu, R., Li, Y., Zhang, Y., Chen, Y., Shields, A. D., & Winder, D. G., et al., (2009). Epitope-tagged receptor knock-in mice reveal that differential desensitization of alpha2-adrenergic responses is because of ligand-selective internalization. *J Biol Chem, 284*, 13233–13243.

Lymperopoulos, A., Rengo, G., Funakoshi, H., Eckhart, A. D., & Koch, W. J. (2007a). Adrenal GRK2 upregulation mediates sympathetic overdrive in heart failure. *Nat Med, 13*, 315–323.

Lymperopoulos, A., Rengo, G., Gao, E., Ebert, S. N., Dorn 2nd, G. W., & Koch, W. J. (2010). Reduction of sympathetic activity via adrenal-targeted *GRK2* gene deletion attenuates heart failure progression and improves cardiac function after myocardial infarction. *J Biol Chem, 285*, 16378–16386.

Lymperopoulos, A., Rengo, G., & Koch, W. J. (2007b). Adrenal adrenoceptors in heart failure: Fine-tuning cardiac stimulation. *Trends Mol Med, 13*, 503–511.

Lymperopoulos, A., Rengo, G., Zincarelli, C., Soltys, S., & Koch, W. J. (2008). Modulation of adrenal catecholamine secretion by in vivo gene transfer and manipulation of G protein-coupled receptor kinase-2 activity. *Mol Ther, 16*, 302–307.

MacDonald, E., Kobilka, B. K., & Scheinin, M. (1997). Gene targeting: Homing in on a2-adrenoceptor-subtype function. *Trends Pharmacol Sci, 18*, 211–219.

MacMillan, L. B., Hein, L., Smith, M. S., Piascik, M. T., & Limbird, L. E. (1996). Central hypotensive effects of the alpha2a-adrenergic receptor subtype. *Science, 273*, 801–803.

Matsuo, S., Jang, I. S., Nabekura, J., & Akaike, N. (2003). alpha 2-Adrenoceptor-mediated presynaptic modulation of GABAergic transmission in mechanically dissociated rat ventrolateral preoptic neurons. *J Neurophysiol, 89*, 1640–1648.

Maze, M., Scarfini, C., & Cavaliere, F. (2001). New agents for sedation in the intensive care unit. *Crit Care Clin, 17*, 881–897.

McClure, M. M., Barch, D. M., Romero, M. J., Minzenberg, M. J., Triebwasser, J., & Harvey, P. D., et al., (2007). The effects of guanfacine on context processing abnormalities in schizotypal personality disorder. *Biol Psychiat, 61*, 1157–1160.

Mizobe, T., Maghsoudi, K., Sitwala, K., Tianzhi, G., Ou, J., & Maze, M. (1996). Antisense technology reveals the alpha2A adrenoceptor to be the subtype mediating the hypnotic response to the highly selective agonist, dexmedetomidine, in the locus coeruleus of the rat. *J Clin Investig, 98*, 1076–1080.

Morrison, S. F., & Nakamura, K. (2011). Central neural pathways for thermoregulation. *Front Biosci, 16*, 74–104.

Morrison, S. F., Nakamura, K., & Madden, C. J. (2008). Central control of thermogenesis in mammals. *Exp Physiol, 93*, 773–797.

Muller, C. S., Haupt, A., Bildl, W., Schindler, J., Knaus, H. G., & Meissner, M., et al., (2010). Quantitative proteomics of the Cav2 channel nano-environments in the mammalian brain. *Proc Natl Acad Sciences USA, 107*, 14950–14957.

Muthig, V., Gilsbach, R., Haubold, M., Philipp, M., Ivacevic, T., & Gessler, M., et al., (2007). Upregulation of soluble vascular endothelial growth factor receptor 1 contributes to angiogenesis defects in the placenta of alpha 2B-adrenoceptor deficient mice. *Circ Res, 101*, 682–691.

Myers, R. D., Beleslin, D. B., & Rezvani, A. H. (1987). Hypothermia: Role of alpha 1- and alpha 2-noradrenergic receptors in the hypothalamus of the cat. *Pharmacol Biochem Behavior, 26*, 373–379.

Nassif, S., Kempf, E., Cardo, B., & Velley, L. (1983). Neurochemical lesion of the locus coeruleus of the rat does not suppress the sedative effect of clonidine. *Eur J Pharmacol, 91*, 69–76.

Nelson, L. E., Guo, T. Z., Lu, J., Saper, C. B., Franks, N. P., & Maze, M. (2002). The sedative component of anesthesia is mediated by GABA(A) receptors in an endogenous sleep pathway. *Nat Neurosci, 5*, 979–984.

Nelson, L. E., Lu, J., Guo, T., Saper, C. B., Franks, N. P., & Maze, M. (2003). The alpha2-adrenoceptor agonist dexmedetomidine converges on an endogenous sleep-promoting pathway to exert its sedative effects. *Anesthesiology, 98*, 428–436.

Philipp, M., Brede, M., & Hein, L. (2002a). Physiological significance of alpha(2)-adrenergic receptor subtype diversity: One receptor is not enough. *Am J Physiol Regul Integr Comp Physiol, 283*, R287–R295.

Philipp, M., Brede, M. E., Hadamek, K., Gessler, M., Lohse, M. J., & Hein, L. (2002b). Placental alpha(2)-adrenoceptors control vascular development at the interface between mother and embryo. *Nat Genet, 31*, 311–315.

Philipp, M., & Hein, L. (2004). Adrenergic receptor knockout mice: distinct functions of 9 receptor subtypes. *Pharmacol Therap, 101*, 65–74.

Ramos, B. P., & Arnsten, A. F. (2007). Adrenergic pharmacology and cognition: Focus on the prefrontal cortex. *Pharmacol Therap, 113*, 523–536.

Rasmussen, S. G., Choi, H. J., Fung, J. J., Pardon, E., Casarosa, P., & Chae, P. S., et al., (2011). Structure of a nanobody-stabilized active state of the beta(2) adrenoceptor. *Nature, 469*, 175–180.

Rasmussen, S. G., Choi, H. J., Rosenbaum, D. M., Kobilka, T. S., Thian, F. S., & Edwards, P. C et al., (2007). Crystal structure of the human beta2 adrenergic G-protein-coupled receptor. *Nature, 450*, 383–387.

Rengo, G., Leosco, D., Zincarelli, C., Marchese, M., Corbi, G., & Liccardo, D., et al., (2010). Adrenal GRK2 lowering is an underlying mechanism for the beneficial sympathetic effects of exercise training in heart failure. *Am J Physiol Heart Circ Physiol, 298*, H2032–H2038.

Rosenbaum, D. M., Cherezov, V., Hanson, M. A., Rasmussen, S. G., Thian, F. S., & Kobilka, T. S et al., (2007). GPCR engineering yields high-resolution structural insights into beta2-adrenergic receptor function. *Science, 318*, 1266–1273.

Rosenbaum, D. M., Zhang, C., Lyons, J. A., Holl, R., Aragao, D., & Arlow, D. H., et al., (2011). Structure and function of an irreversible agonist-beta(2) adrenoceptor complex. *Nature, 469*, 236–240.

Sallinen, J., Haapalinna, A., MacDonald, E., Viitamaa, T., Lahdesmaki, J., & Rybnikova, E., et al., (1999). Genetic alteration of the alpha2-adrenoceptor subtype c in mice affects the development of behavioral despair and stress-induced increases in plasma corticosterone levels. *Mol Psychiat, 4*, 443–452.

Sallinen, J., Haapalinna, A., Viitamaa, T., Kobilka, B. K., & Scheinin, M. (1998). D-amphetamine and L-5-hydroxytryptophan-induced behaviours in mice with genetically-altered expression of the alpha2C-adrenergic receptor subtype. *Neuroscience, 86*, 959–965.

Sallinen, J., Link, R. E., Haapalinna, A., Viitamaa, T., Kulatunga, M., & Sjoholm, B., et al., (1997). Genetic alteration of alpha 2C-adrenoceptor expression in mice: Influence on locomotor, hypothermic, and neurochemical effects of dexmedetomidine, a subtype-nonselective alpha 2-adrenoceptor agonist. *Mol Pharmacol, 51*, 36–46.

Sanders, R. D., & Maze, M. (2007). Adrenergic and cholinergic compounds. *In* ''Handbook of experimental pharmacology: Vol. 177. Analgesia'' (C. Stein ed.), pp. 251–264. Springer, Heidelberg.

Sawamura, S., Kingery, W. S., Davies, M. F., Agashe, G. S., Clark, J. D., & Kobilka, B. K., et al., (2000). Antinociceptive action of nitrous oxide is mediated by stimulation of noradrenergic neurons in the brainstem and activation of alpha2B adrenoceptors. *J Neurosci, 20*, 9242–9251.

Scahill, L., Chappell, P. B., Kim, Y. S., Schultz, R. T., Katsovich, L., & Shepherd, E., et al., (2001). A placebo-controlled study of guanfacine in the treatment of children with tic disorders and attention deficit hyperactivity disorder. *Am J Psychiat, 158*, 1067–1074.

Scheinin, M., Sallinen, J., & Haapalinna, A. (2001). Evaluation of the alpha2C-adrenoceptor as a neuropsychiatric drug target studies in transgenic mouse models. *Life Sci, 68*, 2277–2285.

Scholz, J., & Tonner, P. H. (2000). Alpha2-Adrenoceptor agonists in anaesthesia: A new paradigm. *Curr Opin Anesthesiol, 13*, 437–442.

Spyraki, C., & Fibiger, H. C. (1982). Clonidine-induced sedation in rats: Evidence for mediation by postsynaptic alpha 2-adrenoreceptors. *J Neural Trans, 54*, 153–163.

Starke, K. (2001). Presynaptic autoreceptors in the third decade: Focus on alpha2- adrenoceptors. *J Neurochem, 78*, 685–693.

Starke, K., Endo, T., & Taube, H. D. (1975). Pre- and postsynaptic components in effect of drugs with alpha adrenoceptor affinity. *Nature, 254*, 440–441.

Stone, L. S., MacMillan, L. B., Kitto, K. F., Limbird, L. E., & Wilcox, G. L. (1997). The alpha2a adrenergic receptor subtype mediates spinal analgesia evoked by alpha2 agonists and is necessary for spinal adrenergic-opioid synergy. *J Neurosci, 17*, 7157–7165.

Surprenant, A., Horstman, D. A., Akbarali, H., & Limbird, L. E. (1992). A point mutation of the a2-adrenoceptor that blocks coupling to potassium but not to calcium currents. *Science, 257*, 977–980.

Szot, P., Lester, M., Laughlin, M. L., Palmiter, R. D., Liles, L. C., & Weinshenker, D. (2004). The anticonvulsant and proconvulsant effects of alpha2-adrenoreceptor agonists are mediated by distinct populations of alpha2A-adrenoreceptors. *Neuroscience, 126*, 795–803.

Tan, C. M., Wilson, M. H., MacMillan, L. B., Kobilka, B. K., & Limbird, L. E. (2002). Heterozygous alpha 2A-adrenergic receptor mice unveil unique therapeutic benefits of partial agonists. *Proc Natl Acad Sci USA, 99*, 12471–12476.

Tank, J., Diedrich, A., Szczech, E., Luft, F. C., & Jordan, J. (2004a). Alpha-2 adrenergic transmission and human baroreflex regulation. *Hypertension, 43*, 1035–1041.

Tank, J., Jordan, J., Diedrich, A., Obst, M., Plehm, R., & Luft, F. C., et al., (2004b). Clonidine improves spontaneous baroreflex sensitivity in conscious mice through parasympathetic activation. *Hypertension, 43*, 1042–1047.

Taylor, F. B., & Russo, J. (2001). Comparing guanfacine and dextroamphetamine for the treatment of adult attention-deficit/hyperactivity disorder. *J Clin Psychopharmacol, 21*, 223–228.

Thomas, S. A., Matsumoto, A. M., & Palmiter, R. D. (1995). Noradrenaline is essential for mouse fetal development. *Nature, 374*, 643–646.

Trendelenburg, A. U., Philipp, M., Meyer, A., Klebroff, W., Hein, L., & Starke, K. (2003). All three alpha2-adrenoceptor types serve as autoreceptors in postganglionic sympathetic neurons. *Naunyn Schmiedebergs Arch Pharmacol, 368*, 504–512.

Wang, M., Ramos, B. P., Paspalas, C. D., Shu, Y., Simen, A., & Duque, A., et al., (2007). Alpha2A-adrenoceptors strengthen working memory networks by inhibiting cAMP-HCN channel signaling in prefrontal cortex. *Cell, 129*, 397–410.

Warne, T., Serrano-Vega, M. J., Baker, J. G., Moukhametzianov, R., Edwards, P. C., & Henderson, R., et al., (2008). Structure of a beta1-adrenergic G-protein-coupled receptor. *Nature, 454*, 486–491.

CHAPTER 8

Genetic Variations of α_2-Adrenergic Receptors Illuminate the Diversity of Receptor Functions

Christopher Cottingham, Huaping Chen, Yunjia Chen, Yin Peng, and Qin Wang

Department of Physiology and Biophysics, University of Alabama at Birmingham, AL, USA

I. Overview
II. Introduction
III. Overview of α_2AR Genetic Variants
  A. α_{2A}AR
  B. α_{2B}AR
  C. α_{2C}AR
IV. α_2ARs in Central Nervous System Dysfunction and Disease
  A. Attention Deficit/Hyperactivity Disorder
  B. Neuropsychiatric Disorders
  C. Alzheimer's Disease
  D. Emotional Memory Dysfunction – Possible Novel Role for α_{2B}ARs?
  E. Summary – α_2ARs in the Central Nervous System
V. α_2ARs in Cardiovascular Disease
  A. α_2ARs in Cardiovascular Regulation
  B. Hypertension
  C. Heart Disease
  D. Summary – α_2ARs in the Cardiovascular System
VI. α_2ARs in Metabolism and Type 2 Diabetes
  A. Effects on Insulin Secretion and Glucose Handling
  B. Other Potential Contributions
  C. Summary – α_2ARs as Targets in the Treatment of Type 2 Diabetes
VII. α_2ARs in Other Peripheral Functions
  A. Gastrointestinal System
  B. Renal Functions
  C. Other Peripheral Nervous System-Mediated Functions
VIII. Conclusions
  References

Current Topics in Membranes, Volume 67
0065-230X/10 $35.00

DOI: 10.1016/B978-0-12-384921-2.00008-2

I. OVERVIEW

With advanced genotyping and sequencing technology, tremendous progress has been made in identification of human genome variations and their association with disease risk over the past 10 to 15 years. These studies represent a useful tool for understanding disease etiology and developing effective therapeutic strategies. Meanwhile, the wealth of information obtained in these studies provides valuable insights into endogenous functions of a gene in human physiology. Genetic variants have been identified in genes encoding all subtypes of the α_2 adrenergic receptor (α_2AR) subfamily, which represent an important group of receptors in normal and dysfunctional physiology as well as a significant class of therapeutic drug targets. The study of receptor polymorphisms and their associations with disease states in human populations provides insight complementary to that gained from experimental models on subtype-selective functions in the human body. For example, within the central nervous system (CNS), α_2AR genetic variants have been strongly linked to attention deficit/hyperactivity disorder (ADHD). α_2AR genetic variants have also been associated with various forms of cardiovascular dysfunction including heart disease. A relatively new line of evidence has implicated polymorphisms of the α_{2A}AR subtype specifically in type 2 diabetes. Additional evidence exists linking α_2AR genetic variants with other peripheral disorders, especially those involving autonomic nervous system dysfunction. This chapter will primarily review current knowledge of α_2AR genetic variants and their myriad associations with human disease states.

II. INTRODUCTION

The α_2 subfamily of adrenergic receptors (α_2ARs) consists of three subtypes, α_{2A}, α_{2B}, and α_{2C}, which are products of three distinct genes (Bylund et al., 1994). In native cells, all α_2AR subtypes signal through cognate heterotrimeric G proteins of the $G_{i/o}$ subfamily. Activation of these receptors inhibits adenylyl cyclase and voltage-gated Ca^{2+} channels, and also activates receptor-operated inwardly rectifying K^+ channels (Limbird, 1988; Kobilka, 1992). It has also been reported that stimulation of α_2ARs induces activation of phospholipase C (Gesek, 1996; Dorn, Oswald, McCluskey, Kuhel, & Liggett, 1997), which appears to be required for α_2AR-evoked inhibition of hyperpolarization-activated and cyclic nucleotide-gated channel (HCN) inward currents (Carr, Andrews, Glen, & Lavin, 2007). In addition, activation of α_2ARs leads to signaling propagation through the MAP kinase pathway (Richman & Regan, 1998; Wang et al., 2006a).

α_2ARs are widely distributed throughout the body and mediate a large variety of physiological and pharmacological responses *in vivo*. When activated by their endogenous ligands, epinephrine and norepinephrine (NE), α_2ARs couple to decreased epileptogenesis (Wilson et al., 1998) and anxiety (Schramm, McDonald, & Limbird, 2001), as well as reduction in insulin release (Hiyoshi et al., 1995; Natali et al., 1998). In response to pharmacological α_2AR agonists, activation of α_2ARs lowers blood pressure (MacMillan, Hein, Smith, Piascik, & Limbird, 1996; Altman et al., 1999), evokes sedation (Lakhlani et al., 1997), reduces pain perception (Lakhlani et al., 1997; Stone, MacMillan, Kitto, Limbird, & Wilcox, 1997), and improves working memory (Franowicz, Kessler, Borja, Kobilka, Limbird, & Arnsten, 2002; Ma et al., 2001; Marrs, Kuperman, Avedian, Roth, & Jentsch, 2005; Wang et al., 2007). There are no subtype-selective agonists available for α_2ARs to date, and so subtype-selective functionality has been elucidated primarily through the use of transgenic mouse models. Such studies have revealed *in vivo* functions of individual α_2AR subtypes by characterizing genetically targeted mice. For example, most of the central effects elicited by α_2AR-agonists including sedation and blood pressure reduction can be attributed to the α_{2A}AR subtype, because these responses are lost in mice lacking the functional α_{2A}AR (MacMillan et al., 1996; Altman et al., 1999; Lakhlani et al., 1997). On the other hand, α_2AR agonist-induced peripheral vasoconstriction and hypertensive responses in arteries are mediated by the α_{2B}AR subtype (Link et al., 1996). Functions of individual α_2AR subtypes uncovered by studies in transgenic mice have been extensively reviewed previously (Kable, Murrin, & Bylund, 2000; Philipp, Brede, & Hein, 2002; Knaus et al., 2007), and thus will not be discussed in detail in this review.

With remarkable advances in human genome sequencing and genotyping, identification of genetic variations in human populations and their potential associations with diseases and disorders has become a rapidly growing field. Generally, when a genetic variant appears in more than 1% of a population, it is defined as a polymorphism. Put another way, polymorphisms are (relatively) common genetic variants. In many cases, polymorphisms result in alterations in the expression and/or function of gene products, thereby contributing to disease processes and susceptibility. All α_2AR subtypes have been studied in this regard, and a number of polymorphisms have been identified. These polymorphisms have been investigated for their impact on receptor signaling and pharmacology, and for potential involvement in human diseases and disorders. Pharmacogenetic approaches have also investigated roles for these polymorphisms in relevant drug responses. Given the importance of α_2AR functions in multiple physiological processes and as a therapeutic drug target, it is not surprising that variations at these gene loci have been found to be associated with a number of disease states. Collectively, these genetic association studies illuminate diverse functions of the α_2ARs.

III. OVERVIEW OF α_2AR GENETIC VARIANTS

Before embarking on a review of α_2AR genetic studies in human diseases, we will very briefly introduce α_2AR genetics and outline the common receptor polymorphisms. The three α_2AR subtypes are separate gene products, and all are encoded as intronless genes. Human α_{2A}, α_{2B}, and α_{2C}AR genes are located on chromosomes 10, 2, and 4, respectively. Interestingly, very few polymorphisms have been observed within the coding regions, and all of these reported so far are located within the receptor third intracellular loop (3i loop). Much of the seminal work on adrenergic receptor polymorphisms has been undertaken by the laboratory of Stephen Liggett, and this subject has been extensively reviewed by him and colleagues (e.g., Small, McGraw, & Liggett, 2003). Table I presents a summary by receptor subtype of commonly-studied polymorphisms and their disease associations (which will be detailed in subsequent sections).

A. α_{2A}AR

Although numerous α_{2A}AR genetic variants have been identified, only one nonsynonymous polymorphism has been found within the coding region of the gene. This is the C-to-G substitution at position 753 of the gene which results in an Asn-to-Lys change at residue 251 (N251K), located in the 3i loop (Small, Forbes, Brown, & Liggett, 2000a). The N251K mutant receptor has been shown, at the molecular level, to be a gain-of-function mutant resulting in increased agonist-stimulated G protein coupling to the receptor (Small et al., 2000a). This polymorphism appears to be quite infrequent in the human population (allele frequencies well below 1%), and consequently it has been difficult to study. A number of other polymorphisms have been described in noncoding regions of the gene, including the promoter and 5′-UTR and 3′-UTR regions. Of these, the most-studied is the C-1291G polymorphism in the promoter region, first identified as a restriction fragment length polymorphism (RFLP) creating an Msp1 cleavage site (Lario, Calls, Cases, Oriola, Torras, & Rivera, 1997). The particular effect of this single polymorphism on the receptor itself has not been elucidated. Another well-studied α_{2A}AR polymorphism is the DraI RFLP, originally discovered by Lockette et al. (1995). Subsequent work reported this variant as a G-to-A substitution at position 1780 within the 3′-UTR in combination with a deletion at position 1781 (Finley et al., 2004). This polymorphism (also referred to as rs553668) has since been shown to confer overexpression of the α_{2A}AR in pancreatic islets of carriers of the A allele (Rosengren et al., 2010).

Work from the Liggett laboratory has identified other noncoding polymorphisms and grouped them into 17 haplotypes along with synonymous and

TABLE I

Summary of commonly studied α_2AR polymorphisms and their disease associations arranged by receptor subtype

Receptor	Polymorphism	Association	Reference(s)
α_{2A}AR	C-1291G (MspI RFLP)	ADHD	Roman et al. (2003); Schmitz et al. (2006); Deupree et al. (2006)
		Methylphenidate response	Polanczyk et al. (2007); da Silva et al. (2008)
		Weight gain w/mirtazepine	Lee et al. (2009)
		Weight gain w/olanzapine	Park et al. (2006)
		Suicide	Fukutake et al. (2008)
		Abdominal fat accumulation	Garenc et al. (2002)
		Irritable bowel syndrome	Kim et al. (2004)
	G1780A (DraI RFLP)	ADHD	Park et al. (2005)
		Hypertension	Lockette et al. (1995), Svetkey et al. (1996)
		Type 2 diabetes (risk)	Rosengren et al. (2010)
		Autonomic stress response	Finley et al. (2004)
	C753G (N251K)	Suicide	Sequeira et al. (2004)
α_{2B}AR	Del301-303	Enhanced emotional memory	de Quervain et al. (2007); Rasch et al. (2009); Cousjin et al. (2010)
		Hypertension	von Wowern et al. (2004); Vasudevan et al. (2008)
		Myocardial infarction	Snapir et al. (2001); Laukkanen et al. (2009)
		Sudden cardiac death	Snapir et al. (2003b); Laukkanen et al. (2009)
		Type 2 diabetes (risk)	Siitonen et al. (2004)
		Type 2 diabetes (onset age)	Papazoglou et al. (2006)
		Metabolic rate in obesity	Heinonen et al. (1999)
		Autonomic tone in obesity	Sivenius et al. (2003); Ueno et al. (2006)
α_{2C}AR	Del322-325	Depression	Neumeister et al. (2006)
		Heart failure (risk, in combination w/β1Arg389)	Small et al. (2002)
		Cardiomyopathy (event-free survival)	Regitz-Zagrosek et al. (2006)
		Irritable bowel syndrome	Kim et al. (2004)

nonsynonymous polymorphisms in the coding region (Small, Brown, Seman, Theiss, & Liggett, 2006). Some of these haplotypes were found to affect receptor expression levels in a heterologous system, raising the possibility that combinations of noncoding polymorphisms may affect receptor density endogenously. It should be noted that the position 1781 deletion associated with the DraI RFLP was not observed by the Liggett group in their sample populations (Small et al., 2006).

B. α_{2B}AR

The most well-studied polymorphic α_{2B}AR variant is the deletion of nine base pairs beginning at position 901 in the coding region – this results in a loss of three Glu residues (301–303) in the receptor 3i loop (Small, Brown, Forbes, & Liggett 2001). The deleted residues are part of an acidic stretch in the receptor 3i loop (EDEAEEEEEEEEEEEE) which is known to be necessary for phosphorylation by G protein-coupled receptor kinases (GRKs) (Jewell-Motz & Liggett, 1995). Not surprisingly, then, this variant has been characterized at the molecular level and found to undergo diminished GRK phosphorylation and desensitization in response to agonist (Small et al., 2001). This variant has also been shown to be resistant to chronic agonist-promoted downregulation (Salim, Desai, Taneja, & Eikenburg, 2009). The Del301-303 polymorphism is relatively common, with allele frequencies of 0.31 for Caucasians and 0.12 for African-Americans (Small et al., 2001).

Other relatively common polymorphisms are found in noncoding regions, and include a 12 nucleotide deletion beginning at position -4825 in the 5′-UTR (Crassous et al., 2010), a G-98C SNP in the 5′-UTR (Cayla et al., 2004), a synonymous coding region mutation at position 1182 (Etzel et al., 2005), and a C1776A SNP in the 3′-UTR. Interestingly, these five polymorphisms listed above have been found to be in linkage with each other, especially in the Caucasian population (Cayla et al., 2004; Etzel et al., 2005; Crassous et al., 2010). At the molecular level, the noncoding region polymorphisms may have effects on transcriptional activity at the α_{2B}AR gene, thereby affecting receptor expression levels (Cayla et al., 2004).

C. α_{2C}AR

For the α_{2C}AR, the primary polymorphism that has been identified is a 12 nucleotide deletion beginning at position 964 of the gene. This mutant receptor has a deletion of four residues (Gly–Ala–Gly–Pro) in the receptor 3i loop, and was shown to exhibit deficient coupling to heterotrimeric G proteins and downstream effectors including MAP kinase (Small, Forbes, Rahman, Bridges, & Liggett, 2000b). The Del322-325 variant is infrequent in the Caucasian population (frequency of well below 1%), but is quite common in African-Americans (allele frequency of 0.381). Subsequent work demonstrated that α_{2C}AR polymorphisms (including Del322-325) exist in as many as nine different haplotypes (Small, Mialet-Perez, Seman, Theiss, Brown, & Liggett, 2004). Interestingly, a more recent study indicated that the α_{2C}AR Del322-325 variant does not exhibit any alteration in inhibition of cAMP production compared with wild-type receptor when expressed in HEK293 cells (Montgomery & Bylund, 2010). Thus, the effects of this deletion on α_{2C}AR function may be more complex than initially thought.

IV. α_2ARs IN CENTRAL NERVOUS SYSTEM DYSFUNCTION AND DISEASE

α_2ARs have been well studied with regard to their roles in CNS functions. In particular, α_2ARs are strongly linked with CNS dysfunction involving the noradrenergic system, as has been linked to disorders such as attention deficit/hyperactivity disorder (ADHD) and depression. As well, α_2AR genotypes could potentially serve as predictors of treatment outcomes to therapeutics modulating the brain noradrenergic system. In this section, we will examine available evidence on α_2AR polymorphisms and their roles in various CNS disorders and diseases, including ADHD, mood disorders such as depressive disorders and schizophrenia, Alzheimer's disease (AD), and emotional memory dysfunction.

A. *Attention Deficit/Hyperactivity Disorder*

Dysfunction of the brain noradrenergic system in general has long been implicated in ADHD; this subject has been well-reviewed by others (Biederman, 2005; Brennan & Arnsten, 2008; Prince, 2008). Additionally, genes involved in noradrenergic neurotransmission, including α_2ARs, have often been considered as candidates in genetic studies of ADHD (Banaschewski, Becker, Scherag, Franke, & Coghill, 2010).

Studies of human patients have been carried out to investigate a possible link between α_2AR genetics and ADHD. While some studies have found no link between α_{2A}AR (Xu et al., 2001; Wang, Lu, Zhao, & Limbird, 2006a) and α_{2C}AR (Barr et al., 2001; De Luca, Muglia, Vincent, Lanktree, Jain, & Kennedy, 2004) genetic variants and ADHD, several others have suggested a possible role for α_{2A}AR polymorphisms in ADHD. A pair of studies in an adolescent Brazilian population identified a small contribution of the C-1291G polymorphism to ADHD susceptibility and disorder severity (Roman, Schmitz, Polanczyk, Eizirik, Rohde, & Hutz, 2003), and specifically to susceptibility to the primarily inattentive type of ADHD (Schmitz et al., 2006). Another study in a population of American children found that the DraI RFLP was positively correlated with both the inattentive and hyperactive-impulsive symptoms of ADHD (Park et al., 2005). More recently, Deupree et al. (2006) studied three different polymorphisms (C-1291G and DraI and HhaI RFLPs) and found that certain haplotypes were associated with ADHD. Intriguingly, this study also examined binding characteristics of platelet α_2ARs and found that altered receptor affinity for ligand was correlated with the C-1291G and DraI polymorphisms, thereby providing evidence from human patients that genetic variability can affect α_2AR pharmacological properties. Although not dealing with ADHD specifically, it was recently reported that C-1291G may contribute to inattention and hyperactivity symptoms in adolescents who have experienced maltreatment (Kiive, Kurrikoff, Maestu, & Harro, 2010).

Treatment of ADHD classically relies largely on stimulants, which are thought to work by modulating catecholaminergic neurotransmission (Arnsten, 2006), and, more recently, on α_2AR agonists such as clonidine (Banaschewski, Roessner, Dittmann, Santosh, & Rothenberger, 2004). A pair of studies utilizing a pharmacogenetic approach found a positive association between the C-1291G polymorphism and methylphenidate-induced improvement in inattentive symptoms in cases of ADHD (Polanczyk et al., 2007; da Silva et al., 2008). Collectively, this evidence provides strong support for the general hypothesis that the α_{2A}AR is involved in ADHD, and suggests that continued efforts to target α_{2A}ARs for treatment of this disorder may prove fruitful. Future studies investigating a link between α_2AR genetics and response to clonidine treatment may also be of value.

B. Neuropsychiatric Disorders

α_2ARs, particularly α_{2A}ARs, have been implicated in a number of neuropsychiatric disorders. These receptors may likely be involved in the pathogenesis of these disorders, although details of this are yet to be worked out. Additionally, α_2ARs are valid molecular targets in the therapeutic treatment of these disorders.

1. Depression and Suicide

α_2ARs have been linked quite strongly with mood disorders, in particular depressive disorders including major depressive disorder (MDD), and suicide. A role for α_2ARs was originally suggested by the classical monoamine hypothesis of depression (Belmaker & Agam, 2008). While this hypothesis has proven inadequate to explain the neurobiology of depression, it has nevertheless stimulated a productive line of research into α_2ARs and depressive disorders. In particular, studies on post mortem brain tissue from depressed suicides have consistently yielded results showing upregulation of α_2ARs (Meana, Barturen, & Garcia-Sevilla, 1992; Callado, Meana, Grijalba, Pazos, Sastre, & Garcia-Sevilla, 1998; Garcia-Sevilla et al., 1999; Ordway, Schenk, Stockmeier, May, & Klimek, 2003; Escriba, Ozaita, & Garcia-Sevilla, 2004). Conversely, it has been reported that chronic antidepressant treatment lowers brain α_2AR density in human patients (De Paermentier, Mauger, Lowther, Crompton, Katona, & Horton, 1997) and alters receptor density in experimental models (Barturen & Garcia-Sevilla, 1992; Mateo, Fernandez-Pastor, & Meana, 2001; Subhash, Nagaraja, Sharada, & Vinod, 2003).

Given the evidence outlined above, it seems reasonable to postulate that α_2AR polymorphisms resulting in altered receptor expression levels may underlie the altered receptor expression patterns associated with depressed/suicidal patients. However, to date, there have been very few studies attempting to link α_2AR polymorphisms with depression and few linking the receptors with suicide. With regard to suicide, a 2004 study suggested preliminarily that the N251K variant of the α_{2A}AR was associated with suicide (Sequeira et al., 2004); however, a follow-up study was unable to replicate this result (Martin-Guerrero, Callado, Saitua, Rivero, Garcia-Orad, & Meana, 2006). A potential confound in these studies is the extremely low allele frequency of this coding region polymorphism; indeed, the 2006 study was unable to find any individuals in either the control or suicide group carrying the variant. Therefore, the possibility remains open that the N251K variant may make a contribution to suicidality. Additionally, it is tempting to speculate that this gain-of-function mutant may account for the α_{2A}AR supersensitivity that has been reported in brain tissue from suicide victims (Gonzalez-Maeso, Rodriguez-Puertas, Meana, Garcia-Sevilla, & Guimon, 2002), although a genetic basis for this receptor supersensitivity has not yet been shown. A more recent study uncovered a possible link between the α_{2A}AR C-1291G polymorphism and susceptibility to suicide in Japanese females (Fukutake et al., 2008). With regard to depression itself, a recent study carried out in a Korean sample population found no significant relationship between the α_{2A}AR C-1291G polymorphism and incidence of MDD or response to the anti-α_2-adrenergic antidepressant drug mirtazepine in MDD patients. A weak association was, however, found between the weight gain side effect commonly observed with mirtazepine and the C/C genotype at position -1291 (Lee et al., 2009). A separate study in a Japanese sample population uncovered a potential link between C-1291G and treatment response to the antidepressant milnacipran, although these results are preliminary (Wakeno et al., 2008). Altogether, currently available data suggest that

α_{2A}AR genetics may influence suicidal behavior, but perhaps not depressive disorders in a more general sense, and may have some use in predicting treatment response to therapeutics.

Interestingly, an fMRI-based study in patients with MDD found a positive association between the α_{2C}AR deletion 322-325 variant and abnormal neural responses to facial expressions in the MDD patients (Neumeister et al., 2006). This study raises the possibility that the α_{2C} subtype is involved in depression and may be a potential therapeutic target.

2. Schizophrenia

Schizophrenia is a complex, multifactorial psychiatric disorder with poorly understood etiology. Although the antipsychotic therapeutics used to treat schizophrenia have been designed to primarily target dopaminergic neurotransmission, they are "dirty" drugs with many molecular targets, including α_2ARs (Baldessarini & Tarazi, 2006). Additionally, the enhancement in working memory associated with α_2AR stimulation in the prefrontal cortex has potential therapeutic benefit in schizophrenia (Ramos & Arnsten, 2007). Several pharmacogenetic studies have been carried out investigating a potential role for α_2ARs in the response to antipsychotics. Olanzapine has a well-established side effect of excessive weight gain, which was found to be positively associated with the α_{2A}AR C-1291G polymorphism in a Korean sample population (Park et al., 2006). Other studies have yielded negative results, with no effect of α_{2A}AR C-1291G or a novel 21 bp deletion in the 3′-UTR of the α_{2C}AR on the antipsychotic response (Tsai, Wang, Yu Younger, Lin, Yang, & Hong, 2001a; De Luca et al., 2005). Association studies attempting to link C-1291G (Tsai et al., 2001a; Yamaguchi et al., 2009) and multiple other single α_{2A}AR gene polymorphisms (Clark, Mata, Kerwin, Munro, & Arranz, 2007) with susceptibility to schizophrenia have yielded similarly negative results. Taken together, these results suggest that α_2ARs may be involved in mediating certain clinical effects of antipsychotic drugs, but are perhaps not involved in the underlying disease process. Additionally, these results along with the study from Lee et al. (2009) on the antidepressant mirtazepine indicate that the α_{2A}AR may play a role in mediating the metabolic side effects often associated with psychiatric medications.

3. Other Psychiatric Disorders

Other studies of α_2AR polymorphisms in mood, panic, and personality disorders have yielded largely negative results. A pair of studies by Ohara and colleagues found no association between the C-1291G polymorphism of the α_{2A}AR and a generalized group of mood disorders (Ohara, Nagai, Tani, Tsukamoto, Suzuki, & Ohara, 1998) or panic disorder (Ohara, Suzuki, Ochiai, Terada, & Ohara, 2000). A separate study found no association between the

C-1291G polymorphism and performance on a personality assessment for reward dependence (Tsai, Wang, & Hong, 2001b). It is possible that, as suggested by Small et al. (2006), these and other studies yielding negative results have missed associations by looking at single polymorphisms rather than considering whole haplotypes.

C. Alzheimer's Disease

The loss of noradrenergic input from the locus coeruleus is an early event that often occurs in neurodegenerative diseases, including AD, and this loss has been proposed to play a critical role in the pathogenesis and progression of these diseases (Marien Colpaert, & Rosenquist, 2004). α_2ARs play an essential role in regulating noradrenergic input to the cerebral cortex and the resulting cortical response (Hein, 2006). Although the only study probing for a link between α_2AR polymorphisms and susceptibility to AD yielded a negative result for the α_{2A}AR C-1291G polymorphism (Hong, Wang, Liu, Liu, & Tsai, 2001), the possibility remains that α_2ARs are involved in AD and may be a viable therapeutic target.

D. Emotional Memory Dysfunction - Possible Novel Role for α_{2B}ARs?

Noradrenergic neurotransmission has been consistently implicated in the emotional memory function of the amygdala (McGaugh, 2004; Roozendaal, Barsegyan, & Lee, 2008). This knowledge provides the basis for a recent series of intriguing studies that have suggested a possible novel role for the α_{2B}AR in emotional memory, with implications for post-traumatic stress disorder (PTSD) and other forms of emotional memory dysfunction. These investigators set out to investigate a possible link between the α_{2B}AR deletion 301-303 variant and amygdalar function. The first of these studies found enhanced emotional memory in healthy European individuals and enhanced traumatic memory in African war refugees (with and without a diagnosis of PTSD) carrying the deletion (de Quervain et al., 2007). A follow-up study from the same group used a functional MRI (fMRI) approach to demonstrate enhanced amygdala activity during an emotional memory task in healthy individuals carrying the deletion variant (Rasch et al., 2009). Both of these studies collapsed heterozygous and homozygous carriers of the deletion into a single group, suggesting that just one mutant allele can result in a phenotype. An additional independent study, also using an fMRI approach in health volunteers, discovered that deletion carriers exhibited enhanced amygdala activity during an emotional memory task specifically following exposure to acute stress (Cousjin et al., 2010). This last piece of data provides a further link to the noradrenergic system, which is well-known to be engaged in response to stress, affecting cognition (Sara, 2009).

While these results are indeed interesting, it is important to take some caution in their interpretation. First, these studies have posited that the observed results are due to a loss-of-function effect of the deletion variant (Small et al., 2001), leading to decreased α_{2B}AR presynaptic function and enhanced noradrenergic neurotransmission. However, experimental evidence has suggested that there is little expression of the α_{2B} subtype in the CNS, with no evidence of amygdalar α_{2B} expression (Scheinin et al., 1994; Wang, MacMillan, Fremeau, Magnuson, Lindner, & Limbird 1996), although subtype-specific α_2AR expression is not as well-understood in the human brain. Presynaptic autoreceptor function has been ascribed mainly to the α_{2A} and, to a lesser extent, α_{2C} subtypes (Knaus et al., 2007). Second, the studies rely heavily on fMRI, a technique that relies on changes in blood flow rate to measure neuronal activity, leading to the potential confound of the α_{2B}AR deletion variant affecting fMRI results via direct regulation of vascular function (see Section V) independent of central synaptic transmission. Further studies will be necessary to establish a role for the α_{2B} subtype in noradrenergic neurotransmission in the human brain, and firmly link the α_{2B}AR with the observed abnormalities in amygdalar function. Nevertheless, the studies outlined above have identified the α_{2B}AR as a potential molecular target in emotional memory dysfunction worthy of continued investigation.

E. Summary - α_2ARs in the Central Nervous System

Studies of α_2AR polymorphisms have revealed several potential roles for α_2ARs in the CNS, and have identified α_2ARs as possible molecular targets in therapeutic treatments of CNS diseases and disorders. The α_{2A}AR has been most strongly and consistently linked with ADHD, with the receptor polymorphisms such as C-1291G serving as indicators of disorder severity and/or susceptibility. The receptor may also be a good target for pharmacogenetic studies in ADHD. Within the broad category of neuropsychiatric disorders, α_{2A}ARs have been linked with suicide and metabolic side effects of antidepressant and antipsychotic drugs. Associations with depression and schizophrenia susceptibility have largely not been found; however, the α_{2C}AR deletion 322–325 variant may be linked with MDD. Finally, recent studies have raised the possibility that the α_{2B}AR deletion 301-303 variant may be linked with altered amygdala function in emotional memory, identifying the α_{2B}AR as a possible molecular target in the treatment of disorders such as PTSD. Altogether, the available evidence supports a general role for α_2ARs in cognition and cognitive disorders.

V. α_2ARs IN CARDIOVASCULAR DISEASE

Adrenergic receptors have long been implicated in cardiovascular function and disease. Relevant to our discussion, all three α_2AR subtypes have appreciated roles in the cardiovascular system, through sympathetic regulation and through direct effects in cardiac and vascular tissues. α_2ARs are also known to be targets for a number of sympathomimetic therapeutics utilized in the treatment of cardiovascular dysfunction (Westfall & Westfall, 2006). α_2AR polymorphisms in cardiovascular disease have previously been reviewed (see Flordellis, Manolis, Scheinin, & Paris, 2004). The following sections will briefly discuss roles for α_2AR in cardiovascular regulation, and then focus on α_2AR polymorphisms in hypertension and heart diseases.

A. α_2ARs in Cardiovascular Regulation

Studies from subtype-specific knockout mice have revealed major roles for the α_2ARs in the vascular system with consequences for blood pressure regulation. α_{2A}ARs and α_{2C}ARs have largely been studied in the context of their roles in regulating norepinephrine release from sympathetic terminals. α_{2A}AR-null mice exhibit elevated release of norepinephrine from cardiac sympathetic nerve terminals and higher resting systemic blood pressure and heart rate (Altman et al., 1999). The α_{2A}AR subtype has been demonstrated as the primary presynaptic autoreceptor controlling norepinephrine release from sympathetic terminals, while the α_{2C}AR subtype has been implicated in regulating release specifically at low stimulation frequencies (Hein, Altman, & Kobilka, 1999). Heterozygous and homozygous deletion of the α_{2C}AR resulted in elevated urinary excretion of epinephrine, while heterozygotes were more susceptible to cardiac hypertrophy and heart failure after left-ventricular pressure overload (Gilsbach et al., 2007). Additionally, survival rates of both α_{2A}AR- and α_{2C}AR-null mice have been shown to be decreased following cardiac pressure overload due to heart failure (Brede et al., 2002). In human studies, the α_{2A}AR DraI RFLP has been associated with increased sympathetic drive and elevated blood pressure (Finley et al., 2004), while the α_{2C}AR Del322-325 variant has been associated with elevated basal blood pressure and exaggerated α_2AR antagonist-induced increases in blood pressure/heart rate in healthy volunteer subjects (Neumeister et al., 2005).

The α_{2B}AR subtype has also been extensively studied in this regard. α_{2B}ARs in vascular smooth muscle are known to mediate vasoconstriction contributing to increased blood pressure (Link et al., 1996). The importance of the α_{2B}AR in vascular regulation has since been confirmed in human studies of α_{2B}AR polymorphisms. Work from the laboratory of Mika Scheinin has demonstrated that the α_{2B}AR Del301-303 variant is associated with altered vascular responses to epinephrine. Specifically, the polymorphism was associated with decreased flow-mediated dilation of the brachial artery and increased blunted coronary blood flow following intravenous epinephrine (Heinonen et al., 2002; Snapir et al., 2003a). More recently, it was shown that the haplotype consisting of Del301-303 along with the G-98C, C1182A, and C1176A polymorphisms was associated with resistance to desensitization of the hand vein response to dexmedetomidine (an α_2AR agonist) in Caucasian and African-American patients (Muszkat et al., 2010), although previous studies by this group had yielded negative results (Muszkat et al., 2005a; Muszkat, Sofowora, Xie, Wood, & Stein, 2005b).

B. Hypertension

Despite the promising evidence presented above, studies probing for a link between α_2AR polymorphisms and hypertension have yielded largely mixed results. Positive associations between hypertension and the α_{2A}AR DraI RFLP have been reported in a number of studies. An early study linked the polymorphism with hypertension in a mixed population of hypertensive and normotensive individuals, specifically in Caucasian but not African-American subjects (Svetkey, Timmons, Emovon, Anderson, Preis, & Chen, 1996). Work by Warren Lockette and colleagues demonstrated that the DraI RFLP conferred increased risk of hypertension, particularly in African-Americans (Lockette et al., 1995). These two pieces of evidence illustrate the variability in conclusions commonly seen among different genetic association studies. The largest association study for α_2ARs and hypertension carried out to date by Li, Canham, Vongpatanasin, Leonard, Auchus, and Victor (2006) revealed negative results for both the α_{2A}AR DraI RFLP and α_{2C}AR Del322-325 variant and associations with hypertension and parameters of hypertensive heart disease. Other α_{2A}AR polymorphisms have been studied in the context of hypertension, with no association found between a novel Bsu361 RFLP and hypertension in a Japanese population (Umemura et al., 1994).

The first study investigating a link between the α_{2B}AR Del301-303 variant and hypertension was carried out by Baldwin et al. (1999), and found no genetic linkage between Del301-303 and essential hypertension. Subsequently, several studies have been unable to find a link between α_{2B}AR Del301-303 and hypertension in Western populations (Snapir et al., 2001; Etzel et al., 2005; Iacoviello et al., 2006). One study also found no association between the synonymous coding region C1182A polymorphism and hypertension (Etzel et al., 2005). A pair of studies carried out in a Chinese population showed no difference in frequency of α_{2B}AR Del301-303 genotype in normotensive versus hypertensive groups, but did find an association between the nondeletion allele and elevated blood pressure in Chinese men (Zhang et al., 2005; Li, Li, Y., Wen, Y., & Wang, 2008). On the positive side, studies in a Swedish population identified a hypertension susceptibility locus on chromosome 2 which contains the α_{2B}AR gene and subsequently uncovered an association between α_{2B}AR Del301-303 and early-onset primary hypertension as well as a weak association with nondiabetic primary hypertension (von Wowern et al., 2003; von Wowern, Bengtsson, Lindblad, Ratam, & Melander, 2004). Additionally, a study in a Malaysian population determined that α_{2B}AR Del301-303 was associated with essential hypertension in patients regardless of the presence or absence of type 2 diabetes (Vasudevan, Ismail, Stanslas, Shamsudin, & Ali, 2008).

Collectively, the results outlined above suggest a possible role for the α_{2B}AR as a contributing genetic factor to hypertension, although it is likely not the sole or primary causative factor. The impact of α_{2B}AR Del301-303 observed varies based on a number of factors, particularly the ethnic makeup of the sample population. As well, given that the Del301-303 polymorphism is in linkage with other α_{2B}AR polymorphisms in certain populations (Cayla et al., 2004; Etzel et al., 2005; Crassous et al., 2010), it is possible that there are complex effects of the α_{2B}AR genotype. As an example, the Del301-303 polymorphism may result in enhanced receptor signaling while a noncoding polymorphism such as the 5′-UTR deletion beginning at position -4825 may lead to decreased receptor expression. In such a scenario, the combination of polymorphisms would potentially have offsetting effects, thereby leading to a negative result for either with regard to an association with hypertension; such effects would likely differ among disparate human populations. The hypertension phenotype is also likely affected by the genotype status of many other molecular players, the effects of which may obscure contributions of α_{2B}ARs. Even bearing those caveats in mind, it is clear that the α_{2B}AR is a viable therapeutic target in the treatment of hypertension and other disorders of vascular regulation, although pharmacogenetic studies would likely prove difficult.

C. Heart Disease

Regulation of sympathetic activity by α_{2A}ARs and α_{2C}ARs has provided a rationale for studying the genetics of these receptors in the context of heart disease, particularly heart failure. Additionally, a recent experimental study showed that persistent activation of postsynaptic β1 receptors on cardiomyocytes led to a self-accelerating signaling cycle resulting in heart failure (Dorn, 2010), providing direct evidence that dysregulated sympathetic transmission could contribute to heart failure. Most studies to date have focused on the Del322-325 variant of the α_{2C}AR, and such studies have, unsurprisingly, yielded inconsistent results. An interesting study from Stephen Liggett's group examined the effects of the α_{2C}AR Del322-325 variant in combination with a variant of the β1AR (β1Arg389). The α_{2C}AR variant has decreased function, and the β1AR variant has increased function, and so the rationale was that this combination of variants would lead to both increased norepinephrine release from sympathetic terminals and enhanced adrenergic signaling in the cardiomyocyte, potentially predisposing such individuals to heart failure. Their results demonstrated an increased risk of heart failure for those individuals homozygous for both receptor variants (Small, Wagoner, Levin, Kardia, & Liggett, 2002). However, as pointed out by Dorn, it should be noted that this study had a small number for case-control comparison, and so carries an increased chance of a false-positive association (Dorn, 2010). Indeed, a later study by Savva et al. (2009) was unable to repeat these results, finding no association between α_{2C}AR Del322-325 alone or in combination with β1Arg389 and risk of adverse events in a population of congestive heart failure patients within the MERIT-HF study.

One other positive result was found in a study carried out in patients with dilated cardiomyopathy, which found that the α_{2C}AR Del322-325 polymorphism can protect against adverse outcomes such as death or heart transplant (Regitz-Zagrosek et al., 2006). Several other studies in American, European, African, and Asian populations have found no significant association between the α_{2C}AR Del322-325 variant and heart failure risk or heart failure parameters (Nonen et al., 2005; Metra et al., 2006; Canham et al., 2007; Du Preez, Matolweni, Greenberg, Mentla, Adeyemo, & Mayosi, 2008). While available evidence is mixed, the possibility remains that the α_{2C}AR plays a role in heart failure, although it is likely not the primary causative factor. Future studies on this subject, particularly those which account for haplotypes of α_{2C}AR polymorphisms (Small et al., 2004), seem to be in order. The α_{2C}AR should continue to be considered as a therapeutic target in heart failure moving forward.

α_{2B}ARs have also been studied in the context of heart disease. The Del301-303 variant was first established as a genetic risk factor for acute myocardial infarction in a male Finnish sample population, with an approximately doubled risk of acute coronary events (Snapir et al., 2001). Further studies linked α_{2B}AR Del301-303 with increased risk of sudden cardiac death in a Caucasian population, with particular risk for men under age 50 (Snapir, Mikkelsson, Perola, Penttila, Scheinin, & Karhunen, 2003b). More recently, a separate group confirmed the α_{2B}AR Del301-303 variant as a genetic risk factor for myocardial infarction and sudden cardiac death in middle-aged men (Laukkanen, Makikallio, Kauhanen, & Kurl, 2009). Hence, available evidence strongly links the α_{2B}AR genotype with risk of acute heart disease, suggesting the receptor as a useful potential target for screening.

D. Summary - α_2ARs in the Cardiovascular System

Based upon the evidence outlined in the preceding sections, it is clear that α_2AR genotype can impact the functioning of the cardiovascular system, and potentially predispose one to cardiovascular diseases such as hypertension, heart failure, and coronary artery diseases. A particularly strong link has been found between the α_{2B}AR Del301-303 variant and acute heart disease, including myocardial infarction and sudden cardiac death. More equivocal evidence has linked the α_{2A}AR DraI RFLP and α_{2B}AR Del301-303 variants with hypertension and the α_{2C}AR Del322-325 variant with heart failure. Overall, currently available data suggest that continued study of α_2AR genotypes in the cardiovascular system is warranted, although studies looking at broader genotypes encompassing other relevant receptor types in addition to α_2ARs may be especially informative.

VI. α_2ARS IN METABOLISM AND TYPE 2 DIABETES

α_2ARs have been reported to be expressed in tissues including adipose (Lafontan & Berlan, 1995) and pancreas (Lacey et al., 1996). Within pancreatic islets, α_{2A}ARs are predominantly expressed in β-cells (Lacey et al., 1996), while α_{2C}ARs are expressed in α- and δ-cells (Peterhoff, Sieg, Brede, Chao, Hein, & Ullrich, 2003); α_{2B}AR expression has largely not been reported in islets. Based upon this expression pattern, it can be reasonably assumed that α_{2A}ARs play a role in insulin secretion, and indeed this is the case, with stimulation of β-cell α_{2A}ARs leading to a decrease in insulin secretion. Although dysfunction of the insulin system leading to altered glucose handling has long been understood as an essential component of type 2 diabetes, α_2AR genetics have only very recently begun to be studied in the context of this important human disease.

A. Effects on Insulin Secretion and Glucose Handling

The main premise underlying studies of α_2AR polymorphisms in type 2 diabetes is that dysfunctional α_{2A}AR signaling contributes to altered insulin secretion. Early studies demonstrated that overexpression of α_{2A}ARs in pancreatic β-cells as well as treatment with α_2AR agonists resulted in reduced insulin secretion (Rodriguez-Pena et al., 1997; Hirose, Seto, Maruyama, Dan, Nakamura, & Saruta, 1997). α_2AR antagonists have since been shown to enhance both insulin secretion and the proinsulin effects of a sulfonylurea drug in an α_{2A}AR-dependent fashion (Fagerholm, Scheinin, & Haaparanta, 2008). A more recent series of studies have strongly implicated genetically based alterations in α_{2A}AR expression levels in abnormal insulin secretion and type 2 diabetes. Rosengren et al. (2010) have reported important findings linking the α_{2A}AR DraI RFLP with increased risk of type 2 diabetes in a Scandinavian population. Specifically, the A allele was associated with increased α_{2A}AR expression and diminished insulin secretion in pancreatic islets from human patients. Importantly, the attenuated insulin release in response to glucose was reversed by treatment with α_2AR antagonists, demonstrating dependence on α_2AR signaling.

A separate set of studies have implicated another potential α_{2A}AR polymorphism, referred to as rs10885122, in type 2 diabetes. This polymorphism was identified through a genome-wide association study (GWAS) probing for loci associated with fasting glucose, and is located quite far from the receptor coding region, 202 kilobases downstream of the DraI RFLP in the 5′-UTR (Dupuis et al., 2010; Ingelsson et al., 2010). This polymorphism has been associated with decreased insulin response after oral glucose ingestion in a Danish population (Boesgaard et al., 2010). While this is an interesting finding, given the extreme downstream nature of this polymorphism, it remains to be seen what influence this may have on receptor expression and overall β-cell function.

B. Other Potential Contributions

In addition to direct effects on insulin secretion in pancreatic β-cells, it is possible that α_{2A}AR genetics could influence pancreatic function in a less-direct fashion. Specifically, the α_{2A}AR could regulate insulin secretion via its role in sympathetic neurotransmission impinging on the pancreas (Savoy et al., 2010). α_{2A}AR genetics may also influence metabolism through effects on body fat content, as suggested by data linking the α_{2A}AR C-1291G polymorphism with abdominal fat accumulation in black Canadian subjects (Garenc et al., 2002) and the G1780A polymorphism with BMI and body fat percentage in African Americans (Lima et al., 2007).

Although expression of the α_{2B}AR has largely not been reported in pancreatic islets, the α_{2B}AR Del301-303 variant was nevertheless associated with younger age of onset in a study of type 2 diabetes patients (Papazoglou, Papanas, Papatheodorou, K., Kotsiou, S., Christakidis, D., & Maltezos, 2006). A separate study by a Finnish group linked the Del301-303 allele with risk of type 2 diabetes in patients with impaired glucose tolerance, specifically those who did not receive a lifestyle change intervention (Siitonen et al., 2004). Additionally, the Del301-303 variant has been associated with relatively lower resting metabolic rate in obese patients (Heinonen et al., 1999) and risk of weight gain in nondiabetic patients (Sivenius, Lindi, Niskanen, Laakso, & Uusitupa, 2001), providing evidence of a role for the α_{2B}AR in metabolism.

It has been recently shown that dysregulation of glucagon secretion may play a larger role in both the early development of diabetes and the hyperglycemia associated with later stages than previously thought (Gustavsson, Seah, Lao, Radda, Sudhof, & Han, 2010). Given that α_{2C}ARs are known to be expressed on the pancreatic islet cells involved in glucagon release (α- and δ-cells), the possibility arises that dysfunction of α_{2C}ARs may contribute to type 2 diabetes. Although this has not been demonstrated to date, future investigation of the α_{2C}AR Del322-325 variant for possible association with diabetes may prove fruitful.

C. Summary - α_2ARs as Targets in the Treatment of Type 2 Diabetes

The recent human genetic studies outlined above strongly suggest a role for the α_{2A}AR in type 2 diabetes, likely through its role in regulating insulin secretion from pancreatic β-cells. In particular, the well-established α_{2A}AR DraI RFLP has been linked with receptor overexpression and diminished insulin release as well as heightened risk of type 2 diabetes in human patients. A novel polymorphism downstream of the receptor coding region may also be linked with altered insulin response and glucose handling. In the future, the α_{2A}AR will likely prove to be an important therapeutic target in the treatment of type 2 diabetes and, potentially, other metabolic dysfunctions.

VII. α_2ARs IN OTHER PERIPHERAL FUNCTIONS

In addition to the prominent associations delineated in the previous sections, α_2ARs have also been studied for associations with other peripheral disorders and diseases. These studies are grounded in the knowledge that α_2ARs are ubiquitously expressed in the human body, and that the sympathetic nervous system (in which α_2ARs play a prominent role) enervates the full range of tissue types. Genetic studies have identified roles for α_2ARs in functional bowel disorders, renal functions relating to hypertension, and a handful of other altered autonomic functions.

A. Gastrointestinal System

α_2ARs are increasingly being recognized as important players in the gastro-intestinal (GI) system (for review, see Blandizzi, 2007). It is largely appreciated now that enteric α_2ARs are predominantly of the α_{2A}AR subtype, and these receptors are expressed both in sympathetic terminals and postsynaptically within GI tissues. A role for these α_{2A}ARs in GI pathophysiology, particularly in functional bowel disorders such as irritable bowel syndrome (IBS) was initially suggested by the observation that α_2AR agonists were beneficial in IBS patients, as well as having benefit in improving GI function (Blandizzi, 2007). Subsequently, a genetic association study demonstrated that the α_{2A}AR C-1291G polymorphism was linked to IBS constipation symptoms, while the α_{2C}AR Del322-325 variant was linked with constipation and somatic symptoms (Kim et al., 2004). These results suggest that α_2ARs contribute to the IBS phenotype, and confirm the receptors as viable therapeutic targets in GI dysfunction.

B. *Renal Functions*

A role for α_{2B}ARs in renal functions, particularly with regard to salt-induced hypertension, was suggested by a pair of studies from the Gavras laboratory. These studies demonstrated that α_{2B}AR-deficient mice ($\alpha_{2B}AR^{+/-}$) do not exhibit an elevation in blood pressure in response to dietary salt loading, and that this occurs despite elevated plasma norepinephrine levels; α_{2A}AR- and α_{2C}AR-deficient mice exhibited normal elevations in blood pressure along with elevated plasma norepinephrine levels (Makaritsis, Handy, Johns, Kobilka, Gavras, & Gavras, 1999; Makaritsis, Johns, Gavras, & Gavras, 2000). This data seems to indicate that the salt-induced hypertensive response is mediated by α_{2B}ARs independent of presynaptic autoreceptor function, suggesting a critical role for renal postsynaptic α_{2B}ARs. This role for the α_{2B}AR may contribute to findings outlined above that this receptor is involved as a risk factor in hypertension, in addition to its role in direct regulation of vascular function, and should be considered in the interpretation of such results going forward.

C. *Other Peripheral Nervous System-Mediated Functions*

α_2AR genetics have been linked with abnormal peripheral autonomic nervous system functions. Finley and colleagues demonstrated an association between the α_{2A}AR DraI RFLP and susceptibility to stress-induced motion sickness as well as increased exercise-induced sweat sodium concentrations, both used as readouts of the autonomic stress response (Finley et al., 2004). Other studies have linked the α_{2B}AR Del301-303 polymorphism with altered autonomic function, specifically relatively lower autonomic tone, in obese male and female subjects (Sivenius et al., 2003; Ueno et al., 2006). These studies indicate a contribution of α_2AR genotype in both stress response and autonomic tone in relation to metabolism.

VIII. CONCLUSIONS

Polymorphic variations have been identified for all three subtypes of α_2ARs, and genetic association studies suggest that these receptors are important in mediating diverse physiological functions. These functions appear across a variety of human organ systems, and range from cognitive improvement to central and peripheral blood pressure control to regulation of insulin secretion. Cellular studies in heterologous expression systems have demonstrated that the polymorphic alleles of α_2ARs can result in alterations in receptor signaling and/or pharmacology, by changes in either receptor sequences or expression levels. However, precisely how these polymorphisms cause disease phenotypes *in vivo* remains unclear. An effective way to address such questions would be to generate transgenic mouse models expressing polymorphic α_2ARs, especially in a tissue-specific manner. Such humanized mouse models would complement already-established knockout models to fully illustrate α_2AR functions *in vivo*, and could also reveal novel targets for therapeutic interventions targeting α_2ARs.

References

Altman, J., Trendelenburg, A., Macmillan, L., Bernstein, D., Limbird, L., & Starke, K., et al. (1999). Abnormal regulation of the sympathetic nervous system in alpha2a-adrenergic receptor knock-out mice. *Mol Pharmacol, 56*, 154–161.

Arnsten, A. F. (2006). Stimulants: Therapeutic actions in ADHD. *Neuropsychopharmacology, 31*, 2376–2383.

Baldessarini, R. J., & Tarazi, F. I. (2006). Pharmacotherapy of psychosis and mania. *In* "Goodman & Gilman's the pharmacological basis of therapeutics" (L. L. Brunton., J. S. Lazo, and K. L. Park, eds.),, *eleventh edition. pp. 461–500. McGraw-Hill, New York.*

Baldwin, C. T., Schwartz, F., Baima, J., Burzstyn, M., DeStefano, A. L., & Gavras, I., et al. (1999). Identification of a polymorphic glutamic acid stretch in the alpha2B-adrenergic receptor and lack of linkage with essential hypertension. *Am J Hypertens, 12*, 853–857.

Banaschewski, T., Roessner, V., Dittmann, R. W., Santosh, P. J., & Rothenberger, A. (2004). Non-stimulant medications in the treatment of ADHD. *Eur Child Adolesc Psychiat, 13*(Suppl 1), I102–I116.

Banaschewski, T., Becker, K., Scherag, S., Franke, B., & Coghill, D. (2010). Molecular genetics of attention-deficity/hyperactivity disorder: An overview. *Eur Child Adolesc Psychiat, 19*, 237–257.

Barr, C. L., Wigg, K., Zai, G., Roberts, W., Malone, M., & Schachar, R., et al. (2001). Attention-deficit hyperactivity disorder and the adrenergic receptors alpha-1C and alpha-2C. *Mol Psychiatry, 6*, 334–337.

Barturen, F., & Garcia-Sevilla, J. A. (1992). Long term treatment with desipramine increases the turnover of α_2-adrenoceptors in the rat brain. *Mol Pharmacol, 42*, 846–855.

Belmaker, R. H., & Agam, G. (2008). Mechanisms of disease: Major depressive disorder. *N Engl J Med, 358*, 55–68.

Biederman, J. (2005). Attention-deficit/hyperactivity disorder: A selective overview. *Biol Psychiat, 57*, 1215–1220.

Blandizzi, C. (2007). Enteric alpha-2 adrenoceptors: Pathophysiological implications in functional and inflammatory bowel disorders. *Neurochem Int, 51*, 282–288.

Boesgaard, T. W., Grarup, N., Jorgensen, T., Borch-Johnsen, K., Meta-Analysis of Glucose and Insulin-Related Trait Consortium (MAGIC)., &., & Hansen, T., et al. (2010). Variants at *DGKB/TMEM195, ADRA2A, GLIS3* and *C2CD4B* loci are associated with reduced glucose-stimulated beta cell function in middle-aged Danish people. *Diabetologia, 53*, 1647–1655.

Brede, M., Wiesmann, F., Jahns, R., Hadamek, K., Arnolt, C., & Neubauer, S., et al. (2002). Feedback inhibition of catecholamine release by two different alpha2-adrenoceptor subtypes prevents progression of heart failure. *Circulation, 106*, 2491–2496.

Brennan, A. R., & Arnsten, A. F. (2008). Neuronal mechanisms underlying attention deficit hyperactivity disorder: The influence of arousal on prefrontal cortical function. *Ann NY Acad Sci, 1129*, 236–245.

Callado, L. F., Meana, J. J., Grijalba, B., Pazos, A., Sastre, M., & Garcia-Sevilla, J. A. (1998). Selective increase of alpha2A-adrenoceptor agonist binding sites in the brains of depressed suicide victims. *J Neurochem, 70*, 1114–1123.

Canham, R., Das, S., Leonard, D., Abdullah, S., Mehta, S., & Chung, A., et al. (2007). Alpha2cdel322-325 and beta1arg389 adrenergic polymorphisms are not associated with reduced left ventricular ejection fraction or increased left ventricular volume. *J Am Coll Cardiol, 49*, 274–276.

Carr, D. B., Andrews, G. D., Glen, W. B., & Lavin, A. (2007). Alpha2-noradrenergic receptors activation enhances excitability and synaptic integration in rat prefrontal cortex pyramidal neurons via inhibition of HCN currents. *J Physiol, 584*, 437–450.

Cayla, C., Heinonen, P., Viikari, L., Schaak, S., Snapir, A., & Bouloumie, A., et al. (2004). Cloning, characterization and identification of several polymorphisms in the promoter region of the human alpha2B-adrenergic receptor gene. *Biochem Pharmacol, 67*, 469–478.

Clark, D. A., Mata, I., Kerwin, R. W., Munro, J., & Arranz, M. J. (2007). No association between ADRA2A polymorphisms and schizophrenia. *Am J Med Genet Part B, 144B*, 341–343.

Cousjin, J., Rijpkema, M., Qin, S., van Marle, H. J. F., Franke, B., & Hermans, E. J., et al. (2010). Acute stress modulates genotype effects on amygdala processing in humans. *Proc Natl Acad Sci USA, 107*, 9867–9872.

Crassous, P. A., Blaise, R., Marquette, A., Snapir, A., Scheinin, M., & Paris, H., et al. (2010). Identification of a novel 12-nucleotide insertion polymorphism in the promoter region of ADRA2B: Full linkage with the 9-nucleotide deletion in the coding region and influence on transcriptional activity. *Biochem Pharmacol, 79*, 407–412.

da Silva, T. L., Pianca, T. G., Roman, T., Hutz, M. H., Faraone, S. V., & Schmitz, M., et al. (2008). Adrenergic α2A receptor gene and response to methylphenidate in attention-deficit/hyperactivity disorder-predominantly inattentive type. *J Neural Transm, 115*, 341–345.

Deupree, J. D., Smith, S. D., Kratochvil, C. J., Bohac, D., Ellis, C. R., & Polaha, J., et al. (2006). Possible involvement of alpha-2A adrenergic receptors in attention deficit hyperactivity disorder: Radioligand binding and polymorphism studies. *Am J Med Genet Part B, 141B*, 877–884.

De Luca, V., Muglia, P., Vincent, J. B., Lanktree, M., Jain, U., & Kennedy, J. L. (2004). Adrenergic alpha 2C receptor genomic organization: Association study in adult ADHD. *Am J Med Genet Part B, 127B*, 65–67.

De Luca, V., Vincent, J. B., Muller, D. J., Hwang, R., Shinkai, T., & Volavka, J., et al. (2005). Identification of a naturally occurring 21 bp deletion in alpha 2c noradrenergic receptor gene and cognitive correlates to antipsychotic treatment. *Pharmacol Res, 51*, 381–384.

De Paermentier, F., Mauger, J. M., Lowther, S., Crompton, M. R., Katona, C. L., & Horton, R. W. (1997). Brain alpha-adrenoceptors in depressed suicides. *Brain Res, 757*, 60–68.

de Quervain, D. J., Kolassa, I., Ertl, V., Onyut, P. L., Neuner, F., & Elbert, T., et al. (2007). A deletion variant of the α2b-adrenoceptor is related to emotional memory in Europeans and Africans. *Nat Neurosci, 10*, 1137–1139.

Dorn, G. W. (2010). Adrenergic signaling polymorphisms and their impact on cardiovascular disease. *Physiol Rev, 90*, 1013–1062.

Dorn, G. W., Oswald, K. J., McCluskey, T. S., Kuhel, D. G., & Liggett, S. B. (1997). Alpha 2A-adrenergic receptor stimulated calcium release is transduced by Gi-associated G(Beta Gamma)-mediated activation of phospholipase C. *Biochemistry, 36*, 6415–6423.

Du Pree, J., Matolweni, L. O., Greenberg, J., Mentla, P., Adeyemo, A. A., & Mayosi, B. M. (2008). The alpha 2c del322-325 adrenergic receptor polymorphism is not association with heart failure due to idiopathic dilated cardiomyopathy in black Africans. *Cardiovasc J Àfrica, 19*, 15–16.

Dupuis, J., Langenberg, C., Prokopenko, I.., Saxena, R., Soranzo, N., & Jackson, A. U., et al. (2010). New genetic loci implicated in fasting glucose homeostasis and their impact on type 2 diabetes risk. *Nat Genet, 42*, 105–116.

Escriba, P. V., Ozaita, A., & Garcia-Sevilla, J. A. (2004). Increased mRNA expression of alpha2A-adrenoceptors, serotonin receptors and mu-opioid receptors in the brains of suicide victims. *Neuropyschopharmacology, 29*, 1512–1521.

Etzel, J. P., Rana, B. K., Wen, G., Parmer, R. J., Schork, N. J., & O'Connor, D. T., et al. (2005). Genetic variation at the human alpha2B-adrenergic receptor locus: Role in blood pressure variation and yohimbine response. *Hypertension, 45*, 1207–1213.

Fagerholm, V., Scheinin, M., & Haaparanta, M. (2008). alpha2A-adrenoceptor antagonism increases insulin secretion and synergistically augments the insulinotropic effect of glibenclamide in mice. *Br J Pharmacol, 154*, 1287–1296.

Finley Jr., J. C., O'Leary, M., Wester, D., MacKenzie, S., Shepard, N., & Farrow, S., et al. (2004). A genetic polymorphism of the α_2-adrenergic receptor increases autonomic responses to stress. *J Appl Physiol, 96*, 2231–2239.

Flordellis, C., Manolis, A., Scheinin, M., & Paris, H. (2004). Clinical and pharmacological significance of alpha2-adrenoceptor polymorphisms in cardiovascular diseases. *Int J Cardiol, 97*, 367–372.

Franowicz, J. S., Kessler, L. E., Borja, C. M., Kobilka, B. K., Limbird, L. E., & Arnsten, A. F. (2002). Mutation of the alpha2A-adrenoceptor impairs working memory performance and annuls cognitive enhancement by guanfacine. *J Neurosci, 22*, 8771–8777.

Fukutake, M., Hishimoto, A., Nishiguchi, N., Nushida, H., Ueno, Y., & Shirakawa, O., et al. (2008). Association of alpha2A-adrenergic receptor gene polymorphism with susceptibility to suicide in Japanese females. *Prog Neuro-Psychopharmacol Biol Psychiat, 32*, 1428–1433.

Garcia-Sevilla, J. A., Escriba, P. V., Ozaita, A., La Harpe, R., Walzer, C., & Eytan, A., et al. (1999). Up-regulation of immunolabeled α_{2A}-adrenoceptors, G_i coupling proteins, and regulator receptor kinases in the prefrontal cortex of depressed suicides. *J Neurochem, 72*, 282–291.

Garenc, C., Perusse, L., Chagnon, Y. C., Rankinen, T., Gagnon, J., & Borecki, I. B., et al. (2002). The alpha 2-adrenergic receptor gene and body fat content and distribution: The HERITAGE Family Study. *Mol Med, 8*, 88–94.

Gesek, F. A. (1996). Alpha 2-adrenergic receptors activate phospholipase C in renal epithelial cells. *Mol Pharmacol, 50*, 407–414.

Gilsbach, R., Brede, M., Beetz, N., Moura, E., Muthig, V., & Gerstner, C., et al. (2007). Heterozygous alpha 2C-adrenoceptor-deficient mice develop heart failure after transverse aortic constriction. *Cardivasc Res, 75*, 728–737.

Gonzalez-Maeso, J., Rodriguez-Puertas, R., Meana, J. J., Garcia-Sevilla, J. A., & Guimon, J. (2002). Neurotransmitter receptor-mediated activation of G-proteins in brains of suicide victims with mood disorders: Selective supersensitivity of alpha(2A)-adrenoceptors. *Mol Psychiat, 7*, 755–767.

Gustavsson, N., Seah, T., Lao, Y., Radda, G. K., Sudhof, T. C., & Han, W. (2010). Delayed onset of hyperglycaemia in a mouse model with impaired glucagons secretion demonstrates that dysregulated glucagons secretion promotes hyperglycaemia and type 2 diabetes. *Diabetologia,* published online October 27, 2010.

Hein, L., Altman, J., & Kobilka, B. (1999). Two functionally distinct alpha2-adrenergic receptors regulate sympathetic neurotransmission. *Nature, 402*, 181–184.

Hein, L. (2006). Adrenoceptors and signal transduction in neurons. *Cell Tissue Res, 326*, 541–551.

Heinonen, P., Koulu, M., Pesonen, U., Karvonen, M. K., Rissanen, A., & Laakso, M., et al. (1999). Identification of a three-amino acid deletion in the alpha2B-adrenergic receptor that is associated with reduced basal metabolic rate in obese subjects. *J Clin Endocrinol Metab, 84*, 2429–2433.

Heinonen, P., Jartti, L., Jarvisalo, M. J., Pesonen, U., Kaprio, J. A., & Ronnemaa, T., et al. (2002). Deletion polymorphism in the alpha2B-adrenergic receptor gene is associated with flow-mediated dilation of the brachial artery. *Clin Sci, 103*, 517–524.

Hirose, H., Seto, Y., Maruyama, H., Dan, K., Nakamura, K., & Saruta, T. (1997). Effects of alpha 2-adrenergic agonism, imdizaloines, and G-protein on insulin secretion in beta cells. *Metabolism, 46*, 1146–1149.

Hiyoshi, Y., Miura, H., Uemura, K., Endo, H., Ozawa, K., & Maeda, N., et al. (1995). Effects of imidazoline antagonists of alpha 2-adrenoceptors on endogenous adrenaline-induced inhibition of insulin release. *Eur J Pharmacol, 294*, 117–123.

Hong, C., Wang, Y., Liu, T., Liu, H., & Tsai, S. (2001). A study of alpha-adrenoceptor gene polymorphisms and Alzheimer disease. *J Neural Transm, 108*, 445–450.

Iacoviello, M., Forleo, C., Sorrentino, S., Romito, R., De Tommasi, E., & Lucarelli, K., et al. (2006). Alpha- and beta-adrenergic receptor polymorphisms in hypertensive and normotensive offspring. *J Cardiovasc Med, 7*, 316–321.

Ingelsson, E., Langenberg, C., Hivert, M. F., Prokopenko, I., Lyssenko, V., & Dupuis, J., et al. (2010). Detailed physiologic characterization reveals diverse mechanisms for novel genetic loci regulating glucose and insulin metabolism in humans. *Diabetes, 59*, 1266–1275.

Jewell-Motz, E. A., & Liggett, S. B. (1995). An acidic motif within the third intracellular loop of the alpha2C2 adrenergic receptor is required for agonist-promoted phosphorylation and desensitization. *Biochemistry, 34*, 11946–11953.

Kable, J. W., Murrin, L. C., & Bylund, D. B. (2000). *In vivo* gene modification elucidates subtype-specific functions of alpha(2)-adrenergic receptors. *J Pharmacol Exp Ther, 293*, 1–7.

Kiive, E., Kurrikoff, T., Maestu, J., & Harro, J. (2010). Effect of alpha2A-adrenoceptor C-1291G genotype and maltreatment on hyperactivity and inattention in adolescents. *Prog Neuro-Psychopharmacol Biol Psychiat, 34*, 219–224.

Kim, H. J., Camilleri, M., Carlson, P. J., Cremonini, F., Ferber, I., & Stephens, D., et al. (2004). Association of distinct α_2 adrenoceptor and serotonin transporter polymorphisms with constipation and somatic symptoms in functional gastrointestinal disorders. *Gut, 53*, 829–837.

Knaus, A. E., Muthig, V., Schickinger, S., Moura, E., Beetz, N., & Gilsbach, R., et al. (2007). α_2-adrenoceptor subtypes – unexpected functions for receptors and ligands derived from gene-targeted mouse models. *Neurochem Int, 51*, 277–281.

Kobilka, B. (1992). Adrenergic receptors as models for G protein-coupled receptors. *Annu Rev Neurosci, 15*, 87–114.

Lacey, R. J., Chan, S. L., Cable, H. C., James, R. F., Perett, C. W., & Scarpello, J. H., et al. (1996). Expression of alpha 2- and beta-adrenoceptor subtypes in human islets of Langerhans. *J Endocrinol, 148*, 531–543.

Lafontan, M., & Berlan, M. (1995). Fat cell α_2-adrenoceptors: The regulation of fat cell function and lipolysis. *Endocr Rev, 16*, 716–738.

Lakhlani, P. P., MacMillan, L. B., Guo, T. Z., McCool, B. A., Lovinger, D. M., & Maze, M., et al. (1997). Substitution of mutant alpha2a-adrenergic receptor via "hit and run" gene targeting reveals role of this subtype in sedative, analgesic, and anesthetic-sparing responses *in vivo*. *Proc Natl Acad Sci USA, 94*, 9950–9955.

Lario, S., Calls, J., Cases, A., Oriola, J., Torras, A., & Rivera, F. (1997). MspI identifies an biallelic polymorphism in the promoter region of the alpha 2A-adrenergic receptor gene. *Clin Genet, 51*, 129–130.

Laukkanen, J. A., Makikallio, T. H., Kauhanen, J., & Kurl, S. (2009). Insertion/deletion polymorphism in the alpha2-adrenergic receptor gene is a genetic risk factor for sudden cardiac death. *Am Heart J, 158*, 615–621.

Lee, H., Kang, R., Paik, J., Jeong, Y. J., Chang, H. S., & Han, S., et al. (2009). Association of the adrenergic alpha 2a receptor -1291C/G polymorphism with weight change and treatment response to mirtazepine in patients with major depressive disorder. *Brain Res, 1262*, 1–6.

Li, J., Canham, R., Vongpatanasin, W., Leonard, D., Auchus, R., & Victor, R. (2006). Do allelic variants in alpha2a and alpha2c adrenergic receptors predispose to hypertension in blacks? *Hypertension, 47*, 1140–1146.

Li, L. H., Li, Y., Wen, Y., & Wang, J. G. (2008). Anthropometric and metabolic phenotypes in relation to the ADRA2B deletion/insertion polymorphism in Chinese population. *J Hypertens, 26*, 2161–2167.

Lima, J. J., Feng, H., Duckworth, L., Wang, J., Sylvester, J. E., & Kissoon, N., et al. (2007). Association analyses of adrenergic receptor polymorphisms with obesity and metabolic alterations. *Metabolism, 56*, 757–765.

Limbird, L. E. (1988). Receptors linked to inhibition of adenylate cyclase: Additional signaling mechanisms. *FASEB J, 2*, 2686–2695.

Link, R. E., Desai, K., Hein, L., Stevens, M. E., Chruscinski, A., & Bernstein, D., et al. (1996). Cardiovascular regulation in mice lacking alpha2-adrenergic receptor subtypes b and c. *Science, 273*, 803–805.

Lockette, W., Ghosh, S., Farrow, S., MacKenzie, S., Baker, S., & Miles, P., et al. (1995). Alpha 2-adrenergic receptor gene polymorphism and hypertension in blacks. *Am J Hypertens, 8*, 390–394.

Ma, D., Zerangue, N., Lin, Y. F., Collins, A., Yu, M., & Jan, Y. N., et al. (2001). Role of ER export signals in controlling surface potassium channel numbers. *Science, 291*, 316–319.

MacMillan, L. B., Hein, L., Smith, M. S., Piascik, M. T., & Limbird, L. E. (1996). Central hypotensive effects of the alpha2a-adrenergic receptor subtype. *Science, 273*, 801–803.

Makaritsis, K. P., Handy, D. E., Johns, C., Kobilka, B. K., Gavras, I., & Gavras, H. (1999). Role of the alpha2B-adrenergic receptor in the development of salt-induced hypertension. *Hypertension, 33*, 14–17.

Makaritsis, K. P., Johns, C., Gavras, I., & Gavras, H. (2000). Role of alpha(2)-adrenergic receptor subtypes in the acute hypertensive response to hypertonic saline infusion in anephric mice. *Hypertension, 35*, 609–613.

Marien, M. R., Colpaert, F. C., & Rosenquist, A. C. (2004). Noradrenergic mechanisms in neurodegenerative diseases: A theory. *Brain Res Rev, 45*, 38–78.

Marrs, W., Kuperman, J., Avedian, T., Roth, R. H., & Jentsch, J. D. (2005). Alpha-2 adrenoceptor activation inhibits phencyclidine-induced deficits of spatial working memory in rats. *Neuropsychopharmacology, 30*, 1500–1510.

Martin-Guerrero, I., Callado, L. F., Saitua, K., Rivero, G., Garcia-Orad, A., & Meana, J. J. (2006). The N251K functional polymorphism in the alpha(2A)-adrenoceptor gene is not associated with depression: A study in suicide completers. *Psychopharmacology (Heidelberg Ger), 184*, 82–86.

Mateo, Y., Fernandez-Pastor, B., & Meana, J. J. (2001). Acute and chronic effects of desipramine and clorgyline on α_2-adrenoceptors regulating noradrenergic transmission in the rat brain: A dual-probe microdialysis study. *Br J Pharmacol, 133*, 1362–1370.

McGaugh, J. L. (2004). The amygdala modulates the consolidation of memories of emotionally arousing experiences. *Ann Rev Neurosci, 27*, 1–28.

Meana, J. J., Barturen, F., & Garcia-Sevilla, J. A. (1992). Alpha 2-adrenoceptors in the brain of suicide victims: Increased receptor density associated with major depression. *Biol Psychiat, 31*, 471–490.

Metra, M., Zani, C., Covolo, L., Nodari, S., Pezzali, N., & Gelatti, U., et al. (2006). Role of beta1- and alpha2c-adrenergic receptor polymorphisms and their combination in heart failure: A case control study. *Eur J Heart Fail, 8*, 131–135.

Montgomery, M., & Bylund, D. (2010). Lack of effect of the alpha2c-adrenoceptor del322-325 polymorphism on inhibition of cyclic AMP production in HEK293 cells. *Br J Pharmacol, 159*, 820–830.

Muszkat, M., Kurnik, D., Solus, J., Sofowora, G. G., Xie, H. G., & Jiang, L., et al. (2005a). Variation in the alpha2B-adrenergic receptor gene (ADRA2B) and its relationship to vascular response in vivo. *Pharmacogenet Genomics, 15*, 407–414.

Muszkat, M., Sofowora, G. G., Xie, H. G., Wood, A. J., & Stein, C. M. (2005b). Alpha2B adrenergic receptor 301-303 deletion polymorphism and vascular alpha2 adrenergic receptor response. *Pharmacogenet Genomics, 15*, 23–28.

Muszkat, M., Kurnik, D., Sofowora, G. G., Solus, J., Xie, H. G., & Harris, P. A., et al. (2010). Desensitization of vascular response in vivo: Contribution of genetic variation in the [alpha]2B-adrenergic receptor subtype. *J Hypertens, 28*, 278–284.

Natali, A., Gastaldelli, A., Galvan, A. Q., Sironi, A. M., Ciociaro, D., & Sanna, G., et al. (1998). Effects of acute alpha 2-blockade on insulin action and secretion in humans. *Am J Physiol, 274*, E57–E64.

Neumeister, A., Charney, D., Belfer, I., Geraci, M., Holmes, C., & Sharabi, Y., et al. (2005). Sympathoneural and adrenomedullary functional effects of alpha2c-adrenoceptor gene polymorphism in healthy humans. *Pharmacogenet Genomics, 15*, 143–149.

Neumeister, A., Drevets, W. C., Belfer, I., Luckenbaugh, D. A., Henry, S., & Bonne, O., et al. (2006). Effects of a α_{2C}-adrenoceptor gene polymorphism on neural responses to facial expressions in depression. *Neuropsychopharmacology, 31*, 1750–1756.

Nonen, S., Okamoto, H., Akino, M., Matsui, Y., Fujio, Y., & Yoshiyama, M., et al. (2005). No positive association between adrenergic receptor variants of alpha2cdel322-325, beta1ser49, beta1arg389 and the risk for heart failure in the Japanese population. *Br J Clin Pharmacol, 60*, 414–417.

Ohara, K., Nagai, M., Tani, K., Tsukamoto, T., Suzuki, Y., & Ohara, K. (1998). Polymorphism in the promoter region of the alpha 2A adrenergic receptor gene and mood disorders. *NeuroReport, 9*, 1291–1294.

Ohara, K., Suzuki, Y., Ochiai, M., Terada, H., & Ohara, K. (2000). Polymorphism in the promoter region of the $alpha_{2A}$-adrenergic receptor gene and panic disorders. *Psychiatry Res, 93*, 79–82.

Ordway, G. A., Schenk, J., Stockmeier, C. A., May, W., & Klimek, V. (2003). Elevated agonist binding to α_2-adrenoceptors in the locus coeruleus in major depression. *Biol Psychiat, 53*, 315–323.

Papazoglou, D., Papanas, N., Papatheodorou, K., Kotsiou, S., Christakidis, D., & Maltezos, E. (2006). An insertion/deletion polymorphism in the alpha2B adrenoceptors gene is associated with age at onset of type 2 diabetes mellitus. *Exp Clin Endocrinol Diab, 114*, 424–427.

Park, L., Nigg, J. T., Waldman, I. D., Nummy, K. A., Huang-Pollock, C., & Rappley, M., et al. (2005). Association and linkage of alpha-2A adrenergic receptor gene polymorphisms with childhood ADHD. *Mol Psychiat, 10*, 572–578.

Park, Y., Chung, Y., Lee, S., Lee, K., Kim, H., & Byun, Y., et al. (2006). Weight gain associated with the α_{2A}-adrenergic receptor -1291 C/G polymorphism and olanzapine treatment. *Am J Med Genet Part B, 141B*, 394–397.

Peterhoff, M., Sieg, A., Brede, M., Chao, C. M., Hein, L., & Ullrich, S. (2003). Inhibition of insulin secretion via distinct signaling pathways in alpha2-adrenoceptor knockout mice. *Eur J Endocrinol, 149*, 343–350.

Philipp, M., Brede, M., & Hein, L. (2002). Physiological significance of alpha(2)-adrenergic receptor subtype diversity: One receptor is not enough. *Am J Physiol Regul Integr Comp Physiol, 283*, R287–R295.

Polanczyk, G., Zeni, C., Genro, J. P., Guimaraes, A. P., Roman, T., & Hutz, M. H., et al. (2007). Association of the adrenergic α2A receptor gene with methylphenidate improvement of inattentive symptoms in children and adolescents with attention-deficit/hyperactivity disorder. *Arch Gen Psychiat, 64*, 218–224.

Prince, J. (2008). Catecholamine dysfunction in attention-deficit/hyperactivity disorder: An update. *J Clin Psychopharmacol, 3*(Suppl 2), S39–S45.

Ramos, B. P., & Arnsten, A. F. T. (2007). Adrenergic pharmacology and cognition: Focus on the prefrontal cortex. *Pharmacol Ther, 113*, 523–536.

Rasch, B., Spalek, J., Buholzer, S., Luechinger, R., Boesiger, P., & Papassotiropoulos, A., et al. (2009). A genetic variation of the noradrenergic system is related to differential amygdale activation during encoding of emotional memories. *Proc Natl Acad Sci USA, 106*, 19191–19196.

Regitz-Zagrosek, V., Hocher, B., Bettmann, M., Brede, M., Hadamek, K., & Gerstner, C., et al. (2006). Alpha2c-adrenoceptor polymorphism is associated with improved event-free survival in patients with dilated cardiomyopathy. *Eur Heart J, 27*, 454–459.

Richman, J. G., & Regan, J. W. (1998). Alpha 2-adrenergic receptors increase cell migration and decrease F-actin labeling in rat aortic smooth muscle cells. *Am J Physiol, 283*, C654–C662.

Rodriguez-Pena, M. S., Collins, R., Woodard, C., & Spiegel, A. M. (1997). Decreased insulin content and secretion in RIN 1046-38 cells overexpressing alpha 2-adrenergic receptors. *Endocrine, 7*, 255–260.

Roman, T., Schmitz, M., Polanczyk, G. V., Eizirik, M., Rohde, L. A., & Hutz, M. H. (2003). Is the α-2A adrenergic receptor gene (ADRA2A) associated with attention-deficit/hyperactivity disorder? *Am J Med Genet Part B, 120B*, 116–120.

Roozendaal, B., Barsegyan, A., & Lee, S. (2008). Adrenal stress hormones, amygdala activation, and memory for emotionally arousing experiences. *Prog Brain Res, 167*, 79–97.

Rosengren, A. H., Jokubka, R., Tojjar, D., Granhall, C., Hansson, O., & Li, D., et al. (2010). Overexpression of alpha2A-adrenergic receptors contributes to type 2 diabetes. *Science, 327*, 217–220.

Salim, S., Desai, A. N., Taneja, M., & Eikenburg, D. C. (2009). Chronic adrenaline treatment fails to down-regulate the Del301-303-alpha2B-adrenoceptor in neuronal cells. *Br J Pharmacol, 158*, 314–327.

Sara, S. J. (2009). The locus coeruleus and noradrenergic modulation of cognition. *Nat Rev Neurosci, 10*, 211–223.

Savoy, Y. E., Ashton, M. A., Miller, M. W., Nedza, F. M., Spracklin, D. K., & Hawthorn, M. H., et al. (2010). Differential effects of various typical and atypical antipsychotics on plasma glucose and insulin levels in the mouse: Evidence for the involvement of sympathetic regulation. *Schizophr Bull, 36*, 410–418.

Savva, J., Maqbool, A., White, H., Galloway, S., Yuldasheva, N., & Ball, S., et al. (2009). Polymorphisms of adrenoceptors are not associated with an increased risk of adverse event in heart failure: A MERIT-HF substudy. *J Card Failure, 15*, 435–441.

Scheinin, M., Lomasney, J. W., Hayden-Hixson, D. M., Schambra, U. B., Caron, M. G., & Lefkowitz, R. J., et al. (1994). Distribution of alpha 2-adrenergic receptor subtype gene expression in rat brain. *Mol Brain Res, 21*, 133–149.

Schmitz, M., Denardin, D., Silva, T. L., Pianca, T., Roman, T., & Hutz, M. H., et al. (2006). Association between alpha-2a-adrenergic receptor gene and ADHD inattentive type. *Biol Psychiat, 60*, 1028–1033.

Schramm, N. L., McDonald, M. P., & Limbird, L. E. (2001). The alpha(2a) adrenergic receptor plays a protective role in mouse behavioral models of depression and anxiety. *J Neurosci, 21*, 4875–4882.

Sequeira, A., Mamdani, F., Lalovic, A., Anguelova, M., Lesage, A., & Seguin, M., et al. (2004). Alpha 2A adrenergic receptor gene and suicide. *Psychiat Res, 125*, 87–93.

Siitonen, N., Lindstrom, J., Eriksson, J., Valle, T. T., Hamalainen, H., & Ilanne-Parikka, P., et al. (2004). Association between a deletion/insertion polymorphism in the alpha2B-adrenergic receptor gene and insulin secretion and type 2 diabetes. The Finnish Diabetes Prevention Study. *Diabetologia, 47*, 1416–1424.

Sivenius, K., Lindi, V., Niskanen, L., Laakso, M., & Uusitupa, M. (2001). Effect of a three-amino acid deletion in the alpha2B-adrenergic receptor gene on long-term body weight change in Finnish non-diabetic and type 2 diabetic subjects. *Int J Obes, 25*, 1609–1614.

Sivenius, K., Niskanen, L., Laakso, M., & Uusitupa, M. (2003). A deletion in the alpha2B-adrenergic receptor gene and autonomic nervous function in central obesity. *Obes Res, 11*, 962–970.

Small, K. M., Forbes, S. L., Brown, K. M., & Liggett, S. B. (2000a). An Asn to Lys polymorphism in the third intracellular loop of the human α_{2A}-adrenergic receptor imparts enhanced agonist-promoted G_i coupling. *J Biol Chem, 275*, 38518–38523.

Small, K. M., Forbes, S. L., Rahman, F. F., Bridges, K. M., & Liggett, S. B. (2000b). A four amino acid deletion polymorphism in the third intracellular loop of the human α_{2C}-adrenergic receptor confers impaired coupling to multiple effectors. *J Biol Chem, 275*, 23059–23064.

Small, K. M., Brown, K. M., Forbes, S. L., & Liggett, S. B. (2001). Polymorphic deletion of three intracellular acidic residues of the α_{2B}-adrenergic receptor decreases G protein-coupled receptor kinase-mediated phosphorylation and desensitization. *J Biol Chem, 276*, 4917–4922.

Small, K., Wagoner, L., Levin, A., Kardia, S., & Liggett, S. (2002). Synergistic polymorphisms of beta1- and alpha2c-adrenergic receptors and the risk of congestive heart failure. *N Engl J Med, 347*, 1135–1142.

Small, K. M., McGraw, D. W., & Liggett, S. B. (2003). Pharmacology and physiology of human adrenergic receptor polymorphisms. *Annu Rev Pharmacol Toxicol, 43*, 381–411.

Small, K. M., Mialet-Perez, J., Seman, C. A., Theiss, C. T., Brown, K. M., & Liggett, S. B. (2004). Polymorphisms of cardiac presynaptic alpha2C adrenergic receptors: Diverse intragenic variability with haploptype-specific functional effects. *Proc Natl Acad Sci USA, 101*, 1320–1325.

Small, K. M., Brown, K. M., Seman, C. A., Theiss, C. T., & Liggett, S. B. (2006). Complex haploptypes derived from noncoding polymorphisms of the intronless α_{2A}-adrenergic gene diversify receptor expression. *Proc Natl Acad Sci USA, 103*, 5472–5477.

Snapir, A., Heinonen, P., Tuomainen, T. P., Alhopuro, P., Karvonen, M. K., & Lakka, T. A., et al. (2001). An insertion/deletion polymorphism in the alpha2B-adrenergic receptor gene is a novel genetic risk factor for acute coronary events. *J Am Coll Cardiol, 37*, 1516–1522.

Snapir, A., Koskenvuo, J., Toikka, J., Orho-Melander, M., Hinkka, S., & Saraste, M., et al. (2003a). Effects of common polymorphisms in the alpha1A-, alpha2B-, beta1-, and beta2-adrenoceptors on haemodynamic responses to adrenaline. *Clin Sci, 104*, 509–520.

Snapir, A., Mikkelsson, J., Perola, M., Penttila, A., Scheinin, M., & Karhunen, P. J. (2003b). Variation in the alpha2B-adrenoceptor gene as a risk factor for prehospital fatal myocardial infarction and sudden cardiac death. *J Am Coll Cardiol, 41*, 190–194.

Stone, L. S., MacMillan, L. B., Kitto, K. F., Limbird, L. E., & Wilcox, G. L. (1997). The alpha2a adrenergic receptor subtype mediates spinal analgesia evoked by alpha2 agonists and is necessary for spinal adrenergic-opioid synergy. *J Neurosci, 17*, 7157–7165.

Subhash, M. N., Nagaraja, M. R., Sharada, S., & Vinod, K. Y. (2003). Cortical alpha-adrenoceptor downregulation by tricyclic antidepressants in the rat brain. *Neurochem Int, 7*, 603–609.

Svetkey, L., Timmons, P., Emovon, O., Anderson, N., Preis, L., & Chen, Y. (1996). Association of hypertension with beta2- and alpha2c10-adrenergic receptor genotype. *Hypertension, 27*, 1210–1215.

Tsai, S. J., Wang, Y. C., Yu Younger, W. Y., Lin, C. H., Yang, K. H., & Hong, C. J. (2001a). Association analysis of polymorphism in the promoter region of the alpha2a-adrenoceptor gene with schizophrenia and clozapine response. *Schizophr Res, 49*, 53–58.

Tsai, S. J., Wang, Y. C., & Hong, C. J. (2001b). Allelic variants of the α_{1a} adrenoceptors and the promoter region of the α_{2a} adrenoceptors and temperament factors. *Am J Med Genet (Neuropsychiatr Genet), 105*, 96–98.

Ueno, L. M., Frazzatto, E. S., Batalha, L. T., Trombetta, I. C., do Socorro Brasileiro, M., & Irigoyen, C., et al. (2006). Alpha2B-adrenergic receptor deletion polymorphism and cardiac autonomic nervous system responses to exercise in obese women. *Int J Obes, 30*, 214–220.

Umemura, S., Hirawa, N., Iwamoto, T., Yamaguchi, S., Toya, Y., & Kobayashi, S., et al. (1994). Association analysis of restriction fragment length polymorphism for alpha 2-adrenergic receptor genes in essential hypertension in Japan. *Hypertension, 23*(1 Suppl), I203–I206.

Vasudevan, R., Ismail, P., Stanslas, J., Shamsudin, N., & Ali, A. B. (2008). Association of insertion/deletion polymorphism of alpha-adrenoceptor gene in essential hypertension with or without type 2 diabetes mellitus in Malaysian subjects. *Int J Biol Sci, 4*, 362–367.

von Wowern, F., Bengtsson, K., Lindgren, C. M., Orho-Melander, M., Fyhrquist, F., & Lindblad, U., et al. (2003). A genome wide scan for early onset primary hypertension in Scandinavians. *Hum Mol Genet, 12*, 2077–2081.

von Wowern, F., Bengtsson, K., Lindblad, U., Ratam, L., & Melander, O. (2004). Functional variant in the (alpha)2B adrenoceptors gene, a positional candidate on chromosome 2, associates with hypertension. *Hypertension, 43*, 592–597.

Wakeno, M., Kato, M., Okugawa, G., Fukuda, T., Hosoi, Y., & Takekita, Y., et al. (2008). The alpha 2A-adrenergic receptor gene polymorphism modifies antidepressant responses to milnacipran. *J Clin Psychopharmacol, 28*, 518–524.

Wang, B., Wang, Y., Zhou, R., Li, J., Qian, Q., & Yang, L., et al. (2006a). Possible association of the alpha-2A adrenergic receptor gene (ADRA2A) with symptoms of attention-deficit/hyperactivity disorder. *Am J Med Genet Part B, 141B*, 130–134.

Wang, M., Ramos, B. P., Paspalas, C. D., Shu, Y., Simen, A., & Duque, A., et al. (2007). α2A-adrenoceptors strengthen working memory networks by inhibiting cAMP-HCN channel signaling in prefrontal cortex. *Cell, 129*, 397–410.

Wang, Q., Lu, R., Zhao, J., & Limbird, L. E. (2006b). Arrestin serves as a molecular switch, linking endogenous alpha 2-adrenergic receptors to Src-dependent but not Src-independent ERK activation. *J Biol Chem, 281*, 25948–25955.

Wang, R., MacMillan, L. B., Fremeau, R. T., Magnuson, M. A., Lindner, J., & Limbird, L. E. (1996). Expression of α_2-adrenergic receptor subtypes in the mouse brain: Evaluation of spatial and temporal information imparted by 3 kb of 5' regulatory sequence for the α_{2A}AR-receptor gene in transgenic animals. *Neuroscience, 74*, 199–218.

Westfall, T. C., & Westfall, D. P. (2006). Adrenergic agonists and antagonists. *In* "Goodman & Gilman's The Pharmacological Basis of Therapeutics" (L. L. Brunton., J. S. Lazo, and K. L. Park, eds.),, *Eleventh Edition. pp. 237–296. McGraw-Hill, New York.*

Wilson, M. H., Puranam, R. S., Ottman, R., Gilliam, C., Limbird, L. E., & George Jr., A. L., et al. (1998). Evaluation of the alpha(2A)-adrenergic receptor gene in a heritable form of temporal lobe epilepsy. *Neurology, 51*, 1730–1731.

Xu, C., Schachar, R., Tannock, R., Roberts, W., Malone, M., & Kennedy, J. L., et al. (2001). Linkage study of the α2A adrenergic receptor in attention-deficit hyperactivity disorder families. *Am J Med Genet (Neuropsychiatr Genet), 105*, 159–162.

Yamaguchi, W., Shinkai, T., Inoue, Y., Utsunomiya, K., Sakata, S., & Fukunaka, Y., et al. (2009). Association analysis between the C-1291G polymorphism in the promoter region of the adrenergic alpha2A receptor gene and polydypsia in schizophrenia. *Prog Neuro-Psychopharmacol Biol Psychiat, 33*, 499–502.

Zhang, H., Li, X., Huang, J., Li, Y., Thijs, L., & Wang, Z., et al. (2005). Cardiovascular and metabolic phenotypes in relation to the ADRA2B insertion/deletion polymorphism in a Chinese population. *J Hypertens, 23*, 2201–2207.

CHAPTER 9

β-Adrenergic Receptor Subtype Signaling in the Heart: from Bench to the Bedside

Weizhong Zhu[1], Anthony Yiu-Ho Woo[2], Yan Zhang[2], Chun-Mei Cao,[2] and Rui-Ping Xiao[2]

[1]Center for Translational Medicine, Thomas Jefferson University, Philadelphia, PA, USA
[2] Institute of Molecular Medicine, Peking University, Beijing, China

I. Overview
II. Subtype-Specific Functional Roles of β_1AR and β_2AR in Regulating Cardiomyocyte Survival and Death
   A. Prolonged Stimulation of β_1AR Triggers Cardiomyocyte Apoptosis and Maladaptive Cardiac Remodeling
   B. Cardioprotection by β_2AR Stimulation
III. Mechanisms Underlying β_2AR-Coupled G_i Signaling
IV. RGS2-Mediated Termination of β_2AR-coupled G_i Signaling and Its Potential Pathogenic and Therapeutic Implications
V. Ligand-Directed Selective Activation of β_2AR-Coupling to G_s or G_i
VI. Development of β_2AR Agonists into new Drugs for the Treatment of Heart Failure
   A. Signaling-Selective β_2AR Agonists for the Treatment of Heart Failure
   B. A Combination of β_2AR Activation with β_1AR Blockade Provides a more Effective Therapy for Heart Failure
VII. Future Perspective
References

I. OVERVIEW

Stimulation of β-adrenergic receptor (βAR), a prototypical member of G protein-coupled receptor (GPCR) superfamily, is broadly involved in metabolic regulation, growth control, muscle contraction, cell survival, and cell death. The major βAR subtypes, β_1AR and β_2AR, couple to distinct G proteins and differentially regulate cardiac function and remodeling. Three major discoveries have marked the recent research line with respect to βAR subtype-specific

Current Topics in Membranes, Volume 67 0065-230X/10 $35.00

DOI: 10.1016/B978-0-12-384921-2.00009-4

signal transduction. These include: (1) dual coupling of β_2AR to G_s and G_i proteins in cardiomyocytes; (2) cardioprotection by β_2AR signaling in improving cardiac function and cardiomyocyte viability; (3) PKA-independent, CaMKII-mediated β_1AR signaling in triggering myocyte apoptosis and maladaptive cardiac remodeling. These findings indicate that β_1AR stimulation is cardiac detrimental, while β_2AR stimulation is protective.

Heart failure (HF) is a complex clinical syndrome featured by extensive abnormalities in the βAR system, including elevated circulating catecholamine levels, selective downregulation and desensitization of β_1AR, and increased β_2AR-coupled G_i signaling. In particular, the enhanced G_i signaling negates β_1AR- as well as β_2AR-mediated contractile response, thus contributing to the pathogenesis of HF. Our recent translational studies support the concept that inhibition of the G_i signaling or selective β_2AR-G_s stimulation with fenoterol markedly improves cardiac remodeling and function of the failing heart. In this chapter, we intend to summarize (a) the recent progresses on subtype-specific functional roles of β_1AR and β_2AR, (b) the mechanisms underlying the activation and termination of the unique β_2AR-coupled G_i signaling, and (c) ligand-directed selective activation of β_2AR-coupled G_s or G_i signaling and their potential therapeutic implications.

II. SUBTYPE-SPECIFIC FUNCTIONAL ROLES OF β_1AR AND β_2AR IN REGULATING CARDIOMYOCYTE SURVIVAL AND DEATH

A. *Prolonged Stimulation of β_1AR Triggers Cardiomyocyte Apoptosis and Maladaptive Cardiac Remodeling*

The persistent stimulation of β_1AR and β_2AR exhibits distinct outcomes under certain pathological circumstances such as HF. Specifically, persistent stimulation of β_1AR in mouse cardiomyocytes lacking β_2AR (β_2AR knockout or $\beta_1\beta_2$ double knockout) in conjunction with adenoviral gene transfer of β_1AR triggers cardiomyocyte apoptosis by a CaMKII-dependent mechanism that is independent of PKA signaling (Zhu et al., 2003). Furthermore, β_1AR-activated CaMKII signaling, but not the PKA pathway, is involved in catecholamine-induced neonatal rat cardiomyocyte pathological growth (hypertrophy) (Sucharov, Mariner, Nunley, Long, Leinwand, & Bristow, 2006). These *in vitro* observations have been corroborated by *in vivo* studies using genetic manipulation of either β_1AR or CaMKII signaling components. In particular, cardiac-specific overexpression of the human β_1AR at a modest level (~15-fold) leads to maladaptive cardiac remodeling, severe dilated cardiomyopathy, and premature death of transgenic mice (Bisognano et al., 2000; Engelhardt, Hein, Wiesmann, & Lohse, 1999). This is further supported by the severe HF and premature death of transgenic mice with cardiac-specific overexpression of CaMKII-δC (Zhang et al., 2003). In contrast, the cardiac-specific expression of a peptide inhibitor of CaMKII prevents myocyte apoptosis, maladaptive remodeling, and cardiac contractile dysfunction following excessive βAR stimulation or myocardial infarction (MI; Zhang et al., 2005). In clinical settings, elevations in plasma norepinephrine, an endogenous β_1AR agonist, and a stimulatory anti-β_1AR antibody are directly associated with HF mortality in humans and animal models (Hebert, 2007). Thus, the β_1AR-evoked persistent activation of CaMKII might be a primary etiological mechanism that underlies β_1AR-induced adverse myocardial remodeling and resultant cardiomyopathy.

In contrast to the detrimental effects of persistent CaMKII activation, the consequence of sustained PKA activation in the heart remains controversial. Enhanced AC-cAMP-PKA signaling by overexpressing AC type V or VI does not produce HF and, instead, alleviates HF in some genetic mouse models (Lai et al., 2008; Roth et al., 1999, 2002; Tang, Gao, Roth, Guo, & Hammond, 2004). Likewise, enhanced cAMP-PKA signaling by β_2AR agonist stimulation or by overexpression of the receptor (by $\leq$ 30-fold over its endogenous level) protects cardiomyocytes against apoptosis and exerts beneficial effects in several HF models (Liggett et al., 2000). These findings differ from the HF phenotype of transgenic mice overexpressing either the Gαs subunit or the PKA catalytic subunit (Antos et al., 2001; Iwase et al., 1997). Although the exact mechanism underlying the discrepancy is unknown, it might be attributable to a β_2AR-mediated antiapoptotic effect and the different spatial profile of the cAMP-PKA signaling induced by β_2AR compared with that induced by the overexpression of Gαs or the PKA catalytic subunit (Antos et al., 2001; Iwase et al., 1997). Nevertheless, these paradoxical observations challenge the conventional wisdom that the cAMP-PKA pathway is solely responsible for β_1AR-induced cardiac detrimental effects. Resolving these paradoxes should reveal valuable etiological insights and novel therapeutic targets for the treatment of HF.

B. Cardioprotection by β_2AR Stimulation

In contrast to the cardiotoxic effects of persistent β_1AR activation *in vivo* and *ex vivo,* persistent β_2AR stimulation is cardioprotective. In mice lacking the native β_2AR, the stimulation of endogenous β_1AR with isoproterenol (ISO) triggers more severe myocardial apoptosis *in vivo* compared with wild type control mice (Patterson et al., 2004). This is consistent with earlier pharmacological studies showing that the activation of β_2AR attenuates catecholamine-, hypoxia-, or reactive oxygen species (ROS)-induced apoptosis in both neonatal and adult rat cardiomyocytes (Chesley et al., 2000; Zhu, Zheng, Koch, Lefkowitz, Kobilka, & Xiao, 2001). The cardiac beneficial effects of persistent

β_2AR signaling have been further manifested in recent *in vivo* studies; the selective activation of β_2AR by fenoterol exerts a profound antiapoptotic effect and improves cardiac performance in an ischemic rat HF model (Ahmet et al., 2008). However, further investigation is merited for understanding the exact signaling events that underlie β_2AR-mediated cardiac protection (see following sections).

III. MECHANISMS UNDERLYING β_2AR-COUPLED G_i SIGNALING

The classic view on βAR-coupled G_s signaling involves an agonist-induced change in the receptor conformation that causes the activation of the G_s protein, leading the formation of the second messenger, cAMP, which activates PKA and downstream signaling. The termination of this cascade occurs when GPCR kinases (GRKs) and the second messenger kinase, PKA, phosphorylate the activated receptor and promote the binding of β-arrestins which sterically block the coupling of G_s to the receptor. It is widely accepted that prolonged exposure of βAR to an agonist leads to reduced G_s-mediated responses such as cAMP production and positive inotropic effect. When this occurs in the context of diminished responsiveness to a diverse array of other agonists (heterologous desensitization), it generally results from a negative feedback regulation by PKA-mediated phosphorylation of βARs. Exposure to an agonist also leads to GRKs-dependent, agonist-specific or homologous desensitization, which proceeds the PKA-dependent phosphorylation and constitutes the most efficient means to desensitize GPCRs.

HF is characterized by defective βAR system, including increased circulating catecholamine levels, reduced β_1AR density and signaling efficiency, and enhanced β_2AR-coupled G_i signaling (Rockman, Koch, & Lefkowitz, 2002; Ungerer, Bohm, Elce, Erdmann, & Lohse, 1993). GRK2 (also known as βARK1), the best characterized member of the GRK family, plays a predominant role in desensitizing βARs and has been implicated as a cause factor of HF (Rengo, Lymperopoulos, Leosco, & Koch, 2011). Emerging evidence suggests that activation of GRK2 as well as PKA is essentially involved in the *activation* of the β_2AR-coupled G_i signaling in mammalian cells. First, early work has shown that β_2AR-induced activation of ERK1/2 in HEK293 cells is mediated by a G_i-dependent mechanism, and that phosphorylation of β_2AR by PKA is a prerequisite for the switch of the receptor coupling from G_s to G_i (Daaka, Luttrell, & Lefkowitz, 1997). Second, our recent studies have demonstrated that, in addition to PKA-mediated phosphorylation, elevated β_2AR phosphorylation by GRK2 acerbates the G_i signaling, whereas inhibition of GRK2 activity profoundly suppresses the β_2AR-G_i coupling (Zhu et al., unpublished data). Since GRK2 upregulation occurs prior to the onset of HF and contributes to the development of HF (Choi, Koch, Hunter, & Rockman, 1997; Iaccarino, Tomhave, Lefkowitz, & Koch, 1998; Perrino, Naga Prasad, Schroder, Hata, Milano, & Rockman, 2005; White, Hata, Shah, Glower, Lefkowitz, & Koch, 2000) enhanced GRK2 activation may play an important role in the exacerbated β_2AR-coupled G_i signaling in the failing heart. Indeed, disruption of G_i signaling with PTX or inhibition of GRK2 with a peptide inhibitor, βARK-ct, can restore cardiac contractile response to βAR stimulation in multiple HF models (Chakir et al., 2009; Koch et al., 1995; Tachibana, Naga Prasad, Lefkowitz, Koch, & Rockman, 2005; Xiao et al., 2003). It is also noteworthy that, while the β_1AR subtype cannot couple to G_i under normal conditions, β_1AR contractile response is cross-inhibited by enhanced β_2AR-G_i signaling in the failing heart. Thus, the enhanced β_2AR-G_i signaling contributes to the dysfunction of both β_1AR and β_2AR in the failing heart (Lokuta et al., 2005; Sato, Gong, Terracciano, Ranu, & Harding, 2004; Xiao & Balke, 2004; Zhu et al., 2005).

Taken together, these recent studies have defined GRK2 as an important etiological link between enhanced β_2AR-G_i signaling and the development of HF. These studies also suggest that the previously reported beneficial effects of βARK-ct in improving the function of the failing heart (Choi et al., 1997; Iaccarino et al., 1998; Perrino et al., 2005; White et al., 2000) is mediated, at least in part, by attenuating GRK2-dependent β_2AR-G_i signaling.

IV. RGS2-MEDIATED TERMINATION OF β_2AR-COUPLED G_i SIGNALING AND ITS POTENTIAL PATHOGENIC AND THERAPEUTIC IMPLICATIONS

As discussed above, in contrast to the βAR-G_s signaling, the β_2AR-G_i signaling is enhanced by PKA- and GRK2-mediated phosphorylation of the receptor (Daaka et al., 1997; Hausdorff, Lohse, Bouvier, Liggett, Caron, & Lefkowitz, 1990; Liu, Ramani, Soto, De Arcangelis, & Xiang, 2009). The next fundamental question is what is the mechanism underlying the termination of the β_2AR-G_i signaling. In this regard, it has been shown that upon GPCR activation, GDP is exchanged for GTP on the Gα subunit, resulting in dissociation of the Gα from Gβγ subunits and the activation of downstream effectors. The intrinsic GTPase activity of the α subunit of G proteins serves as a molecular clock, turning down GPCR signaling via returning G proteins to the GDP-bond heterotrimeric form. Regulator of G protein signaling (RGS) proteins are GTPase-activating proteins (GAPs) which accelerate GTPase-mediated hydrolysis of GTP to GDP on Gα, thus reconstituting the heterotrimeric G protein complex and terminating G protein signaling (Hollinger & Hepler, 2002; Ross & Wilkie, 2000).

The small RGS B/R4 subfamily members, RGS2-5, are the predominant RGS proteins expressed in the cardiovascular system (Hao, Michalek, Zhang, Zhu, Xu, & Mende, 2006; Riddle, Schwartzman, Bond, & Insel, 2005). In the heart of mammalian species, RGS2-5 are abundantly expressed (Riddle et al., 2005). RGS3-5 equally regulate both $G\alpha_{i/o}$ and $G\alpha_{q/11}$(Hao et al., 2006), whereas cardiomyocyte RGS2 displays selectivity for $G_{\alpha q/11}$ (Hao et al., 2006; Heximer, Watson, Linder, Blumer, & Hepler, 1997; Zhang, Watson, Zahner, Rottman, Blumer, & Muslin, 1998). In particular, RGS2 can negatively regulate the signaling of G_q-coupled receptors, including α_{1A}-adrenergic receptor, angiotensin II receptor 1A, and interleukin receptor (Hao et al., 2006; Zou, Roy, Zhao, Kirshenbaum, Karmazyn, & Chidiac, 2006). Deregulation of RGS2 has been implicated in the pathogenesis of cardiac hypertrophy and hypertension (Wieland, Lutz, & Chidiac, 2007). It has been recently reported that RGS2 can physically interact with β_2AR (Roy, Lemberg, & Chidiac, 2003), in addition to G_q-coupled GPCRs such as M_1 muscarinic receptor (Bernstein et al., 2004) and α_{1A}AR (Hague, Bernstein, Ramineni, Chen, Minneman, & Hepler, 2005). More importantly, our recent studies have provided multiple lines of evidence to demonstrate that RGS2 is a primary terminator of the β_2AR-G_i signaling. These include (a) prolonged absence of agonist stimulation for 24 h impairs the β_2AR-G_i signaling, resulting in enhanced β_2AR- but not β_1AR-mediated contractile response in cultured adult mouse cardiomyocytes; (b) increased β_2AR contractile response is accompanied by a selective upregulation of RGS2 in the absence of alterations in other major cardiac RGS proteins (RGS3-5) or G_s, G_i or βAR subtypes; (c) administration of a βAR agonist, ISO, prevents RGS2 upregulation and restores the β_2AR-G_i signaling in cultured cells; (d) RGS2 ablation, similar to βAR agonist stimulation, sustains the β_2AR-G_i signaling in cultured cells, whereas adenoviral overexpression of RGS2 suppresses agonist-activated β_2AR-G_i signaling in cardiomyocytes and HEK293 cells (Chakir et al., 2011). Thus, RGS2 functions as a novel negative regulator of the β_2AR-G_i signaling. Since RGS2 constitutively binds to β_2AR (Roy et al., 2003), it is reasonable to assume that the β_2AR-coupled G_i signaling is constitutively suppressed by RGS2 under physiological conditions, leading to apparent G_s-predominant β_2AR signaling in cardiac myocytes of most mammalian species except mouse (Xiao, Cheng, Zhou, Kuschel, & Lakatta, 1999; Xiao et al., 2006). These findings have revealed a novel negative regulation of β_2AR-activated G_i signaling by RGS2, and suggest that βAR agonist-induced switch of β_2AR signaling from G_s to G_i pathway is mediated, at least in part, by agonist-induced downregulation of RGS2 protein.

The selective upregulation of RGS2 and the concurrent augmentation of β_2AR contractile response induced by the lack of βAR stimulation in cultured myocytes are intriguing given the fact that β-blocker therapy can resensitize βAR-mediated contractile support and improve cardiac function in patients with

HF. The beneficial effects of β-blockers might be mediated, in part, by increasing RGS2 expression. In contrast, adrenergic overdriving, as is the case in hypertension and cardiac hypertrophy in different animal models, is accompanied by a selective downregulation of RGS2 (Heximer et al., 2003; Wang et al., 2008; Zhang et al., 2006). In this regard, recent studies have shown that RGS2 gene silencing blocks α_1AR-induced cardiac myocyte hypertrophy (Hao et al., 2006), and that pressure overload by trans-aortic constriction results in enhanced G_q signaling, exacerbated cardiac hypertrophy, HF, and premature death in RGS2-deficient mice as compared to wild type counterparts (Takimoto et al., 2009), implying that RGS2 may play a central role in protecting the heart against stress-induced maladaptive remodeling. Thus, under various pathological circumstances, downregulation or malfunction of RGS2 is expected to enhance G_q- and G_i-mediated signaling, constituting a pathogenic element for the development of HF in addition to its known role in the pathogenesis of hypertension (Tsang, Woo, Zhu, & Xiao, 2010).

V. LIGAND-DIRECTED SELECTIVE ACTIVATION OF β_2AR-COUPLING TO G_s OR G_i

It is now well established that any given ligand for a GPCR does not simply possess a single defined efficacy. Rather, a ligand possesses multiple efficacies, depending on the specific down-stream signal transduction pathway analyzed. This diversity reflects ligand-specific GPCR conformations and is often referred to as "Functional Selectivity." It has been known for a century that stereoisomers of catecholamines differ in their potency and efficacy. However, the molecular basis for differences in efficacy of GPCR ligand stereoisomers has remained poorly defined till now. β_2AR couples dually to G_s and PTX-sensitive G_i proteins, resulting in functionally opposing effects on cardiomyocyte contractility. A fundamental question regarding receptor–G protein interaction is, therefore, whether different agonists can traffic a receptor to different intracellular signaling pathways. Recent studies have demonstrated, while most β_2AR agonists activate both G_s and G_i, fenoterol selectively activates G_s (Xiao et al., 2003). Furthermore, we have synthesized all fenoterol stereoisomers, including *R,R*-, *R,S*-, *S,R*-, and *S,S*-forms, and found that the *R,R*-fenoterol fails to activate G_i signaling, as evidenced by the absence of PTX-sensitivity of its contractile response and its inability to activate G_i-dependent ERK1/2 signaling, but *S,R*-fenoterol exhibits a robust PTX-sensitivity in these responses, suggesting that the *S,R*-isomer enables β_2AR to activate both G_s and G_i (Woo et al., 2009). The same conclusion holds true for some fenoterol derivatives. For instance, *S,R*-methoxy-fenoterol, but not *R,R*-methoxy-fenoterol, activates β_2AR-coupled G_i signaling in cardiomyocytes (Woo et al., 2009). Thus, in addition to β_2AR phosphorylation, stereoisomers of an agonist can direct β_2AR to different G protein(s). This finding is important because it is the first account to show that even the subtle chemical differences within a ligand stereoisomer pair are sufficient to stabilize GPCR conformations with distinct G-protein coupling properties, highlighting how important it is to carefully examine both the "active" and the "inactive" stereoisomer to understand the exact mechanism of action and cellular effects of a GPCR ligand. This finding may also have broad reaching implications in GPCR biology and signaling pathway-targeted drug development (Seifert & Dove, 2009).

VI. DEVELOPMENT OF β_2AR AGONISTS INTO NEW DRUGS FOR THE TREATMENT OF HEART FAILURE

A hallmark of HF is the desensitization of βAR signaling, characterized by downregulation of βAR, reduced signaling efficient of remaining receptors, increased G_i signaling, and elevated circulating catecholamine levels. However, HF-associated loss of βAR is selective for β_1AR, with little change in β_2AR. Previous studies have demonstrated that (a) β_2AR dually couples to the G_i and the G_s signaling pathways in the heart with the G_i coupling negating the G_s-mediated contractile response, whereas β_1AR couples solely to the G_s signaling cascade (Kilts et al., 2000; Xiao, 2001; Xiao et al., 1994, 2003); (b) β_1AR and β_2AR stimulation oppositely regulate cardiomyocyte viability and myocardial remodeling with β_1AR detrimental and β_2AR protective (Bisognano et al., 2000; Engelhardt et al., 1999; Liggett et al., 2000; Patterson et al., 2004; Zhu et al., 2001); and (c) β_1AR blockade exhibits salutary effects on patients with HF, thus becoming a major therapy in the treatment of HF (Bristow, 2000; Sabbah, 1999). Our recent research focus is to translate βAR subtype signaling principles into drug development and novel therapies to improve the structure and function of the failing heart. Specifically, we have tested the hypothesis that G_s-selective activation of β_2AR alone or in combination with clinically used β_1AR blockers might provide a potential novel therapy with greater efficacy and fewer side effects for HF.

A. *Signaling-Selective β_2AR Agonists for the Treatment of Heart Failure*

In HF, impaired βAR response is often associated with increased G_i signaling and selective downregulation of β_1AR (higher β_2/β_1 ratio) (Bristow et al., 1986; Eschenhagen et al., 1992). We have demonstrated that inhibition of G_i with PTX restores the markedly depressed β_2AR contractile response in myocytes from the failing heart of the rat (Xiao et al., 2003). Further, we have identified a unique β_2AR agonist, fenoterol, which selectively activates β_2AR-G_s signaling, bypassing the G_i pathway, fully restores the diminished β_2AR inotropic effect in myocytes from failing spontaneously hypertensive rat (SHR) hearts in the absence of PTX (Xiao et al., 2003). This suggests that selective activation of the β_2AR-G_s signaling may provide a useful therapeutic target for the treatment of HF.

Follow-up *in vivo* studies have further demonstrated that prolonged use of fenoterol not only improves cardiac function, but also retards cardiac maladaptive remodeling, and that the overall beneficial effects of fenoterol are greater than the salutary effects of β_1AR blockade in a rat HF model (Ahmet, Krawczyk, Heller, Moon, Lakatta, & Talan, 2004; Ahmet, Lakatta, & Talan, 2005). Specifically, the effectiveness of the β_2AR agonist in attenuating left ventricle (LV) dilatation, functional decline, and myocyte apoptosis significantly exceeds that of the clinically sued β_1AR blocker, metoprolol (Ahmet et al., 2004). Since fenoterol and its derivatives can increase myocardial contractility *in vivo* and delay the development of HF in the rat ischemic HF model, we envision some of the promising new G_s-signaling selective β_2AR agonists may be developed into drugs to improve the structure and function of the failing heart. The therapeutic effect of β_2AR stimulation has been recently confirmed in a rat model of autoimmune myocarditis (Nishii et al., 2006).

B. A Combination of β_2AR Activation with β_1AR Blockade Provides A More Effective Therapy for Heart Failure

In the final analysis a main measure of therapeutic efficacy of the treatment for HF is its effect on mortality. In recent *in vivo* studies, we have compared the efficacy of a combined therapy with a β_1AR blocker and a β_2AR agonist (fenoterol plus metoprolol) with that of a single therapy of either the β_1AR blocker or the β_2AR agonist, in terms of animal mortality as a primary outcome. We have found that the beneficial effects of the combined therapy on animal survival, cardiac remodeling and function exceed the effects of either single therapy (Ahmet et al., 2005, 2008).

Since standard therapy for advanced HF includes RAS inhibition, we have also compared the combined (fenoterol plus metoprolol) therapy with a combination of β_1AR blocker and angiotensin-converting enzyme inhibitor, and found that the combined therapy of the β_1AR blocker and the β_2AR agonist is equally effective compared to the standard therapy for HF with respect to mortality and exceeds the latter with respect to cardiac remodeling and myocardial infarction (MI) expansion (Ahmet et al., 2008).

Altogether, recent *in vivo* studies have revealed that in the rat HF model, β_2AR activation alone or in combination with a β_1AR blocker or a RAS inhibitor is superior to β_1AR blockade alone in preventing MI expansion, improving animal survival, and attenuating LV maladaptive remodeling, contractile dysfunction, myocyte apoptosis, and arrhythmia (Ahmet et al., 2008). These *in vivo* studies have provided a proof of principle for the combined therapy of β_2AR agonists with the clinically used β_1AR blockers or RAS inhibitors for the treatment of HF.

VII. FUTURE PERSPECTIVE

Studies over the past decade have greatly enriched our understanding of βAR subtype-specific signal transduction and biological functions in normal and disease conditions, but also raised many perplexing questions regarding β_2AR signaling and the potential interaction between βAR subtypes in the failing heart. First, if the β_2AR-coupled G_i signaling is inactive during β_2AR stimulation with *R,R*-isomers of fenoterol and its derivatives, what is the mechanism underlying their prosurvival effects in cardiomyocytes and the beneficial effects of fenoterol *in vivo*? Second, what is the mechanism by which enhanced β_2AR-G_i signaling cross-inhibits β_1AR-mediated contractile support in the failing heart? In other words, how do β_1AR and β_2AR interact with each other in the failing heart? To address these important questions, it is necessary to systematically characterize βAR subtype signaling complexes and dissect their signaling circuitries at the gene, protein, and cellular levels using interdisciplinary approaches, and integrate what we have learned into better understanding of how they function in the whole animal level. These mechanistic studies will greatly expand our understanding of subtype-specific βAR signaling in special and GPCR biology in general. As another frontline of future directions, extensive efforts should be continuously invested into translating bench discoveries into clinic use.

References

Ahmet, I., Krawczyk, M., Heller, P., Moon, C., Lakatta, E. G., & Talan, M. I. (2004). Beneficial effects of chronic pharmacological manipulation of β-adrenoreceptor subtype signaling in rodent dilated ischemic cardiomyopathy. *Circulation, 110*(9), 1083–1090.

Ahmet, I., Krawczyk, M., Zhu, W., Woo, A. Y., Morrell, C., & Poosala, S., et al., (2008). Cardioprotective and survival benefits of long-term combined therapy with β_2-adrenoreceptor (AR) agonist and beta1 AR blocker in dilated cardiomyopathy postmyocardial infarction. *J Pharmacol Exp Ther, 325*(2), 491–499.

Ahmet, I., Lakatta, E. G., & Talan, M. I. (2005). Pharmacological stimulation of β_2-adrenergic receptors (β_2AR) enhances therapeutic effectiveness of β_1AR blockade in rodent dilated ischemic cardiomyopathy. *Heart Fail Rev, 10*(4), 289–296.

Antos, C. L., Frey, N., Marx, S. O., Reiken, S., Gaburjakova, M., & Richardson, J. A., et al., (2001). Dilated cardiomyopathy and sudden death resulting from constitutive activation of protein kinase a. *Circ Res, 89*(11), 997–1004.

Bernstein, L. S., Ramineni, S., Hague, C., Cladman, W., Chidiac, P., & Levey, A. I., et al., (2004). RGS2 binds directly and selectively to the M1 muscarinic acetylcholine receptor third intracellular loop to modulate Gq/11alpha signaling. *J Biol Chem, 279*(20), 21248–21256.

Bisognano, J. D., Weinberger, H. D., Bohlmeyer, T. J., Pende, A., Raynolds, M. V., & Sastravaha, A., et al., (2000). Myocardial-directed overexpression of the human β_1-adrenergic receptor in transgenic mice. *J Mol Cell Cardiol, 32*(5), 817–830.

Bristow, M. R. (2000). Mechanistic and clinical rationales for using β-blockers in heart failure. *J Card Fail, 6*(2 Suppl 1), 8–14.

Bristow, M. R., Ginsburg, R., Umans, V., Fowler, M., Minobe, W., & Rasmussen, R., et al., (1986). β_1- and β_2-adrenergic-receptor subpopulations in nonfailing and failing human ventricular myocardium: Coupling of both receptor subtypes to muscle contraction and selective beta 1-receptor down-regulation in heart failure. *Circ Res, 59*(3), 297–309.

Chakir, K., Daya, S. K., Aiba, T., Tunin, R. S., Dimaano, V. L., & Abraham, T. P., et al., (2009). Mechanisms of enhanced β-adrenergic reserve from cardiac resynchronization therapy. *Circulation, 119*(9), 1231–1240.

Chakir, K., Zhu, W., Tsang, S., Woo, A. Y. H., Yang, D., & Wang, X., et al., (2011). RGS2 Is a primary terminator of β_2-adrenergic receptor-mediated G_i signaling. *J Mol Cell Cardiol,* in press.

Chesley, A., Lundberg, M. S., Asai, T., Xiao, R. P., Ohtani, S., & Lakatta, E. G., et al., (2000). The β_2-adrenergic receptor delivers an antiapoptotic signal to cardiac myocytes through G_i-dependent coupling to phosphatidylinositol 3'-kinase. *Circ Res, 87*(12), 1172–1179.

Choi, D. J., Koch, W. J., Hunter, J. J., & Rockman, H. A. (1997). Mechanism of β-adrenergic receptor desensitization in cardiac hypertrophy is increased β-adrenergic receptor kinase. *J Biol Chem, 272*(27), 17223–17229.

Daaka, Y., Luttrell, L. M., & Lefkowitz, R. J. (1997). Switching of the coupling of the β_2-adrenergic receptor to different G proteins by protein kinase A. *Nature, 390*(6655), 88–91.

Engelhardt, S., Hein, L., Wiesmann, F., & Lohse, M. J. (1999). Progressive hypertrophy and heart failure in β_1-adrenergic receptor transgenic mice. *Proc Natl Acad Sci USA, 96*(12), 7059–7064.

Eschenhagen, T., Mende, U., Nose, M., Schmitz, W., Scholz, H., & Haverich, A., et al., (1992). Increased messenger RNA level of the inhibitory G protein alpha subunit Giα-2 in human end-stage heart failure. *Circ Res, 70*(4), 688–696.

Hague, C., Bernstein, L. S., Ramineni, S., Chen, Z., Minneman, K. P., & Hepler, J. R. (2005). Selective inhibition of alpha1A-adrenergic receptor signaling by RGS2 association with the receptor third intracellular loop. *J Biol Chem, 280*(29), 27289–27295.

Hao, J., Michalek, C., Zhang, W., Zhu, M., Xu, X., & Mende, U. (2006). Regulation of cardiomyocyte signaling by RGS proteins: Differential selectivity towards G proteins and susceptibility to regulation. *J Mol Cell Cardiol, 41*(1), 51–61.

Hausdorff, W. P., Lohse, M. J., Bouvier, M., Liggett, S. B., Caron, M. G., & Lefkowitz, R. J. (1990). Two kinases mediate agonist-dependent phosphorylation and desensitization of the β_2-adrenergic receptor. *Symp Soc Exp Biol, 44*, 225–240.

Hebert, T. E. (2007). Anti-beta1AR antibodies in dilated cardiomyopathy: Are these a new class of receptor agonists? *Cardiovasc Res, 76*(1), 5–7.

Heximer, S. P., Knutsen, R. H., Sun, X., Kaltenbronn, K. M., Rhee, M. H., & Peng, N., et al., (2003). Hypertension and prolonged vasoconstrictor signaling in RGS2-deficient mice. *J Clin Invest, 111*(4), 445–452.

Heximer, S. P., Watson, N., Linder, M. E., Blumer, K. J., & Hepler, J. R. (1997). RGS2/G0S8 is a selective inhibitor of Gqalpha function. *Proc Natl Acad Sci USA, 94*(26), 14389–14393.

Hollinger, S., & Hepler, J. R. (2002). Cellular regulation of RGS proteins: Modulators and integrators of G protein signaling. *Pharmacol Rev, 54*(3), 527–559.

Iaccarino, G., Tomhave, E. D., Lefkowitz, R. J., & Koch, W. J. (1998). Reciprocal *in vivo* regulation of myocardial G protein-coupled receptor kinase expression by β-adrenergic receptor stimulation and blockade. *Circulation, 98*(17), 1783–1789.

Iwase, M., Uechi, M., Vatner, D. E., Asai, K., Shannon, R. P., & Kudej, R. K., et al., (1997). Cardiomyopathy induced by cardiac Gsα overexpression. *Am J Physiol, 272*(1 Pt 2), H585–H589.

Kilts, J. D., Gerhardt, M. A., Richardson, M. D., Sreeram, G., Mackensen, G. B., & Grocott, H. P., et al., (2000). β_2-adrenergic and several other G protein-coupled receptors in human atrial membranes activate both G_s and G_i. *Circ Res, 87*(8), 705–709.

Koch, W. J., Rockman, H. A., Samama, P., Hamilton, R. A., Bond, R. A., & Milano, C. A., et al., (1995). Cardiac function in mice overexpressing the β-adrenergic receptor kinase or a βARK inhibitor. *Science, 268*(5215), 1350–1353.

Lai, N. C., Tang, T., Gao, M. H., Saito, M., Takahashi, T., & Roth, D. M., et al., (2008). Activation of cardiac adenylyl cyclase expression increases function of the failing ischemic heart in mice. *J Am Coll Cardiol, 51*(15), 1490–1497.

Liggett, S. B., Tepe, N. M., Lorenz, J. N., Canning, A. M., Jantz, T. D., & Mitarai, S., et al., (2000). Early and delayed consequences of β_2-adrenergic receptor overexpression in mouse hearts: critical role for expression level. *Circulation, 101*(14), 1707–1714.

Liu, R., Ramani, B., Soto, D., De Arcangelis, V., & Xiang, Y. (2009). Agonist dose-dependent phosphorylation by protein kinase A and G protein-coupled receptor kinase regulates β_2-adrenoceptor coupling to G_i proteins in cardiomyocytes. *J Biol Chem, 284*(47), 32279–32287.

Lokuta, A. J., Maertz, N. A., Meethal, S. V., Potter, K. T., Kamp, T. J., & Valdivia, H. H., et al., (2005). Increased nitration of sarcoplasmic reticulum Ca^{2+}-ATPase in human heart failure. *Circulation, 111*(8), 988–995.

Nishii, M., Inomata, T., Niwano, H., Takehana, H., Takeuchi, I., & Nakano, H., et al., (2006). β_2-Adrenergic agonists suppress rat autoimmune myocarditis: Potential role of beta2-adrenergic stimulants as new therapeutic agents for myocarditis. *Circulation, 114*(9), 936–944.

Patterson, A. J., Zhu, W., Chow, A., Agrawal, R., Kosek, J., & Xiao, R. P., et al., (2004). Protecting the myocardium: A role for the β_2-adrenergic receptor in the heart. *Crit Care Med, 32*(4), 1041–1048.

Perrino, C., Naga Prasad, S. V., Schroder, J. N., Hata, J. A., Milano, C., & Rockman, H. A. (2005). Restoration of beta-adrenergic receptor signaling and contractile function in heart failure by disruption of the βARK1/phosphoinositide 3-kinase complex. *Circulation, 111*(20), 2579–2587.

Rengo, G., Lymperopoulos, A., Leosco, D., & Koch, W. J. (2011). GRK2 as a novel gene therapy target in heart failure. *J Mol Cell Cardiol., 50*(5), 785–792.

Riddle, E. L., Schwartzman, R. A., Bond, M., & Insel, P. A. (2005). Multi-tasking RGS proteins in the heart: the next therapeutic target? *Circ Res, 96*(4), 401–411.

Rockman, H. A., Koch, W. J., & Lefkowitz, R. J. (2002). Seven-transmembrane-spanning receptors and heart function. *Nature, 415*(6868), 206–212.

Ross, E. M., & Wilkie, T. M. (2000). GTPase-activating proteins for heterotrimeric G proteins: Regulators of G protein signaling (RGS) and RGS-like proteins. *Annu Rev Biochem, 69*, 795–827.

Roth, D. M., Bayat, H., Drumm, J. D., Gao, M. H., Swaney, J. S., & Ander, A., et al., (2002). Adenylyl cyclase increases survival in cardiomyopathy. *Circulation, 105*(16), 1989–1994.

Roth, D. M., Gao, M. H., Lai, N. C., Drumm, J., Dalton, N., & Zhou, J. Y., et al., (1999). Cardiac-directed adenylyl cyclase expression improves heart function in murine cardiomyopathy. *Circulation, 99*(24), 3099–3102.

Roy, A. A., Lemberg, K. E., & Chidiac, P. (2003). Recruitment of RGS2 and RGS4 to the plasma membrane by G proteins and receptors reflects functional interactions. *Mol Pharmacol, 64*(3), 587–593.

Sabbah, H. N. (1999). The cellular and physiologic effects of beta blockers in heart failure. *Clin Cardiol, 22*(Suppl 5), V16–V20.

Sato, M., Gong, H., Terracciano, C. M., Ranu, H., & Harding, S. E. (2004). Loss of β-adrenoceptor response in myocytes overexpressing the Na^+/Ca^{2+}-exchanger. *J Mol Cell Cardiol, 36*(1), 43–48.

Seifert, R., & Dove, S. (2009). Functional selectivity of GPCR ligand stereoisomers: New pharmacological opportunities. *Mol Pharmacol, 75*(1), 13–18.

Sucharov, C. C., Mariner, P. D., Nunley, K. R., Long, C., Leinwand, L., & Bristow, M. R. (2006). A β_1-adrenergic receptor CaM kinase II-dependent pathway mediates cardiac myocyte fetal gene induction. *Am J Physiol Heart Circ Physiol, 291*(3), H1299–H1308.

Tachibana, H., Naga Prasad, S. V., Lefkowitz, R. J., Koch, W. J., & Rockman, H. A. (2005). Level of β-adrenergic receptor kinase 1 inhibition determines degree of cardiac dysfunction after chronic pressure overload-induced heart failure. *Circulation, 111*(5), 591–597.

Takimoto, E., Koitabashi, N., Hsu, S., Ketner, E. A., Zhang, M., & Nagayama, T., et al., (2009). Regulator of G protein signaling 2 mediates cardiac compensation to pressure overload and antihypertrophic effects of PDE5 inhibition in mice. *J Clin Invest, 119*(2), 408–420.

Tang, T., Gao, M. H., Roth, D. M., Guo, T., & Hammond, H. K. (2004). Adenylyl cyclase type VI corrects cardiac sarcoplasmic reticulum calcium uptake defects in cardiomyopathy. *Am J Physiol Heart Circ Physiol, 287*(5), H1906–H1912.

Tsang, S., Woo, A. Y., Zhu, W., & Xiao, R. P. (2010). Deregulation of RGS2 in cardiovascular diseases. *Front Biosci (Schol Ed), 2*, 547–557.

Ungerer, M., Bohm, M., Elce, J. S., Erdmann, E., & Lohse, M. J. (1993). Altered expression of β-adrenergic receptor kinase and β_1-adrenergic receptors in the failing human heart. *Circulation, 87*(2), 454–463.

Wang, X., Adams, L. D., Pabon, L. M., Mahoney Jr., W. M., Beaudry, D., & Gunaje, J., et al., (2008). RGS5, RGS4, and RGS2 expression and aortic contractibility are dynamically co-regulated during aortic banding-induced hypertrophy. *J Mol Cell Cardiol, 44*(3), 539–550.

White, D. C., Hata, J. A., Shah, A. S., Glower, D. D., Lefkowitz, R. J., & Koch, W. J. (2000). Preservation of myocardial β-adrenergic receptor signaling delays the development of heart failure after myocardial infarction. *Proc Natl Acad Sci USA, 97*(10), 5428–5433.

Wieland, T., Lutz, S., & Chidiac, P. (2007). Regulators of G protein signalling: A spotlight on emerging functions in the cardiovascular system. *Curr Opin Pharmacol, 7*(2), 201–207.

Woo, A. Y., Wang, T. B., Zeng, X., Zhu, W., Abernethy, D. R., & Wainer, I. W., et al., (2009). Stereochemistry of an agonist determines coupling preference of β_2-adrenoceptor to different G proteins in cardiomyocytes. *Mol Pharmacol, 75*(1), 158–165.

Xiao, R. P. (2001). β-adrenergic signaling in the heart: dual coupling of the β_2-adrenergic receptor to G_s and G_i proteins. *Sci STKE, 2001*(104), re15.

Xiao, R. P., & Balke, C. W. (2004). Na^+/Ca^{2+} exchange linking β_2-adrenergic G_i signaling to heart failure: Associated defect of adrenergic contractile support. *J Mol Cell Cardiol, 36*(1), 7–11.

Xiao, R. P., Cheng, H., Zhou, Y. Y., Kuschel, M., & Lakatta, E. G. (1999). Recent advances in cardiac beta(2)-adrenergic signal transduction. *Circ Res, 85*(11), 1092–1100.

Xiao, R. P., Hohl, C., Altschuld, R., Jones, L., Livingston, B., & Ziman, B., et al., (1994). β_2-adrenergic receptor-stimulated increase in cAMP in rat heart cells is not coupled to changes in Ca^{2+} dynamics, contractility, or phospholamban phosphorylation. *J Biol Chem, 269*(29), 19151–19156.

Xiao, R. P., Zhang, S. J., Chakir, K., Avdonin, P., Zhu, W., & Bond, R. A., et al., (2003). Enhanced G_i signaling selectively negates β_2-adrenergic receptor (AR) – but not β_1-AR-mediated positive inotropic effect in myocytes from failing rat hearts. *Circulation, 108*(13), 1633–1639.

Xiao, R. P., Zhu, W., Zheng, M., Cao, C., Zhang, Y., & Lakatta, E. G., et al., (2006). Subtype-specific α_1- and β-adrenoceptor signaling in the heart. *Trends Pharmacol Sci, 27*(6), 330–337.

Zhang, R., Khoo, M. S., Wu, Y., Yang, Y., Grueter, C. E., & Ni, G., et al., (2005). Calmodulin kinase II inhibition protects against structural heart disease. *Nat Med, 11*(4), 409–417.

Zhang, S., Watson, N., Zahner, J., Rottman, J. N., Blumer, K. J., & Muslin, A. J. (1998). RGS3 and RGS4 are GTPase activating proteins in the heart. *J Mol Cell Cardiol, 30*(2), 269–276.

Zhang, T., Maier, L. S., Dalton, N. D., Miyamoto, S., Ross Jr., J., & Bers, D. M., et al., (2003). The deltaC isoform of CaMKII is activated in cardiac hypertrophy and induces dilated cardiomyopathy and heart failure. *Circ Res, 92*(8), 912–919.

Zhang, W., Anger, T., Su, J., Hao, J., Xu, X., & Zhu, M., et al., (2006). Selective loss of fine tuning of Gq/11 signaling by RGS2 protein exacerbates cardiomyocyte hypertrophy. *J Biol Chem, 281*(9), 5811–5820.

Zhu, W. Z., Chakir, K., Zhang, S., Yang, D., Lavoie, C., & Bouvier, M., et al., (2005). Heterodimerization of β_1- and β_2-adrenergic receptor subtypes optimizes β-adrenergic modulation of cardiac contractility. *Circ Res, 97*(3), 244–251.

Zhu, W. Z., Wang, S. Q., Chakir, K., Yang, D., Zhang, T., & Brown, J. H., et al., (2003). Linkage of β_1-adrenergic stimulation to apoptotic heart cell death through protein kinase A-independent activation of Ca^{2+}/calmodulin kinase II. *J Clin Invest, 111*(5), 617–625.

Zhu, W. Z., Zheng, M., Koch, W. J., Lefkowitz, R. J., Kobilka, B. K., & Xiao, R. P. (2001). Dual modulation of cell survival and cell death by β_2-adrenergic signaling in adult mouse cardiac myocytes. *Proc Natl Acad Sci USA, 98*(4), 1607–1612.

Zou, M. X., Roy, A. A., Zhao, Q., Kirshenbaum, L. A., Karmazyn, M., & Chidiac, P. (2006). RGS2 is upregulated by and attenuates the hypertrophic effect of alpha1-adrenergic activation in cultured ventricular myocytes. *Cell Signal, 18*(10), 1655–1663.

CHAPTER 10

β-Adrenergic Receptor, Amyloid β Peptide, and Alzheimer's Disease

Dayong Wang[1] **and Yang K. Xiang**[2]

[1]Molecular and Integrative Physiology, University of Illinois at Urbana-Champaign, Urbana, IL, USA

[2]Molecular and Integrative Physiology and Neuroscience Program, University of Illinois at Urbana-Champaign, Urbana, IL, USA

I. Overview
II. Introduction
III. Adrenergic Signaling and APP Expression
IV. Activation of βAR Promotes Amyloid Process
V. Interaction of Adrenergic Receptor with Amyloid β Peptide
VI. CNS Adrenergic System in AD
VII. Clinic Implication of Activation of Downstream Targets by Aβ/βAR Signaling Axis in CNS
VIII. Adrenergic System in Peripheral Tissues
IX. Potential Drug Targets on Aβ-β_2AR Signaling Axis for AD Therapy
X. Conclusion and Future Direction
Acknowledgments
References

I. OVERVIEW

Studies *in vitro* and *in vivo* have shown that soluble oligomeric amyloid beta (Aβ) peptides are cytotoxic and induce synaptic degeneration and dysfunction that contributes to the pathophysiology of Alzheimer's Disease (AD) in early stage. Accumulating evidence indicates that CNS noradrenergic system plays an essential role in AD, and may mediate and facilitate the detrimental cellular effects of soluble Aβ peptides. Recent studies have revealed direct regulation between CNS noradrenergic system and amyloidogenesis. A reinforced reciprocal regulation between Aβ and β_2 adrenergic receptor (β_2AR) may not only contribute to synaptic and neuronal degeneration in AD but also to its psychological stress-dependent symptoms

Current Topics in Membranes, Volume 67 0065-230X/10 $35.00

DOI: 10.1016/B978-0-12-384921-2.00010-0

including agitation, aggression, irritability, and anxiety. In this chapter, we discuss these progresses and publications showing (patho-)physiological roles of Aβ peptides and β_2AR in AD, and their implications for preventing the development of AD.

II. INTRODUCTION

AD is a progressive neurodegenerative disease and the most common form of dementia affecting more than 35 million people worldwide (Prince & Jackson, 2009). AD is characterized by cognitive failure associated with pathology including cerebral senile plaques containing Aβ peptide, cerebral intraneuronal neurofibrillary tangle of tau protein, neuron and synapse loss, and neurotransmission changes. Genetically, AD is divided into two forms, about 5–10% of familial cases with mendelian inheritance of predominantly early-onset (<60 years), and 90–95% sporadic cases with less apparent or no familial aggregation and usually of later onset age (>60 years). Overall, AD affects more than 5% of the population over the age of 65, and up to 40% of those over the age of 85. Despite the first diagnosis of AD more than 100 years ago, the causes of the late-onset disease remain unclear. The long-term care of AD patients is of enormous social and economic burden. The increasing age of the population generates an urgent need in the search for understanding the pathogenesis and effective treatment strategies.

The works on familial AD form the bedrock of the amyloid cascade hypothesis, which holds that an increase in Aβ may trigger AD (Goedert & Spillantini, 2006). Although it offers a framework to explain AD pathogenesis, it is still an evolving hypothesis (Hardy & Selkoe, 2002). The early studies suggested that plaque numbers were directly related to quantitative measures of cognitive decline in the aged population. However, subsequent studies cast doubt on the predicative value of plaque numbers, suggesting instead that neurofibrillary tangles and synapse loss are more reliable predicators of cognitive decline (Terry, 2006). Two lines of evidence support this argument. First, the correlation between the burden of Aβ and neuronal loss or cognitive impairment is weak (Giannakopoulos et al., 2003; Lee, Zhu, Castellani, Nunomura, Perry, & Smith, 2007). Second, high Aβ loads or deposits can be seen in cognitively intact old people (Crystal et al., 1988; Lee et al., 2007). In animal studies, different groups have reported that knockout of presenilin 1 (PS1), a subunit of γ-secretases that is responsible for amyloidogenic process of amyloid precursor protein (APP), does not improve cognition in APP-transgenic animals. In contrast, PS1 knockout-APP transgenic animals lacking Aβ perform worse than APP-transgenic animals with high levels of Aβ, which even argue that Aβ is beneficial in certain circumstances (Dewachter et al., 2002; Lee, Casadesus, Zhu, Joseph, Perry, & Smith 2004a; Lee, Casadesus, Zhu, Takeda, Perry, & Smith, 2004b; Lee et al., 2007). Accordingly, Aβ is secreted under normal physiological conditions in healthy persons (Goedert & Spillantini, 2006; Selkoe, 2006). Until now, ectopic application of Aβ has generated highly heterogeneous results, ranging from acute increase in spontaneous synaptic activity accompanied by neurotoxicity and intrinsic excitability of neurons (Hartley et al., 1999; Ye, Walsh, Selkoe, & Hartley, 2004) to a lack of effect on synaptic transmission (Shankar et al., 2008) or depression (Kamenetz et al., 2003; Nimmrich et al., 2008). These observations suggest that the oligomeric states of Aβ and treating time differentially affect synaptic function. One of the key questions that remains to be addressed is which cellular targets are responsible for a myriad cellular signaling and physiological and pathological conditions induced by different oligomeric Aβ species.

Increasing evidence indicates that β adrenergic receptor (βAR), a prototype G protein-coupled receptor (GPCR) in central nervous system (CNS), may play an important role in sporadic late-onset AD (Yu et al., 2008). Several polymorphisms of β_2AR are linked to sporadic late-onset AD (Rosenberg et al., 2008; Yu et al., 2008). CNS adrenergic system is significantly altered in specific brain regions (Raskind, Cyrus, Ruzicka, & Gulanski, 1999a). In peripheral system, βAR response is blunted in fibroblasts and lymphocytes isolated from AD patients as well as in cardiac contractile function (Huang & Gibson, 1993; Garlind, Johnston, Algotsson, Winblad, & Cowburn, 1997; Turdi, Guo, Huff, Wolf, Culver, & Ren, 2009). Recent epidemiological studies have shown that βAR antagonists delay the rate of functional decline and the incidence of AD pathogenesis (Khachaturian et al., 2006; Rosenberg et al., 2008). One of our studies has identified β_2AR as a cellular membrane target for soluble dimer of Aβ, and the binding of Aβ to β_2AR triggers the receptor signaling cascades targeting a variety of intracellular substrates. This signaling paradigm may offer an explanation of some cellular effects of Aβ on neurons contributing to the development of AD. Moreover, another study has shown that stimulation of β_2AR leads to the activation of γ-secretases, which enhances the amyloidogenic process of APP in HEK293 cells and hippocampal neurons (Ni et al., 2006). Inhibition of β_2AR with antagonist ICI118551 significantly reduces the amyloid plaque formation and sizes in AD animals (Ni et al., 2006). These studies together indicate that reciprocal regulations between Aβ and β_2AR may potentially form a reinforcing pathological cycle as the activation of β_2AR increases Aβ production, and the increased Aβ levels in turn stimulate β_2AR signaling to induce synaptic dysfunction in AD (Fig. 1).

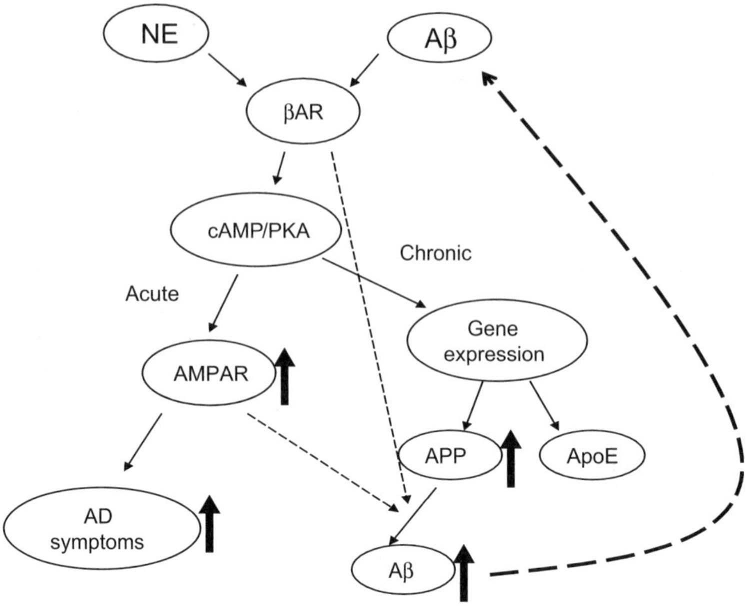

FIGURE 1 Working models on the reciprocal interaction between CNS adrenergic system and amyloid β peptide. Both norepinephrine (NE) and Aβ activate β_2 adrenergic receptor, which leads to cAMP and PKA activity in CNS. Acute PKA activities promote phosphorylation and activities of α-amino-2,3-dihydro-5-methyl-3-oxo-4-isoxazolepropanoic acid receptor (AMPAR), contributing to psychological stress-dependent symptoms in AD. Chronic PKA activities promote gene expression of amyloid precursor protein (APP) and apolipoprotein E (ApoE). In addition, activation of β_2 adrenergic receptor and AMPAR activity (dash line) promote amyloidogenic APP process by increasing activities of γ-secretases.

III. ADRENERGIC SIGNALING AND APP EXPRESSION

APP belongs to a family of genes encoding type I transmembrane proteins, with a single membrane-spanning domain, a large extracellular N-terminal region, and a shorter cytoplasmic C-terminal region (Kang et al., 1987). There are three major isoforms of mammalian APP: APP695, APP751, and APP770 (Wasco, Bupp, Magendantz, Gusella, Tanzi, & Solomon, 1992; Wasco et al., 1993). The function of APP and its homologues are under intensive study since the characterization of APP as the precursor of Aβ more than 20 years ago (Kang et al., 1987). Numerous studies *in vivo* and *in vitro* have yielded evidence for different roles of APP in the developing and adult nervous systems. APPs are involved in cell adhesion, neuronal survival, neurite outgrowth, synaptogenesis, vesicle trafficking, and neuronal migration (Jacobsen & Iverfeldt, 2009). They also modulate synaptic plasticity and insulin and glucose homeostasis (Jacobsen & Iverfeldt, 2009). The expression of APP is usually higher in CNS than in peripheral tissues (Lahiri & Ge, 2004). The levels of APP expression display regional and temporal patterns in CNS with highest expression observed between 2 and 3 weeks after birth, suggesting potential roles of APP in development (Jacobsen & Iverfeldt, 2009). Studies have also shown that different signaling pathways can stimulate the expression of APP in adulthood, which may be linked to amyloid pathogenesis (Lee, Araki, & Wurtman, 1997; Rossner, Ueberham, Schliebs, Perez-Polo, & Bigl, 1998a, 1998b; Lee, Knapp, & Wurtman, 1999).

Activation of βAR induces coupling to Gs protein, which turns on adenylyl cyclases to increase cAMP for subsequent PKA activation. Stimulation of cAMP/PKA activities with norepinephrine or isoproterenol for 24 h increases both APP mRNA and holoprotein levels in fibroblast cells and primary astrocytes, and these increases are blocked by the β-adrenergic antagonist propranolol (Lee et al., 1997). In addition, stimulation of βAR leads to activation of MAPK, which also contributes to expression in APP. Directly activating cAMP/PKA pathway with 8-bromo-adenosine 3′,5′-cyclic monophosphate or forskolin for 24 h similarly increases APP holoprotein levels (Lee et al., 1997). The same stimulation also induces transformation and activation of astrocytes with increased expression of glial fibrillary acidic protein, suggesting that these cultured astrocytes resemble reactive astrocytes found *in vivo* (Lee et al., 1997, 1999). Similar to adrenergic stimulation, activation of prostaglandin E2 (PGE2) receptors increases cAMP formation and stimulates expression of APP mRNA and holoprotein as well as glial fibrillary acidic protein in primary cortical astrocytes. These results suggest that catecholamines and prostaglandins produced by brain injury or inflammation can activate APP transcription in astrocytes for the pathophysiological processes underlying AD (Lee, Jimenez, Cox, & Wurtman, 1996; Lee et al., 1999; Lee & Wurtman, 2000). Besides neurotransmitters and hormones, neurotrophins that stimulate growth, differentiation, and survival of neurons also promote APP expression in brain tissues. Nerve growth factor (NGF) activates the high affinity receptor TrkA to regulate APP transcription, whereas NGF activates the low affinity receptor p75 to regulate APP translation in brain tissues (Rossner et al., 1998a, 1998b).

Among adrenergic receptors, several studies have indicated the links of different polymorphisms with AD. Genetically, two polymorphisms in the β_2AR gene, Gly16Arg and Gln27Glu, have been linked to an increased risk of sporadic late-onset AD in a Han Chinese population (Yu et al., 2008). Both 16Gly and 27Glu polymorphisms promote stronger cAMP/PKA activation (Johnson, 1998; Oostendorp et al., 2005), and are associated with an increased risk of late onset AD. The Gly16/Glu27 haplotype displays an even stronger increase in the risk of late-onset AD. Moreover, they also show a highly significant interaction with the apolipoprotein E gene epsilon4 allele, a gene with high risk factor to AD (Yu et al., 2008). In another study with a case-control sample of AD, two functional polymorphisms in β_1AR (ADRB1 C) and G protein β_3 subunit (GNB3 T) genes, together produce a significant risk for AD, suggesting that the combined effect of both polymorphisms influences AD susceptibility (Bullido et al., 2004). Accordingly, the coexpression of GNB3 T and ADRB1 C alleles, compared with GNB3 C and ADRB1 G alone, produces higher cAMP levels and MAPK activation following adrenergic stimulation of transfected human cell lines. The co-expression of these alleles also promotes APP expression. These data again strongly indicate that the combination of G protein and adrenergic receptor polymorphisms produces AD susceptibility by changing responsiveness of the cell to adrenergic stimulation. In the case of β_2AR, it also interacts with the ApoE epsilon4 allele to markedly increase the risk of AD.

IV. ACTIVATION OF βAR PROMOTES AMYLOID PROCESS

Aβ peptides generated by the proteolysis of APP in the TGN/endosome compartments play an important role in the pathogenesis of AD. The proteolytic process is carried by two proteases β- and γ-secretases in the secretory pathways. In addition, APP can also be processed by α-secretases, which yields nonamyloidogenic fragments. The nature and regulation of APP process by different enzymes is not entirely clear. The process of APP to Aβ can be promoted in raft microdomain since γ-secretases are enriched with flotillin, a lipid rafts marker (Ehehalt, Keller, Haass, Thiele, & Simons, 2003; Abad-Rodriguez et al., 2004). Optimal activity of β-secretases at pH4.5 *in vitro* implicates acidic organelles such as endosome as major sites of β-secretase cleavage of APP. The efficiency of APP processing is also greatly affected by its subcellular trafficking, and therefore the regulators of intracellular trafficking and subcellular localization of APP and the secretases have been extensively examined. Indeed, endocytosis of APP has been shown to be critical for Aβ production both in cultured cells and *in vivo* (Koo & Squazzo, 1994; Cirrito et al., 2008). In transfected non-neuronal cells such as CHO and HEK293 cells, Aβ is mainly generated in the TGN and endosomes as APP is trafficked through the secretory and recycling pathways, consistent with predominant localization of β- and γ-secretases. Accordingly, synaptic activation increases transportation of membrane proteins between cell surface and endosome/TGN compartments, and enhances the process of APP (Lesne et al., 2005; Cirrito et al., 2008; Hoey, Williams, & Perkinton, 2009).

Aβ is constantly produced in brain and cerebrospinal fluid (CSF), which is also rapidly metabolized (Li et al., 2009). The levels of cerebrospinal fluid (CSF) Aβ levels display a 24 h clock cycle (Kang et al., 2009). Despite that the extensive characterization of biological processes of APP to yield Aβ, the physiological implication of the peptide production is not clear. The levels of CSF Aβ are also increased in aged population (Lambert et al., 2009; Robakis, 2009). Under stressful condition, the CSF Aβ level is significantly increased (Yu et al., 2010). This is consistent with the hypothesis that environment factors like stress, inflammation, cardiovascular diseases, and metabolic disorders are true causal risk factors driving the pathogenic amyloidogenic processes leading to the development of AD. βARs are activated during stress, exerting response to norepinephrine released in CNS to influence different aspects of cognitive function. A recent study in a mouse model shows that acute restraint stress for 2 weeks significantly increases Aβ peptides production in the brain (Yu et al., 2010). This increase is completely blocked by injection of the β_2AR-selective antagonist ICI118551. Meanwhile, injection of the β_2AR-selective agonist clenbuterol hydrochloride enhances the production of acute stress-induced Aβ peptides production (Yu et al., 2010). The data suggested for the first time that stress induces abnormal activation of β_2AR for enhancing Aβ production, which may play a role in onset and/or pathogenesis of AD.

A recent study clearly illustrates the mechanism by which stimulation of adrenergic receptor leads to increased process of APP to yield Aβ. Norepinephrine can induce β_2AR-dependent activation of γ-secretases activation for APP process to yield Aβ in hippocampal neurons (Ni et al., 2006). Stimulation of β_2AR leads to transportation of the receptor and γ-secretases into late endosomal/lysosomal compartments, in which γ-secretases are optimally activated in acidic environment to process APP to yield Aβ peptide in hippocampal neurons. This event is mediated by activation of β_2AR in a G-protein activation-independent manner, but requires arrestin-dependent receptor internalization (Ni et al., 2006). Inhibition of both Gs and Gi signaling pathways does not affect the norepinephrine-induced APP process, however, inhibition of receptor internalization with sucrose and concanavalin A abolishes the stimulation of APP process by norepinephrine. Meanwhile, dominant-negative arrestin 3 V54D and dynamin K44E block the agonist-induced receptor internalization; they also block the activation of γ-secretases and process of APP. Moreover, in an AD mouse model in which APP is overexpressed, selective inhibition of β_2AR with ICI118551, but not inhibition of β_1AR with CGP20712A, significantly reduces the formation and size of amyloid plaques in brain tissues (Ni et al., 2006).

Studies have indicated that stimulation of other GPCRs can promote amyloidogenic process of APP *in vitro* and *in vivo* via similar mechanisms; these include delta opioid receptor (Teng, Zhao, Wang, Ma, & Pei, 2010), prostaglandin receptors EP2 and EP4 (Qin et al., 2003; Hoshino et al., 2007, 2009), the serotonin receptor HTR2C (Nitsch, Deng, Growdon, & Wurtman, 1996), and orphan G protein-coupled receptor 3 (GPR3) (Thathiah et al., 2009). These data indicate that amyloid genesis is broadly associated with a variety of neurohormonal stimuli in CNS, including adrenergic stress and inflammation. In the case of PGE2, a strong inducer of inflammation, it stimulates the production of Aβ through EP2 and EP4 receptors (Qin et al., 2003; Hoshino et al., 2007, 2009). Interestingly, inhibitors of adenylyl cyclase and PKA suppress EP2, not EP4, receptor-mediated stimulation of the Aβ production. In contrast, activation of EP4 receptor leads to activation of γ-secretases via the agonist-induced and receptor-mediated internalization. Similarly, activation of delta-opioid receptor (DOR) leads to internalization of corresponding receptor together with PS-1, a catalytic subunit of γ-secretases (Teng et al., 2010). In all cases, the internalized PS-1 displays colocalization with internalized GPCR in late endosome and lysosome, which promotes the process of APP by γ-secretases. Disruption of endocytosis of GPCRs also retards the endocytosis of BACE1 and γ-secretases, and the receptor-mediated stimulation of Aβ production (Qin et al., 2003; Hoshino et al., 2007, 2009; Teng et al., 2010). In another study, the authors have screened a library of adenoviruses encoding nearly 2000 potential drug targets for modulators of Aβ production, and found GPR3 as a novel regulator of Aβ secretion in HEK293 cells expressing APP (Thathiah et al., 2009). The *GPR3* gene has been mapped to a chromosomal locus that is associated with increased risk for AD and its expression modulates activities of γ-secretases via promoting formation and cell-surface localization of the mature γ-secretase complex (Thathiah et al., 2009). GPR3 is elevated in the sporadic AD brain, whereas genetic ablation of GPR3 prevents accumulation of Aβ *in vitro* and *in vivo* of an AD mouse model (Thathiah et al., 2009). Thus, GPR3 represents a potential therapeutic target for the treatment of AD.

The ubiquitous effects of different GPCRs on activation of PS-1 are dependent on the association of these receptors to γ-secretases (Hoshino et al., 2009; Teng et al., 2010). Physical interaction has been revealed between EP4 receptor and PS-1, and between DOR and BACE1 and γ-secretases by immunoprecipitation assays (Hoshino et al., 2009; Teng et al., 2010). It is not clear whether individual receptors form direct protein–protein interaction with γ-secretases and neither is it clear how activation of GPCRs induces the internalization of γ-secretases. One possible scenario is that both GPCRs and γ-secretases are enriched in lipid rafts domains, which are cointernalized together when the GPCRs are stimulated by ligands. If this is true, the regulation of amyloidogenic process of APP would not be specific to individual GPCR activation, which further argues the heterogeneity of sporadic late-onset AD and the complexity in targeting and treatment strategies in individual AD patients. Nevertheless, the GPCR-induced proteolytic activities are specific for processing of APP, but not for Notch, N-cadherin, or APLP (Thathiah et al., 2009; Teng et al., 2010). Together, these data indicates that a variety of GPCR signaling under different pathophysiological condition could selectively modulate activities of γ-secretases in CNS to enhance processing of APP, but not affecting the process of other γ-secretase substrates such as Notch (Thathiah et al., 2009; Teng et al., 2010). These studies together not only uncover molecular mechanisms for the formation of different GPCR/secretase complexes that regulate the specificity of secretase for Aβ production but also suggests that intervention of either formation or trafficking of these GPCR/secretase complexes could lead to a new strategy against AD, potentially with fewer side effects. This offers unique opportunities to modulate Aβ production in AD patients without affecting other physiological function of secretases, which is a common problem in therapies targeting the enzymes directly.

V. INTERACTION OF ADRENERGIC RECEPTOR WITH AMYLOID β PEPTIDE

Ecotopic application of Aβ induces a variety of intracellular signaling in neurons that modulates different cellular targets for distinct responses. Aβ induces activation of CaMKII, AKT, PKC, PKA, GSK3, JNK, and MAPK, and phosphatases such PP2 and STEP in either fibroblast cells or neurons (Thathiah & De Strooper, 2009). The substrates, such as AMPAR, NMDAR, and tau play a critical role in synaptic dysfunction, dendritic generation, as well as neuronal degeneration. In recent years, emerging evidence indicates that Aβ may activate cAMP/PKA signaling through β_2AR (Echeverria, Ducatenzeiler, Chen, & Cuello, 2005; Igbavboa, Johnson-Anuna, Rossello, Butterick, Sun, & Wood, 2006; Iijima-Ando et al., 2008). Our recent studies have revealed that β_2AR functions as a receptor for Aβ and activates the class G protein-dependent activation for a variety of cellular targets in neurons (Wang, Govindaiah, Liu, De Arcangelis, Cox, & Xiang, 2010).

Many GPCRs can bind to both small molecules and peptide ligands at different sites to modulate synaptic function (Neubig & Siderovski, 2002; Gainetdinov, Premont, Bohn, Lefkowitz, & Caron, 2004). We have recently found that β_2AR has a capacity to bind to $A\beta_{1\text{-}42}$ (Wang et al., 2010). Replacement of the N-terminus of β_2AR with that of β_1AR greatly decreases its binding to $A\beta_{1\text{-}42}$, indicating that the N-terminus of β_2AR is required for the binding, and different strategies are used by the peptide and small molecular ligands in binding to the receptor. βAR antagonist alprenolol does not block the binding, and Aβ does not compete with antagonist dihydroalprenolol in binding to β_2AR (Wang et al., 2010). These data together suggest that Aβ does not exclusively occupy the catechol-binding sites, but binds to β_2AR via an allosteric site that requires the N-terminus of the receptor. Although the monomer is more abundant than dimer in the $A\beta_{1\text{-}42}$ solution, β_2AR are mostly coimmunoprecipitated with $A\beta_{1\text{-}42}$ dimer (Wang et al., 2010). Since the oligomerization of $A\beta_{1\text{-}42}$ happens rather quickly, even during the SDS-PAGE (Chromy et al., 2003), it is therefore possible that β_2AR can bind to the monomer of $A\beta_{1\text{-}42}$.

Further study showed that the binding of $A\beta_{1\text{-}42}$ to β_2AR leads to a receptor-dependent G protein activation in a reconstituted β_2AR–G_s system (Whorton et al., 2007). Activation of β_2AR/Gs protein turns on adenylyl cyclases to increase cAMP/PKA activities, one of the key components of the β_2AR/Gs protein-induced signaling cascades. $A\beta_{1\text{-}42}$-induced cAMP accumulation is selectively blocked by βAR antagonist alprenolol, possibly indicating an allosteric antagonism of alprenolol on the $A\beta_{1\text{-}42}$ effect. The role of β_2AR in the effect of $A\beta_{1\text{-}42}$ on cAMP signaling is further confirmed with cells isolated from mice lacking the β_2AR gene (Wang et al., 2010). Accordingly, several other reports have shown that $A\beta_{1\text{-}42}$ but not $A\beta_{1\text{-}40}$ induces β_2AR-dependent cAMP signaling in PC12 cells (Echeverria et al., 2005), primary astrocytes (Igbavboa et al., 2006), and hippocampal neurons (Prapong, Uemura, & Hsu, 2001). Moreover, cAMP level increases in CSF of AD patients (Martinez, Fernandez, Frank, Guaza, de la Fuente, & Hernanz, 1999). Although alprenolol does not block the binding of $A\beta_{1\text{-}42}$ to β_2AR, it blocks the increase of $A\beta_{1\text{-}42}$-induced intracellular cAMP. Chekman and colleagues found that the binding of alprenolol to the receptor increases the entropy of receptor-membrane complex, which is energetically favorable for the inactive receptor conformation (Chekman, Budarin, Nurishchenko, Ivanov, & Pogrebnoi, 1985).

Upon binding to cAMP, PKA is activated to phosphorylate a variety of downstream substrates to regulate gene expression, cell differentiation, and function of receptors and ion channels (Prapong et al., 2001). Using a Fluorescence Resonance Energy Transfer (FRET)-based PKA activity indicator *AKAR2.2* (Zhang, Hupfeld, Taylor, Olefsky, & Tsien, 2005), we have found that Aβ increases the β_2AR-dependent instantaneous PKA activity in prefrontal cortical neurons and MEF cells (Wang et al., 2010).

In neurons, one of the primary functions of βAR is to modulate postsynaptic effects of synaptic transmission. Activation of cAMP/PKA signaling cascades can directly lead to phosphorylation of ion channels such as AMPA receptor and L-type calcium channel to increase synaptic transmission and long-term potentiation (LTP) in hippocampal neurons. Activation of β_2AR by Aβ induces an acute increase in PKA-dependent phosphorylation of GluR1 in prefrontal cortical neurons (Wang et al. 2010). Meanwhile, $A\beta_{1\text{-}42}$ also increase the phosphorylation of β_2AR at PKA-dependent site S262, further confirming the activation of β_2AR-cAMP/PKA signaling pathway (Wang et al., 2010). Moreover, β_2AR mediates acute $A\beta_{1\text{-}42}$ (60 s) treatment-induced increase in both mEPSC frequency and amplitude, and inhibition of PKA abolishes the A$\beta_{1\text{-}42}$-induced response in AMPAR activity (Wang et al., 2010), consistent with that PKA phosphorylation of GluR1 at S845 increases AMPAR opening probability (Banke, Bowie, Lee, Huganir, Schousboe, & Traynelis, 2000).

Interestingly, Aβ-induced PKA activities are modest, less than one-third of that induced by βAR agonist isoproterenol. The weak PKA activities induced by Aβ suggest that the effect may be localized. In agreement, a recent study shows that the locally produced soluble Aβ reduces dendritic spine number and neuroplasticity in the neurons within the range of 5–10 μM distance (Wei, Nguyen, Kessels, Hagiwara, Sisodia, & Malinow, 2009). Moreover, β_2AR forms a protein complex with AMPAR GluR1, containing PKA, AKAP150, which is a protein anchoring PKA holoenzyme to discrete locations in neurons, PP2A, which is a ubiquitously expressed serine/threonine phosphatase, and PSD-95, which is a specialized scaffold protein with multiple protein interaction domains that forms the backbone of extensive postsynaptic protein complexes at synaptic contact zone (Joiner et al., 2010; Wang et al., 2010). Further analysis shows that both the i3 and C-terminus of β_2AR are required for the formation of β_2AR–GluR1 complex (Wang et al., 2010), probably via binding to PSD-95 (Joiner et al., 2010). The complex between β_2AR/AMPAR supports a local impact of soluble Aβ on synaptic activity via a β_2AR-dependent signaling pathway.

PKA activation under adrenergic stimulation is involved in learning and memory consolidation in the hippocampus, mnemonic functions of the amygdala, and reward-motivated action and appetitive associative learning in the nucleus accumbens (Ramos, Birnbaum, Lindenmayer, Newton, Duman, & Arnsten, 2003; Arnsten, Ramos, Birnbaum, & Taylor, 2005). The observation of β_2AR activation by Aβ for phosphorylation of AMPAR supports that Aβ has a modulator role in synaptic transmission. Recently, it is shown that enhancing Aβ level increases synaptic activities in young mice (Abramov, Dolev, Fogel, Ciccotosto, Ruff, & Slutsky, 2009). In contrast, deletion of APP gene leads to reduced synaptic activity in mice brain (Abramov et al., 2009). Therefore, it is possible that Aβ modulates the CNS adrenergic system to influence synaptic function. However, unlike neurotransmitters, Aβ may not be rapidly catabolized at synapses. Therefore, under AD pathophysiology, an elevated Aβ level may lead to a persistent activation of β_2AR by Aβ for AMPAR hyperactivity. On the other hand, a persistent stimulation could also induce receptor desensitization and degradation, which functionally impairs normal neurotransmission following presynaptic release of the receptor's agonists.

Although the mechanism is not clear, Aβ is shown to activate angiotensin AT1R to induce Gq activation (Thathiah & De Strooper, 2009), which may contribute to cholinergic dysfunction. While Aβ has been reported to directively induce neurotoxicity, it also causes indirect neuronal damage by activation of microglia that induces inflammatory responses near amyloid plaques in the CNS. This is may be due to Aβ activation on FPR-1, a G-protein-coupled chemoattractant receptor. The activation of FPR-1 leads to accumulation and aggregation of microglia around the amyloid plaques, and facilitates inflammatory responses that damages neurons, induces synaptic and neuronal degeneration, and impairs learning and memory (Le et al., 2001; Tiffany et al., 2001; Yazawa et al., 2001). Aβ is also known to bind specifically to the scavenger receptors expressed in microglia and macrophages (El Khoury, Hickman, Thomas, Cao, Silverstein, & Loike, 1996; Paresce, Ghosh, & Maxfield, 1996) that may induce cell death by generating free radicals (Munch et al., 1998). Aβ also binds to the ubiquitous receptor for advanced glycation end products (RAGE) to promote Aβ clearance (Yan et al., 1996; Du Yan et al., 1997). However the mechanisms of these binding are not clear.

In comparison, oligomeric Aβ can also bind a variety of cellular and membrane proteins including α_7 nicotinic receptor (Wang, Lee, D'Andrea, Peterson, Shank, & Reitz, 2000) and glyceraldehyde phosphodehydrogenase (Verdier et al., 2008). Recently, a study shows that oligomeric Aβ binds to hippocampal neurons in a cellular prion protein (PrP^C) expression dependent fashion (Lauren, Gimbel, Nygaard, Gilbert, & Strittmatter, 2009). PrPs form very toxic beta sheet structures resistant to proteolytic digestion, and cause a lethal neurodegenerative Creutzfeldt-Jakob disease or mad cow disease (Prusiner, 1987). In other studies, oligomeric Aβ binds to hippocampal neurons at postsynaptic sites, which immobilizes mGluR5 through clustering of the receptors on distinct membrane sites (Renner et al., 2010). The immobilization of mGluR5 leads to increased intracellular calcium concentration and neurodegeneration (Renner et al., 2010). The ability of oligomeric Aβ in binding different proteins may be relevant to their β-sheet structure, which enhances hydrophobic protein–protein interactions.

VI. CNS ADRENERGIC SYSTEM IN AD

The central noradrenergic system projected from the locus coeruleus (LC) is thought to determine the brain's global orientation concerning events in the external world and within the viscera (Cooper, Bloo, & Roth, 1991). It plays a critical role in arousal, which is important in regulating consciousness, attention, and information processing, and is crucial for promoting certain behaviors such as mobility, learning, and pursuit of food (Csikszentmihalyi, 1997). Neurotransmitter norepinephrine is released from adrenergic neurons projected from the LC into different brain regions and tissues, and activates noradrenergic receptors throughout CNS. Noradrenergic receptors are classified in three subfamilies (α_1, α_2, and β), and with three genes in each subfamily (α_{1A}, α_{1B}, α_{1C},

α_{2A}, α_{2B}, α_{2C}, β_1, β_2, and β_3) (Ramos & Arnsten, 2007). βARs are widely distributed in different tissues, including the frontal, parietal, piriform, and retrosplenial cortices, medial septal nuclei, olfactory tubercle, midbrain, striatum, amygdala, hippocampus, and thalamic nuclei (Asanuma, Ogawa, Mizukawa, Haba, Hirata, & Mori, 1991), regulating working memory and other basic brain functions (Satler, Garrido, Sarmiento, Leme, Conde, & Tomaz, 2007; Ramos, Colgan, Nou, & Arnsten, 2008; Wang et al., 2010).

Clinically studies have long pointed out alterations of CNS noradrenergic system in AD patients (Raskind, Peskind, Holmes, & Goldstein, 1999b). Significant losses of LC neurons are observed in AD patients in comparison to healthy adults (Hoogendijk, Pool, Troost, van Zwieten, & Swaab, 1995). However, other studies have indicated that CSF norepinephrine levels are significantly increased in AD patients (Raskind, Peskind, Halter, & Jimerson, 1984). This difference may be explained by the hyperactivity of remaining neurons to compensate the losses of neurons in the LC (Hoogendijk et al., 1999). Studies have also shown alterations of noradrenergic receptor expression in AD brain tissues. The βAR density in the LC projection areas is increased in AD postmortem brain tissue (Shimohama, Taniguchi, Fujiwara, & Kameyama 1987; Kalaria, Andorn, Tabaton, Whitehouse, Harik, & Unnerstall, 1989). In another study, increased βAR density in AD cerebellar cortex tissue was specific to a subgroup of AD subjects with an antemortem history of agitation, aggression, and disruptive behaviors (Russo-Neustadt & Cotman, 1997). In comparison, no significant differences were found in adrenergic receptor concentrations within the frontal cortex or hypothalamus. Meanwhile, aggressive AD patients had markedly increased (by approximately 70%) concentrations of α_2ARs in the cerebellar cortex compared with nonaggressive patients with similar levels of cognitive deficit (Russo-Neustadt & Cotman, 1997). AD subjects also manifest substantially greater agitation following α_2ARs antagonist yohimbine than did healthy old or young subjects (Peskind et al., 1995). Because the increases of CNS norepinephrine following yohimbine did not differ between healthy old subjects and AD subjects, these findings suggest an increased sensitivity to the activation of CNS norepinephrine, which is consistent with that CNS βARs may be responsible for the increased behavioral sensitivity (agitation) to noradrenergic stimulation in AD (Peskind et al., 1995, 1998). Together, these data indicates that at least, CNS adrenergic system contributes to agitation, aggression, and disruptive behaviors associated with AD.

Consistent with the effects of Aβ on activation of βARs and AT1Rs, recent studies suggest that receptor function in AD may be compromised due to disrupted postreceptor signal transduction, in particular in the G-protein regulated phosphoinositide hydrolysis and adenylyl cyclase pathways (Cowburn, O'Neill, Bonkale, Ohm, & Fastbom, 2001). In postmortem cerebral cortex tissues from AD, βAR-mediated activation of G proteins is significantly reduced despite the intact expression and function of the G alpha proteins in AD brains. In addition, the stimulation of muscarinic cholinergic receptors is also reduced in brain tissues from AD patients. This leads to impaired agonist and G-protein regulation of phospholipase C, decreased protein kinase C (PKC) levels and activity (Cowburn et al., 2001). Although the mechanism is not clear, a chronic activation of β_2AR (Wang et al., 2010) and AT1R (Thathiah & De Strooper, 2009) as well as the alteration of mGluR5 distribution (Renner et al., 2010) may contribute to the disruption of receptor/G-protein regulation of adenylyl cyclase and PLC–PKC pathways.

VII. CLINIC IMPLICATION OF ACTIVATION OF DOWNSTREAM TARGETS BY Aβ/βAR SIGNALING AXIS IN CNS

Under emotional stress, adrenergic activation increases PKA phosphorylation of GluR1 (Satler et al., 2007) for ionotropic GluR1 AMPAR activation (Banke et al., 2000). In AD patients, there are stress-dependent symptoms, for example, anxiety, irritability, agitation, and aggression, which involve the activation of noradrenergic and glutamatergic systems (Russo-Neustadt & Cotman, 1997; Kapus et al., 2008; Francis, 2009). Recent studies show that AMPAR contributes to the hyperactivities observed near amyloid plaques in cortical tissues of AD animal models. These spontaneous activities occur at the proximity of the amyloid plaque and decrease with increasing distance. The spontaneous activities also facilitate the activation waves from the plaques into distance. These observations have spurred new theories on the imbalance of CNS neuronal network activities that may explain epilepsy and seizure observed in AD patients. Therefore, a chronic activation of β_2AR and AMPAR by Aβ in AD may contribute to AMPAR-dependent hyperactivities in cortical tissues, thus linking to common psychiatric symptoms such as agitation, aggression, and anxiety in AD patients (Kornischka, Cordes, & Agelink, 2007; Vekovischeva, Aitta-aho, Verbitskaya, Sandnabba, & Korpi, 2007). While directly blocking AMPAR activities poses significant risk of compromising other CNS behaviors and normal physiology, increased behavioral sensitivity to CNS noradrenergic stimulation and increased brain tissue βAR binding in AD provide rationale for the hypothesis that a β-adrenergic antagonist that readily enters the CNS could reduce disruptive agitation in AD. Two uncontrolled trials of the β-adrenergic antagonist propranolol provide support for this hypothesis. Shankle reported that low dose propranolol (10–80 mg/day) reduced agitation and aggression in AD outpatients with minimal adverse effects (Shankle, Nielson, & Cotman, 1995). Weiler reported that propanolol in doses ranging from 60 to 560 mg/day decreases agitation in all six dementia patients treated in a nursing home (Weiler,

Mungas, & Bernick, 1988). These preliminary clinical studies suggest that the β-adrenergic antagonist propanolol reduces disruptive agitation in AD and is well tolerated. This is true even in persons over age 80 and has prompted recommendations for broader use of β antagonists in late life (Gottlieb, McCarter, & Vogel, 1998). In a pilot placebo-controlled propranolol study, very elderly (mean age 84, range 73–101 years) nursing home residents with disruptive agitation were randomized to propranolol (mean dose 108 mg/day) or placebo. Propranolol was significantly more effective than placebo for reducing agitation and was well tolerated (Raskind et al., 1999a, 1999b).

On the other hand, chronic stress is associated with hippocampal damage and impaired memory in animals and humans (Esteban, Shi, Wilson, Nuriya, Huganir, & Malinow, 2003). Several recent clinical studies have proved that the clinical application of β-blockers is associated with a delayed rate of functional decline (Rosenberg et al., 2008), and a decreased incidence of AD pathogenesis (Khachaturian et al., 2006). In one study of about 800 older Catholic clergy members, people who often feel negative emotions are twice as likely to develop AD compared to those least prone to negative emotions (Esteban et al., 2003).

VIII. ADRENERGIC SYSTEM IN PERIPHERAL TISSUES

βARs widely expressed in peripheral tissues are involved in regulation of cardiovascular functions, pulmonary function, metabolism, immune systems. They also facilitate fibroblast growth and secretion of a host of growth factors and cytokines. Both hypothalamus–pituitary–adrenal (HPA) axis and sympathetic ganglia nerves can modulate the peripheral adrenergic system. Studies have linked both AD and affective disorders to reduced serotonergic (5-HT) activity and hyperactivity of the HPA axis (de Leon et al., 1988; O'Brien, Ames, Schweitzer, Colman, Desmond, & Tress, 1996). In AD patients, the hyperactive HPA axis may contribute to constant stimulation of βAR in peripheral tissues. Indeed, in comparison to aged healthy controls, AD patients show a significantly reduced response to the adrenergic agonist when using iontophoresis to measure cutaneous responses (erythema and blanching) (Hornqvist, Henriksson, Back, Bucht, & Winblad, 1987). In contrast, the methacholine response did not differ between AD and age-matched controls. Thus, a reduced peripheral adrenergic reactivity is selective in AD patients (Hornqvist et al., 1987).

Later, adrenergic signaling was also evaluated with skin fibroblast isolated from AD patients. The β-adrenergic-stimulated increase in cAMP in fibroblasts from AD was reduced approximately to 80% of that from age-matched controls (Huang & Gibson, 1993). Further analysis reveals that function of downstream signaling components such as Gs protein and adenylyl cyclases are not affected in AD fibroblasts (Huang & Gibson, 1993). The results suggest either that the coupling of the G-protein(s) to β-adrenergic receptor is abnormal or that the sensitivity of receptor is altered with AD. Similar observations of desensitized adrenergic responses have also been made on peripheral blood lymphocytes in patients (Garlind et al., 1997) and cardiac tissues in AD animal models (Turdi et al., 2009). These studies indicate that there is a widespread disruption of β-adrenoceptor–G-protein–enzyme coupling in different tissues from AD patients, and that adenylyl cyclase disturbances previously reported in AD brain do not occur as a consequence of disease pathology or of terminal illness (Bonkale, Fastbom, Wiehager, Ravid, Winblad, & Cowburn, 1996; Garlind et al., 1997). The impaired adrenergic signaling is probably in part due to the elevated HPA axis activity. Alternatively, since Aβ is also generated in peripheral tissues, a constantly elevated plasma Aβ level could also contribute to βAR desensitization in peripheral tissues. Moreover, chronic stimulation of βAR can lead to insulin insensitivity; thus chronic stimulation of βARs under elevated plasma catecholamines and/or Aβ may contribute to insulin resistance commonly observed in AD patients (Sabayan, Foroughinia, Mowla, & Borhanihaghighi, 2008).

IX. POTENTIAL DRUG TARGETS ON Aβ-β_2AR SIGNALING AXIS FOR AD THERAPY

Mounting evidence now indicates that soluble oligomers of Aβ are the main neurotoxins that lead to early neuronal dysfunction and memory deficits in AD. The unique requirement of the N-terminal domain in binding of Aβ to β_2AR may offer opportunities to screen ligands and drugs specifically blocking the peptide binding to the receptor with minimal effect on catecholamine binding. It is an essential consideration that β-adrenergic activity can enhance LTP and memory, and potential depression and memory impairments may associate with chronic β-blocker therapies (Ko, Hebert, Coffey, Sedrakyan, Curtis, & Krumholz, 2002; van Melle & de Jonge, 2009). While most orthosteric βAR ligands bind to the intratransmembrane domain pocket involving the TM3, TM5, and TM6, the nature of Aβ binding to the receptor is still not clear. Aβ does not compete against catecholamines in binding to the β_2AR and essentially rules out the possibility that the peptide may interact with the orthosteric pocket in β_2AR with the help of the N-terminus. More likely, the peptide binding requires both the N-terminal domain and the groove formed from the TM3, 5, and 6 on the extracellular surface (Rasmussen et al., 2007), which may together form a binding pocket for the peptide. It is therefore important to probe the structural–functional properties of the peptide binding to β_2AR. Unlike traditional β_2AR antagonists that block the catecholamine stimulation and β_2AR-mediated Aβ pathological effects in CNS, a compound selectively blocking the peptide binding may have minimal effect on catecholamine stimulation. Such a component will be desirable to abolish the pathological effect of Aβ-β_2AR axis for treatment of AD as well as AD-associated anxiety symptoms.

Furthermore, the impaired CNS noradrenergic system also suggests that targeting the downstream cAMP pathway may be a potential effective strategy in treating AD (De Felice, Wasilewska-Sampaio, Barbosa, Gomes, Klein, & Ferreira, 2007). cAMP is a central component of intracellular signaling pathways that regulate a wide range of biological functions, including memory. Notably, cAMP triggers the phosphorylation and activation of the cAMP-responsive element binding protein (CREB), a transcription factor that regulates the expression of genes that are important for long-term memory (De Felice et al., 2007).

X. CONCLUSION AND FUTURE DIRECTION

Recent discovery of the interaction between Aβ and β_2AR of CNS noradrenergic system offer a new approach to dissect (patho-)physiological effects of Aβ *in vivo*. Since Aβ is not metabolized and taken up by neurons like norepinephrine, the stimulation of β_2AR by Aβ modulates synaptic activities quite differently from that by norepinephrine. This may play a critical role in initiating a cascade of events that eventually lead to the onset of AD pathogenesis under specific conditions. Several critical questions remain to be addressed. Is there a physiological implication of the activation of β_2AR by Aβ? How do (nor-) epinephrine and Aβ cooperate under chronic stresses or with elevated (nor-) epinephrine levels associated with cardiovascular and metabolic disorders? At which stage does the activation of β_2AR by Aβ intervene in onset and development of sporadic AD? A detailed structural analysis of the Aβ receptor binding is also needed for finding medicines targeting Aβ and central noradrenergic system for the purpose of preventing AD development and treating its psychiatric symptoms.

Acknowledgments

This study is supported by NIH grant and an Alzheimer's association NIRG grant to Yang K. Xiang. Dayong Wang is a recipient of an award from the Illinois Department of Public Health and a NARSAD young investigator award.

References

Abad-Rodriguez, J., Ledesma, M. D., Craessaerts, K., Perga, S., Medina, M., & Delacourte, A., et al. (2004). Neuronal membrane cholesterol loss enhances amyloid peptide generation. *J Cell Biol, 167*(5), 953–960.

Abramov, E., Dolev, I., Fogel, H., Ciccotosto, G. D., Ruff, E., & Slutsky, I. (2009). Amyloid-beta as a positive endogenous regulator of release probability at hippocampal synapses. *Nat Neurosci, 12*(12), 1567–1576.

Arnsten, A. F., Ramos, B. P., Birnbaum, S. G., & Taylor, J. R. (2005). Protein kinase A as a therapeutic target for memory disorders: Rationale and challenges. *Trends Mol Med, 11*(3), 121–128.

Asanuma, M., Ogawa, N., Mizukawa, K., Haba, K., Hirata, H., & Mori, A. (1991). Distribution of the beta-2 adrenergic receptor messenger RNA in the rat brain by in situ hybridization histochemistry: Effects of chronic reserpine treatment. *Neurochem Res, 16*(12), 1253–1256.

Banke, T. G., Bowie, D., Lee, H., Huganir, R. L., Schousboe, A., & Traynelis, S. F. (2000). Control of GluR1 AMPA receptor function by cAMP-dependent protein kinase. *J Neurosci, 20*(1), 89–102.

Bonkale, W. L., Fastbom, J., Wiehager, B., Ravid, R., Winblad, B., & Cowburn, R. F. (1996). Impaired G-protein-stimulated adenylyl cyclase activity in Alzheimer's disease brain is not accompanied by reduced cyclic-AMP-dependent protein kinase A activity. *Brain Res, 737* (1–2), 155–161.

Bullido, M. J., Ramos, M. C., Ruiz-Gomez, A., Tutor, A. S., Sastre, I., & Frank, A., et al. (2004). Polymorphism in genes involved in adrenergic signaling associated with Alzheimer's. *Neurobiol Aging, 25*(7), 853–859.

Chekman, I. S., Budarin, L. I., Nurishchenko, N. E., Ivanov, L. V., & Pogrebnoi, P. V. (1985). Interaction of beta-adrenergic agents with the plasma membranes of the myocardium. *Biull Eksp Biol Med, 100*(8), 208–210.

Chromy, B. A., Nowak, R. J., Lambert, M. P., Viola, K. L., Chang, L., & Velasco, P. T., et al. (2003). Self-assembly of Abeta(1-42) into globular neurotoxins. *Biochemistry, 42*(44), 12749–12760.

Cirrito, J. R., Kang, J. E., Lee, J., Stewart, F. R., Verges, D. K., & Silverio, L. M., et al. (2008). Endocytosis is required for synaptic activity-dependent release of amyloid-beta *in vivo*. *Neuron, 58*(1), 42–51.

Cooper, J. R., Bloom, F. E., & Roth, R. H. (1991). The biochemical basis of neuropharmacology. Oxford UP, New York.

Cowburn, R. F., O'Neill, C., Bonkale, W. L., Ohm, T. G., & Fastbom, J. (2001). Receptor-G-protein signalling in Alzheimer's disease. *Biochem Soc Symp,* (67):163–175.

Crystal, H., Dickson, D., Fuld, P., Masur, D., Scott, R., & Mehler, M., et al. (1988). Clinicopathologic studies in dementia: nondemented subjects with pathologically confirmed Alzheimer's disease. *Neurology, 38*(11), 1682–1687.

Csikszentmihalyi, M. (1997). Finding flow: The psychology of engagement with everyday life. BasicBooks, New York.

De Felice, F. G., Wasilewska-Sampaio, A. P., Barbosa, A. C., Gomes, F. C., Klein, W. L., & Ferreira, S. T. (2007). Cyclic AMP enhancers and Abeta oligomerization blockers as potential therapeutic agents in Alzheimer's disease. *Curr Alzheimer Res, 4*(3), 263–271.

de Leon, M. J., McRae, T., Tsai, J. R., George, A. E., Marcus, D. L., & Freedman, M., et al. (1988). Abnormal cortisol response in Alzheimer's disease linked to hippocampal atrophy. *Lancet, 2* (8607), 391–392.

Dewachter, I., Reverse, D., Caluwaerts, N., Ris, L., Kuiperi, C., & Van den Haute, C., et al. (2002). Neuronal deficiency of presenilin 1 inhibits amyloid plaque formation and corrects hippocampal long-term potentiation but not a cognitive defect of amyloid precursor protein [V717I] transgenic mice. *J Neurosci, 22*(9), 3445–3453.

Du Yan, S., Zhu, H., Fu, J., Yan, S. F., Roher, A., & Tourtellotte, W. W., et al. (1997). Amyloid-beta peptide-receptor for advanced glycation endproduct interaction elicits neuronal expression of macrophage-colony stimulating factor: A proinflammatory pathway in Alzheimer disease. *Proc Natl Acad Sci USA, 94*(10), 5296–5301.

Echeverria, V., Ducatenzeiler, A., Chen, C. H., & Cuello, A. C. (2005). Endogenous beta-amyloid peptide synthesis modulates cAMP response element-regulated gene expression in PC12 cells. *Neuroscience, 135*(4), 1193–1202.

Ehehalt, R., Keller, P., Haass, C., Thiele, C., & Simons, K. (2003). Amyloidogenic processing of the Alzheimer beta-amyloid precursor protein depends on lipid rafts. *J Cell Biol, 160*(1), 113–123.

El Khoury, J., Hickman, S. E., Thomas, C. A., Cao, L., Silverstein, S. C., & Loike, J. D. (1996). Scavenger receptor-mediated adhesion of microglia to beta-amyloid fibrils. *Nature, 382*(6593), 716–719.

Esteban, J. A., Shi, S. H., Wilson, C., Nuriya, M., Huganir, R. L., & Malinow, R. (2003). PKA phosphorylation of AMPA receptor subunits controls synaptic trafficking underlying plasticity. *Nat Neurosci, 6*(2), 136–143.

Francis, P. T. (2009). Altered glutamate neurotransmission and behaviour in dementia: Evidence from studies of memantine. *Curr Mol Pharmacol, 2*(1), 77–82.

Gainetdinov, R. R., Premont, R. T., Bohn, L. M., Lefkowitz, R. J., & Caron, M. G. (2004). Desensitization of G protein-coupled receptors and neuronal functions. *Annu Rev Neurosci, 27*, 107–144.

Garlind, A., Johnston, J. A., Algotsson, A., Winblad, B., & Cowburn, R. F. (1997). Decreased beta-adrenoceptor-stimulated adenylyl cyclase activity in lymphocytes from Alzheimer's disease patients. *Neurosci Lett, 226*(1), 37–40.

Giannakopoulos, P., Herrmann, F. R., Bussiere, T., Bouras, C., Kovari, E., & Perl, D. P., et al. (2003). Tangle and neuron numbers, but not amyloid load, predict cognitive status in Alzheimer's disease. *Neurology, 60*(9), 1495–1500.

Goedert, M., & Spillantini, M. G. (2006). A century of Alzheimer's disease. *Science, 314*(5800), 777–781.

Gottlieb, S. S., McCarter, R. J., & Vogel, R. A. (1998). Effect of beta-blockade on mortality among high-risk and low-risk patients after myocardial infarction. *N Engl J Med, 339*(8), 489–497.

Hardy, J., & Selkoe, D. J. (2002). The amyloid hypothesis of Alzheimer's disease: Progress and problems on the road to therapeutics. *Science, 297*(5580), 353–356.

Hartley, D. M., Walsh, D. M., Ye, C. P., Diehl, T., Vasquez, S., & Vassilev, P. M., et al. (1999). Protofibrillar intermediates of amyloid beta-protein induce acute electrophysiological changes and progressive neurotoxicity in cortical neurons. *J Neurosci, 19*(20), 8876–8884.

Hoey, S. E., Williams, R. J., & Perkinton, M. S. (2009). Synaptic NMDA receptor activation stimulates alpha-secretase amyloid precursor protein processing and inhibits amyloid-beta production. *J Neurosci, 29*(14), 4442–4460.

Hoogendijk, W. J., Feenstra, M. G., Botterblom, M. H., Gilhuis, J., Sommer, I. E., & Kamphorst, W., et al. (1999). Increased activity of surviving locus ceruleus neurons in Alzheimer's disease. *Ann Neurol, 45*(1), 82–91.

Hoogendijk, W. J., Pool, C. W., Troost, D., van Zwieten, E., & Swaab, D. F. (1995). Image analyser-assisted morphometry of the locus coeruleus in Alzheimer's disease, Parkinson's disease and amyotrophic lateral sclerosis. *Brain, 118*(Pt 1), 131–143.

Hornqvist, R., Henriksson, R., Back, O., Bucht, G., & Winblad, B. (1987). Iontophoretic study of adrenergic and cholinergic skin vessel reactivity in normal ageing and Alzheimer's disease. *Gerontology, 33*(6), 374–379.

Hoshino, T., Nakaya, T., Homan, T., Tanaka, K., Sugimoto, Y., & Araki, W., et al. (2007). Involvement of prostaglandin E2 in production of amyloid-beta peptides both *in vitro* and *in vivo*. *J Biol Chem, 282*(45), 32676–32688.

Hoshino, T., Namba, T., Takehara, M., Nakaya, T., Sugimoto, Y., & Araki, W., et al. (2009). Prostaglandin E2 stimulates the production of amyloid-beta peptides through internalization of the EP4 receptor. *J Biol Chem, 284*(27), 18493–18502.

Huang, H. M., & Gibson, G. E. (1993). Altered beta-adrenergic receptor-stimulated cAMP formation in cultured skin fibroblasts from Alzheimer donors. *J Biol Chem, 268*(20), 14616–14621.

Igbavboa, U., Johnson-Anuna, L. N., Rossello, X., Butterick, T. A., Sun, G. Y., & Wood, W. G. (2006). Amyloid beta-protein1-42 increases cAMP and apolipoprotein E levels which are inhibited by beta1 and beta2-adrenergic receptor antagonists in mouse primary astrocytes. *Neuroscience, 142*(3), 655–660.

Iijima-Ando, K., Hearn, S. A., Granger, L., Shenton, C., Gatt, A., & Chiang, H. C., et al. (2008). Overexpression of neprilysin reduces alzheimer amyloid-beta42 (Abeta42)-induced neuron loss and intraneuronal Abeta42 deposits but causes a reduction in cAMP-responsive element-binding protein-mediated transcription, age-dependent axon pathology, and premature death in *Drosophila*. *J Biol Chem, 283*(27), 19066–19076.

Jacobsen, K. T., & Iverfeldt, K. (2009). Amyloid precursor protein and its homologues: A family of proteolysis-dependent receptors. *Cell Mol Life Sci, 66*(14), 2299–2318.

Johnson, M. (1998). The beta-adrenoceptor. *Am J Respir Crit Care Med, 158*(5 Pt 3), S146–S153.

Joiner, M. L., Lise, M. F., Yuen, E. Y., Kam, A. Y., Zhang, M., & Hall, D. D., et al. (2010). Assembly of a beta2-adrenergic receptor – GluR1 signalling complex for localized cAMP signalling. *Embo J, 29*(2), 482–495.

Kalaria, R. N., Andorn, A. C., Tabaton, M., Whitehouse, P. J., Harik, S. I., & Unnerstall, J. R. (1989). Adrenergic receptors in aging and Alzheimer's disease: Increased beta 2-receptors in prefrontal cortex and hippocampus. *J Neurochem, 53*(6), 1772–1781.

Kamenetz, F., Tomita, T., Hsieh, H., Seabrook, G., Borchelt, D., & Iwatsubo, T., et al. (2003). APP processing and synaptic function. *Neuron, 37*(6), 925–937.

Kang, J., Lemaire, H. G., Unterbeck, A., Salbaum, J. M., Masters, C. L., & Grzeschik, K. H., et al. (1987). The precursor of Alzheimer's disease amyloid A4 protein resembles a cell-surface receptor. *Nature, 325*(6106), 733–736.

Kang, J. E., Lim, M. M., Bateman, R. J., Lee, J. J., Smyth, L. P., & Cirrito, J. R., et al. (2009). Amyloid-beta dynamics are regulated by orexin and the sleep-wake cycle. *Science, 326*(5955), 1005–1007.

Kapus, G. L., Gacsalyi, I., Vegh, M., Kompagne, H., Hegedus, E., & Leveleki, C., et al. (2008). Antagonism of AMPA receptors produces anxiolytic-like behavior in rodents: Effects of GYKI 52466 and its novel analogues. *Psychopharmacol (Berl), 198*(2), 231–241.

Khachaturian, A. S., Zandi, P. P., Lyketsos, C. G., Hayden, K. M., Skoog, I., & Norton, M. C., et al. (2006). Antihypertensive medication use and incident Alzheimer disease: The Cache County Study. *Arch Neurol, 63*(5), 686–692.

Ko, D. T., Hebert, P. R., Coffey, C. S., Sedrakyan, A., Curtis, J. P., & Krumholz, H. M. (2002). Beta-blocker therapy and symptoms of depression, fatigue, and sexual dysfunction. *J Am Med Assoc, 288*(3), 351–357.

Koo, E. H., & Squazzo, S. L. (1994). Evidence that production and release of amyloid beta-protein involves the endocytic pathway. *J Biol Chem, 269*(26), 17386–17389.

Kornischka, J., Cordes, J., & Agelink, M. W. (2007). 40 years beta-adrenoceptor blockers in psychiatry. *Fortschr Neurol Psychiatr, 75*(4), 199–210.

Lahiri, D. K., & Ge, Y. W. (2004). Role of the APP promoter in Alzheimer's disease: Cell type-specific expression of the beta-amyloid precursor protein. *Ann N Y Acad Sci, 1030*, 310–316.

Lambert, J. C., Schraen-Maschke, S., Richard, F., Fievet, N., Rouaud, O., & Berr, C., et al. (2009). Association of plasma amyloid beta with risk of dementia: The prospective Three-City Study. *Neurology, 73*(11), 847–853.

Lauren, J., Gimbel, D. A., Nygaard, H. B., Gilbert, J. W., & Strittmatter, S. M. (2009). Cellular prion protein mediates impairment of synaptic plasticity by amyloid-beta oligomers. *Nature, 457* (7233), 1128–1132.

Le, Y., Gong, W., Tiffany, H. L., Tumanov, A., Nedospasov, S., & Shen, W., et al. (2001). Amyloid (beta)42 activates a G-protein-coupled chemoattractant receptor, FPR-like-1. *J Neurosci, 21*(2), RC123.

Lee, H. G., Casadesus, G., Zhu, X., Joseph, J. A., Perry, G., & Smith, M. A. (2004a). Perspectives on the amyloid-beta cascade hypothesis. *J Alzheimers Dis, 6*(2), 137–145.

Lee, H. G., Casadesus, G., Zhu, X., Takeda, A., Perry, G., & Smith, M. A. (2004 b). Challenging the amyloid cascade hypothesis: Senile plaques and amyloid-beta as protective adaptations to Alzheimer disease. *Ann N Y Acad Sci, 1019*, 1–4.

Lee, H. G., Zhu, X., Castellani, R. J., Nunomura, A., Perry, G., & Smith, M. A. (2007). Amyloid-beta in Alzheimer disease: The null versus the alternate hypotheses. *J Pharmacol Exp Ther, 321*(3), 823–829.

Lee, R. K., Araki, W., & Wurtman, R. J. (1997). Stimulation of amyloid precursor protein synthesis by adrenergic receptors coupled to cAMP formation. *Proc Natl Acad Sci USA, 94*(10), 5422–5426.

Lee, R. K., Jimenez, J., Cox, A. J., & Wurtman, R. J. (1996). Metabotropic glutamate receptors regulate APP processing in hippocampal neurons and cortical astrocytes derived from fetal rats. *Ann N Y Acad Sci, 777*, 338–343.

Lee, R. K., Knapp, S., & Wurtman, R. J. (1999). Prostaglandin E2 stimulates amyloid precursor protein gene expression: Inhibition by immunosuppressants. *J Neurosci, 19*(3), 940–947.

Lee, R. K., & Wurtman, R. J. (2000). Regulation of APP synthesis and secretion by neuroimmunophilin ligands and cyclooxygenase inhibitors. *Ann N Y Acad Sci, 920*, 261–268.

Lesne, S., Ali, C., Gabriel, C., Croci, N., MacKenzie, E. T., & Glabe, C. G., et al. (2005). NMDA receptor activation inhibits alpha-secretase and promotes neuronal amyloid-beta production. *J Neurosci, 25*(41), 9367–9377.

Li, G., Peskind, E. R., Millard, S. P., Chi, P., Sokal, I., & Yu, C. E., et al. (2009). Cerebrospinal fluid concentration of brain-derived neurotrophic factor and cognitive function in non-demented subjects. *PLoS One, 4*(5), e5424.

Martinez, M., Fernandez, E., Frank, A., Guaza, C., de la Fuente, M., & Hernanz, A. (1999). Increased cerebrospinal fluid cAMP levels in Alzheimer's disease. *Brain Res, 846*(2), 265–267.

Munch, G., Schinzel, R., Loske, C., Wong, A., Durany, N., & Li, J. J., et al. (1998). Alzheimer's disease – synergistic effects of glucose deficit, oxidative stress and advanced glycation end-products. *J Neural Transm, 105*(4–5), 439–461.

Neubig, R. R., & Siderovski, D. P. (2002). Regulators of G-protein signalling as new central nervous system drug targets. *Nat Rev Drug Discov, 1*(3), 187–197.

Ni, Y., Zhao, X., Bao, G., Zou, L., Teng, L., & Wang, Z., et al. (2006). Activation of beta2-adrenergic receptor stimulates gamma-secretase activity and accelerates amyloid plaque formation. *Nat Med, 12*(12), 1390–1396.

Nimmrich, V., Grimm, C., Draguhn, A., Barghorn, S., Lehmann, A., & Schoemaker, H., et al. (2008). Amyloid beta oligomers (A beta(1–42) globulomer) suppress spontaneous synaptic activity by inhibition of P/Q-type calcium currents. *J Neurosci, 28*(4), 788–797.

Nitsch, R. M., Deng, M., Growdon, J. H., & Wurtman, R. J. (1996). Serotonin 5-HT2a and 5-HT2c receptors stimulate amyloid precursor protein ectodomain secretion. *J Biol Chem, 271*(8), 4188–4194.

O'Brien, J. T., Ames, D., Schweitzer, I., Colman, P., Desmond, P., & Tress, B. (1996). Clinical and magnetic resonance imaging correlates of hypothalamic–pituitary–adrenal axis function in depression and Alzheimer's disease. *Br J Psychiatry, 168*(6), 679–687.

Oostendorp, J., Postma, D. S., Volders, H., Jongepier, H., Kauffman, H. F., & Boezen, H. M., et al. (2005). Differential desensitization of homozygous haplotypes of the beta2-adrenergic receptor in lymphocytes. *Am J Respir Crit Care Med, 172*(3), 322–328.

Paresce, D. M., Ghosh, R. N., & Maxfield, F. R. (1996). Microglial cells internalize aggregates of the Alzheimer's disease amyloid beta-protein via a scavenger receptor. *Neuron, 17*(3), 553–565.

Peskind, E. R., Elrod, R., Dobie, D. J., Pascualy, M., Petrie, E., & Jensen, C., et al. (1998). Cerebrospinal fluid epinephrine in Alzheimer's disease and normal aging. *Neuropsychopharmacology, 19*(6), 465–471.

Peskind, E. R., Wingerson, D., Murray, S., Pascualy, M., Dobie, D. J., & Le Corre, P., et al. (1995). Effects of Alzheimer's disease and normal aging on cerebrospinal fluid norepinephrine responses to yohimbine and clonidine. *Arch Gen Psychiatry, 52*(9), 774–782.

Prapong, T., Uemura, E., & Hsu, W. H. (2001). G protein and cAMP-dependent protein kinase mediate amyloid beta-peptide inhibition of neuronal glucose uptake. *Exp Neurol, 167*(1), 59–64.

Prince, M., & Jackson, J. (2009). World Alzheimer's Report 2009. Alzheimer's Disease International, London.

Prusiner, S. B. (1987). Prions causing degenerative neurological diseases. *Annu Rev Med, 38*, 381–398.

Qin, W., Ho, L., Pompl, P. N., Peng, Y., Zhao, Z., & Xiang, Z., et al. (2003). Cyclooxygenase (COX)-2 and COX-1 potentiate beta-amyloid peptide generation through mechanisms that involve gamma-secretase activity. *J Biol Chem, 278*(51), 50970–50997.

Ramos, B. P., & Arnsten, A. F. (2007). Adrenergic pharmacology and cognition: Focus on the prefrontal cortex. *Pharmacol Ther, 113*(3), 523–536.

Ramos, B. P., Birnbaum, S. G., Lindenmayer, I., Newton, S. S., Duman, R. S., & Arnsten, A. F. (2003). Dysregulation of protein kinase a signaling in the aged prefrontal cortex: New strategy for treating age-related cognitive decline. *Neuron, 40*(4), 835–845.

Ramos, B. P., Colgan, L. A., Nou, E., & Arnsten, A. F. (2008). Beta2 adrenergic agonist, clenbuterol, enhances working memory performance in aging animals. *Neurobiol Aging, 29* (7), 1060–1069.

Raskind, M. A., Cyrus, P. A., Ruzicka, B. B., & Gulanski, B. I. (1999a). The effects of metrifonate on the cognitive, behavioral, and functional performance of Alzheimer's disease patients. Metrifonate Study Group. *J Clin Psychiatry, 60*(5), 318–325.

Raskind, M. A., Peskind, E. R., Halter, J. B., & Jimerson, D. C. (1984). Norepinephrine and MHPG levels in CSF and plasma in Alzheimer's disease. *Arch Gen Psychiatry, 41*(4), 343–346.

Raskind, M. A., Peskind, E. R., Holmes, C., & Goldstein, D. S. (1999b). Patterns of cerebrospinal fluid catechols support increased central noradrenergic responsiveness in aging and Alzheimer's disease. *Biol Psychiatry, 46*(6), 756–765.

Rasmussen, S. G., Choi, H. J., Rosenbaum, D. M., Kobilka, T. S., Thian, F. S., & Edwards, P. C., et al. (2007). Crystal structure of the human beta2 adrenergic G-protein-coupled receptor. *Nature, 450*(7168), 383–387.

Renner, M., Lacor, P. N., Velasco, P. T., Xu, J., Contractor, A., & Klein, W. L., et al. (2010). Deleterious effects of amyloid beta oligomers acting as an extracellular scaffold for mGluR5. *Neuron, 66*(5), 739–754.

Robakis, N. K. (2011). Mechanisms of AD neurodegeneration may be independent of Abeta and its derivatives. *Neurobiol Aging, 32*(3), 372–379.

Rosenberg, P. B., Mielke, M. M., Tschanz, J., Cook, L., Corcoran, C., & Hayden, K. M., et al. (2008). Effects of cardiovascular medications on rate of functional decline in Alzheimer disease. *Am J Geriatr Psychiatry, 16*(11), 883–892.

Rossner, S., Ueberham, U., Schliebs, R., Perez-Polo, J. R., & Bigl, V. (1998a). p75 and TrkA receptor signaling independently regulate amyloid precursor protein mRNA expression, isoform composition, and protein secretion in PC12 cells. *J Neurochem, 71*(2), 757–766.

Rossner, S., Ueberham, U., Schliebs, R., Perez-Polo, J. R., & Bigl, V. (1998b). The regulation of amyloid precursor protein metabolism by cholinergic mechanisms and neurotrophin receptor signaling. *Prog Neurobiol, 56*(5), 541–569.

Russo-Neustadt, A., & Cotman, C. W. (1997). Adrenergic receptors in Alzheimer's disease brain: Selective increases in the cerebella of aggressive patients. *J Neurosci, 17*(14), 5573–5580.

Sabayan, B., Foroughinia, F., Mowla, A., & Borhanihaghighi, A. (2008). Role of insulin metabolism disturbances in the development of Alzheimer disease: Mini review. *Am J Alzheimers Dis Other Demen, 23*(2), 192–199.

Satler, C., Garrido, L. M., Sarmiento, E. P., Leme, S., Conde, C., & Tomaz, C. (2007). Emotional arousal enhances declarative memory in patients with Alzheimer's disease. *Acta Neurol Scand, 116*(6), 355–360.

Selkoe, D. J. (2006). Amyloid beta-peptide is produced by cultured cells during normal metabolism: a reprise. *J Alzheimers Dis, 9*(3 Suppl), 163–168.

Shankar, G. M., Li, S., Mehta, T. H., Garcia-Munoz, A., Shepardson, N. E., & Smith, I., et al. (2008). Amyloid-beta protein dimers isolated directly from Alzheimer's brains impair synaptic plasticity and memory. *Nat Med, 14*(8), 837–842.

Shankle, W. R., Nielson, K. A., & Cotman, C. W. (1995). Low-dose propranolol reduces aggression and agitation resembling that associated with orbitofrontal dysfunction in elderly demented patients. *Alzheimer Dis Assoc Disord, 9*(4), 233–237.

Shimohama, S., Taniguchi, T., Fujiwara, M., & Kameyama, M. (1987). Changes in beta-adrenergic receptor subtypes in Alzheimer-type dementia. *J Neurochem, 48*(4), 1215–1221.

Teng, L., Zhao, J., Wang, F., Ma, L., & Pei, G. (2010). A GPCR/secretase complex regulates beta- and gamma-secretase specificity for Abeta production and contributes to AD pathogenesis. *Cell Res, 20*(2), 138–153.

Terry, R. D. (2006). My own experience in early research on Alzheimer disease. *J Alzheimers Dis, 9* (3 Suppl), 117–119.

Thathiah, A., & De Strooper, B. (2009). G protein-coupled receptors, cholinergic dysfunction, and Abeta toxicity in Alzheimer's disease. *Sci Signal, 2*(93), re8.

Thathiah, A., Spittaels, K., Hoffmann, M., Staes, M., Cohen, A., & Horre, K., et al. (2009). The orphan G protein-coupled receptor 3 modulates amyloid-beta peptide generation in neurons. *Science, 323*(5916), 946–951.

Tiffany, H. L., Lavigne, M. C., Cui, Y. H., Wang, J. M., Leto, T. L., & Gao, J. L., et al. (2001). Amyloid-beta induces chemotaxis and oxidant stress by acting at formylpeptide receptor 2, a G protein-coupled receptor expressed in phagocytes and brain. *J Biol Chem, 276*(26), 23645–23652.

Turdi, S., Guo, R., Huff, A. F., Wolf, E. M., Culver, B., & Ren, J. (2009). Cardiomyocyte contractile dysfunction in the APPswe/PS1dE9 mouse model of Alzheimer's disease. *PLoS One, 4*(6), e6033.

van Melle, J. P., & de Jonge, P. (2009). Beta-blocker use and the development of depression. *Am J Cardiol, 103*(9), 1331–1332.

Vekovischeva, O. Y., Aitta-aho, T., Verbitskaya, E., Sandnabba, K., & Korpi, E. R. (2007). Acute effects of AMPA-type glutamate receptor antagonists on intermale social behavior in two mouse lines bidirectionally selected for offensive aggression. *Pharmacol Biochem Behav, 87*(2), 241–249.

Verdier, Y., Foldi, I., Sergeant, N., Fulop, L., Penke, Z., & Janaky, T., et al. (2008). Characterization of the interaction between Abeta 1-42 and glyceraldehyde phosphodehydrogenase. *J Pept Sci, 14* (6), 755–762.

Wang, D., Govindaiah, G., Liu, R., De Arcangelis, V., Cox, C. L., & Xiang, Y. K. (2010). Binding of amyloid beta peptide to beta2 adrenergic receptor induces PKA-dependent AMPA receptor hyperactivity. *Faseb J, 24*(9), 3511–3521.

Wang, H. Y., Lee, D. H., D'Andrea, M. R., Peterson, P. A., Shank, R. P., & Reitz, A. B. (2000). beta-Amyloid(1-42) binds to alpha7 nicotinic acetylcholine receptor with high affinity. Implications for Alzheimer's disease pathology. *J Biol Chem, 275*(8), 5626–5632.

Wasco, W., Bupp, K., Magendantz, M., Gusella, J. F., Tanzi, R. E., & Solomon, F. (1992). Identification of a mouse brain cDNA that encodes a protein related to the Alzheimer disease-associated amyloid beta protein precursor. *Proc Natl Acad Sci USA, 89*(22), 10758–10762.

Wasco, W., Gurubhagavatula, S., Paradis, M. D., Romano, D. M., Sisodia, S. S., & Hyman, B. T., et al. (1993). Isolation and characterization of APLP2 encoding a homologue of the Alzheimer's associated amyloid beta protein precursor. *Nat Genet, 5*(1), 95–100.

Wei, W., Nguyen, L. N., Kessels, H. W., Hagiwara, H., Sisodia, SMalinow, R. (2009). Amyloid beta from axons and dendrites reduces local spine number and plasticity. *Nat Neurosci, 13*(2), 190–196.

Weiler, P. G., Mungas, D., & Bernick, C. (1988). Propranolol for the control of disruptive behavior in senile dementia. *J Geriatr Psychiatry Neurol, 1*(4), 226–230.

Whorton, M. R., Bokoch, M. P., Rasmussen, S. G., Huang, B., Zare, R. N., & Kobilka, B., et al. (2007). A monomeric G protein-coupled receptor isolated in a high-density lipoprotein particle efficiently activates its G protein. *Proc Natl Acad Sci USA, 104*(18), 7682–7687.

Yan, S. D., Chen, X., Fu, J., Chen, M., Zhu, H., & Roher, A., et al. (1996). RAGE and amyloid-beta peptide neurotoxicity in Alzheimer's disease. *Nature, 382*(6593), 685–691.

Yazawa, H., Yu, Z. X., Takeda., Le, Y., Gong, W., & Ferrans, V. J., et al. (2001). Beta amyloid peptide (Abeta42) is internalized via the G-protein-coupled receptor FPRL1 and forms fibrillar aggregates in macrophages. *Faseb J, 15*(13), 2454–2462.

Ye, C., Walsh, D. M., Selkoe, D. J., & Hartley, D. M. (2004). Amyloid beta-protein induced electrophysiological changes are dependent on aggregation state: *N*-methyl-D-aspartate (NMDA) versus non-NMDA receptor/channel activation. *Neurosci Lett, 366*(3), 320–325.

Yu, J. T., Tan, L., Ou, J. R., Zhu, J. X., Liu, K., & Song, J. H., et al. (2008). Polymorphisms at the beta2-adrenergic receptor gene influence Alzheimer's disease susceptibility. *Brain Res, 1210*, 216–222.

Yu, N. N., Wang, X. X., Yu, J. T., Wang, N. D., Lu, R. C., & Miao, D., et al. (2010). Blocking beta2-adrenergic receptor attenuates acute stress-induced amyloid beta peptides production. *Brain Res, 1317*, 305–310.

Zhang, J., Hupfeld, C. J., Taylor, S. S., Olefsky, J. M., & Tsien, R. Y. (2005). Insulin disrupts beta-adrenergic signalling to protein kinase A in adipocytes. *Nature, 437*(7058), 569–573.

Index

A

A_{2A}-adenosine receptor, 33
α_{1A}-Adrenergic receptor
  mRNA, murine bone marrow, 115
  transgenic mice overexpressing, 133
α_{2A}-Adrenergic receptor
  C-1291G polymorphism, 178
  chemical structures of, 109
  DraI RFLP, 179, 180
  FRET signals, action of ligand, 108
  in functional bowel disorders, 179
  human embryonic cells, 107
  μ-opioid receptor, 109
  recording and watching, 104
  stabilization/retention of, 88
  sulfonylurea drug, 177
α_{2A}-Adrenoceptor pathway
  C-1291G polymorphism of, 170
  polymorphism, 164, 165
  working memory, regulating, 151
α_{1A}AR. *See* α_{1A}-Adrenergic receptor
α_{2A}AR. *See* α_{2A}-Adrenergic receptor
α_{2A}-D79N mice
  dexmedetomidine anesthetic-sparing effect on, 145
  electrophysiological experiments in, 144
α_1-Adrenergic receptor
  on adaptive immune system
    B-lymphocytes, 129
    peripheral blood mononuclear cells, 126
    T-lymphocytes, 127–129
  $CD68^+$ cells, 131
  dendritic cells, 122
  in disease states, 133–134
  electrical stimulation, 130
  enhanced green fluorescent protein, 124
    immunohistochemistry of, 125
  expression in immune system, 115–117
  immune cell function, 113, 115
    blood/circulating cytokines, 131–132
    catecholamines *in vivo,* 114
    dendritic cells, 122
    Epi/NE, 114
    macrophages, 119–122
    mast cells, 123
    monocytes, 117–119
    natural killer cells, 124–126
    neutrophils, 122–123
    nonimmune tissue, 132–133
    spleen, 129–130
    thymus, 116, 130–131
  on isolated PMBCs, 129
  mRNA, 117, 118
  on PMBC, immunocytochemistry analysis, 116
  pre- *vs.* postsynaptic signaling, 139
  in spleen, 130
  T cells, 127
  on thymocytes, 131
α_2-Adrenergic receptor, 162
  antidepressant treatment, 169
  attention deficit/hyperactivity disorder and, 162
  in cardiovascular disease, 173–177
    genotype, 177
    heart disease, 175–176
    hypertension, 174–175
    regulation of, 173–174
  in central nervous system, 162
    and Alzheimer's disease, 171
    attention deficit/hyperactivity disorder, 167–168
    dysfunction of, 172
    emotional memory dysfunction, 171–172
    neuropsychiatric disorders, 168–171
  first intracellular loop (ICL1), 82
  gastrointestinal system, 179–180

genes encoding, 162
genetic variants, overview of, 164
α_{2A}AR, 164–166
α_{2B}AR, 166
α_{2C}AR, 166–167
in major depressive disorder, 168
in metabolism and type 2 diabetes, 177–179
glucose handling, effects, 177–179
insulin secretion, effects, 177–178
pancreatic islets, 178
treatment of, 179
mutating specific residues, effect of, 84
nervous system-mediated functions, 180
and norepinephrine, 163
phosphorylation, arrestin-dependent stabilization of, 70
renal functions, 180
subfamily of, 162
third intracellular loop (ICL3), 81
α_2-Adrenoceptors
blockade of, 153
cellular localization of, 140
G protein-coupled receptors, 139
neuronal precursor cell, 153
in non-adrenergic cells and neurons, 13
pre/post-synaptic, 140, 141
α_2-adrenoceptor-mediated antinociception, 148
α_2-adrenoceptor-mediated sedation and hypnosis, 145–147
α_2-agonists, cardiovascular effects of, 153–154
behavior and depression, 152–153
cognitive functions, 149–152
hypothermia, 148–149
transgenic dissection of, 144
subtypes, functions of, 141
α_{2A}-adrenoceptors, 142
α_{2B}-adrenoceptors, 142–143
α_{2C}-adrenoceptors, 143
α_1AR. *See* α_1-Adrenergic receptor
α_2AR. *See* α_2-Adrenergic receptor
α_{2B}-Adrenergic receptor, 79
colocalization of, 86
Del301-303 variant, 174, 175, 180
ER-to-Golgi transport of, 92
GTPases role in, 94
ARF, cell surface, 90–91
Rab, in ER–Golgi-cell surface transport, 91–93
Sar1, 89–90
heart disease, 176, 177
L48 residue, 85
stabilization/retention of, 88
structural basis of, 82
C-terminal F(x)$_6$IL motif, 82–85
ICL3, basolateral targeting of, 88
L48 residue in, 85–86
N-terminal Y12/S13 motif in, 86–88
in vascular smooth muscle, 173
α_{2B}-AR. *See* α_{2B}-Adrenergic receptor
α_{2C}-Adrenergic receptor
Del322-325 polymorphism, 176
heart failure, 143
α_{2C}AR. *See* α_{2C}-Adrenergic receptor
ACTH receptors, 3
AD. *See* Alzheimer's disease
Adaptive immune system, α_1-Adrenergic receptor on
B-lymphocytes, 129
peripheral blood mononuclear cells, 126
T-lymphocytes, 127–129
Adenylyl cyclase (AC), 21
ADHD. *See* Attention-deficient hyperactivity disorder
Adrenergic receptors, 1. *See also* α_{10A}-Adrenergic receptor; α_1-Adrenergic receptor; α_2-Adrenergic receptor; β-Adrenergic receptor; β_2-Adrenergic receptor
activation in live cells
kinetics of, 105–106
ligand efficacy modulation, 107–110
ligand intrinsic efficacy, 106–107
activation of, 113
α_2-adrenergic receptors, pre- *vs.* postsynaptic roles of, 11–12
β_2-adrenergic receptor, structure for, 10
identification of, 3–4
interaction with amyloid β peptide, 213–216
protein–protein interactions, 12–13
rhodopsin, natural enrichment of, 4–5
specificity/therapeutic selectivity, 13
in vivo settings, gene targeting studies, 10–12
Adrenergic signaling, 208–210
Adrenoceptor subtypes, cloning of, 142

Alzheimer's disease, 206
CNS adrenergic system in, 216–218
pathophysiology of, 205
Amino acids glutamic acid/aspartic acid–arginine–tyrosine, 26
Amyloid beta (Aβ) peptides, 205
and CNS adrenergic system, 208
Amyloid precursor protein, 206
and adrenergic signaling, 208–210
amyloidogenic process of, 206
family of, 208
in HEK293 cells, 207
involved in, 208–209
proteolysis of, 210
Angiotensin II type 1 receptor, 83
Antagonist *vs.* agonist, 4
APP. *See* Amyloid precursor protein
Arg389, 108
Arrestins
as adapters for endocytosis, via clathrin-coated pits, 6–7
homologous desensitization, by adrenergic receptors, 5–6
molecular cloning permits structure–function analysis of, 8–10
β_2-adrenergic receptor, structure for, 10
scaffolding of non-G protein-dependent signaling, 7–8
ARs. *See* Adrenergic receptors
AT1R. *See* Angiotensin II type 1 receptor
Attention-deficient hyperactivity disorder, 152, 162, 167

B

β-Adrenergic ligands, 25
β-Adrenergic receptor, 19, 51, 52, 191, 192, 207
activation of, 209
amyloid process, 210–213
in AD, 218
asymmetric GPCR complexes, 33–35
β-arrestin-biased signaling, 60
biased signaling, 56
crystal structures, 24–27
CaMKII signaling, 192
cardiomyocytes, 59
cell surface receptor, 35–39
clinic implication of, 218–219
CNS, 217, 218–219, 221
desensitizing receptors, 35–39
G_i signaling, 194
G protein-coupled receptor signal, 36
signal transduction systems, complexity of, 20–21
GRK-mediated phosphorylation, 62
G_s signaling, 194, 209
GTP-binding proteins and adenylyl cyclase, 4
heterologous desensitization, 5
with Nup-62, 38
ontogeny of, 31–33
peripheral tissues, adrenergic system in, 219–220
polymorphism, in heart failure patients, 107
radioligand binding strategies, 3, 4
receptor arrays
heterodimerization, 28–29
homodimers, 27–28
interactions with G proteins, 30–31
interactions with receptor classes, 29
seven-transmembrane receptors, 51
signaling diversity in, 21–24
signaling pathways, diversity of, 22
subtype-specific signal transduction, in heart, 192
β_2AR agonists development, 198–200
β_1AR/β_2AR roles, 192
β_2AR, ligand-directed selective activation, 197–198
β_2AR, RGS2-mediated termination of, 195–197
β_2AR stimulation, cardioprotection, 193–194
β_1AR triggers cardiomyocyte apoptosis, 192–193
G_i signaling, 194–195
G_s signaling, 194–195
in vitro studies, 31
β_2-Adrenergic receptor, 207
with antagonist ICI118551, 207
β-arrestin complex, 63
cAMP-PKA signaling, 193, 214
cardioprotection, stimulation of, 193–194
chimeric receptors, 8
conformational changes, 102
crystallization of, 10
cysteine-scanning mutagenesis of, 9
desensitization, 7
future direction, 221

genomic/cDNA cloning, 8
GRK-mediated phosphorylation, 62
GRK5/6 phosphorylation, 60
G_s/G_i proteins, 197–198
Gs heterotrimer, 33
G_s to G_i pathway, 196
heart failure treatment and, 198–200
HEK-293, 57
immune system, 115
ionic lock, 10
PDZ binding motif, 70
potential drug targets on, 220–221
structure for, 10
7TMRs, 55
TM-VI of, 26
in vivo studies, 194
β_3-Adrenoceptors, neuronal precursor cell, 153
βAR. *See* β-Adrenergic receptor
β-Arrestin
S-nitrosylation of β-Arrestin2, 67–68
ubiquitination of, 66
β-Arrestin-biased signaling, 52
by β_1-adrenergic receptor
ligand bias, 59–61
receptor bias, 61
by β_2-adrenergic receptor, 54
ligand bias, 55–58
receptor bias, 58–59
reciprocal regulation of, 67
β-Arrestin-dependent ERK activation, 57
β-Arrestin-dependent signals, 23, 64, 72
β-arrestin2, S-nitrosylation of, 67–68
β_1/β_2 -ARs, GRK-mediated phosphorylation of, 62–63
cardioprotective effects of, 72
conformational changes in, 68–69
desensitize, 69
future perspectives, 72
inhibitors of
NHERF 1 and 2, 70–71
spinophilin, 69–70
ubiquitin specific protease 33, 71–72
pathway of, 59
phosphorylation of, 65–66
receptor endocytosis, 63–64
scaffolding properties of, 64–65
temporal/spatial features, 61–62
ubiquitination of, 66–67
β-Arrestin–USP33 binding, 71
β-Blockers, 102
B cell receptor, 129
B cells, 124, 125, 126
BCR. *See* B cell receptor
Binding affinity, 26
Biochemical responses, 107
B lymphocytes. *See* B cells
Bone marrow-derived lymphocytes cells. *See* B cells
Brain noradrenergic system, dysfunction of, 167
BRET biosensor, 69

C

cAMP, biosensors for, 57
cAMP-dependent phosphorylation, 5
cAMP/PKA activities, 214
cAMP-responsive element binding protein, 221
Cardiac muscle contraction, beta responses of, 2
Cardiomyocytes, 54
survival, β_1AR/β_2AR role in, 192
Cardiovascular disease, α_2-Adrenergic receptor in, 173–177
genotype, 177
heart disease, 175–176
hypertension, 174–175
regulation of, 173–174
Cardiovascular homeostasis, 114
Catecholamines, 2
Epi and NE, 114
inhibition of, 149
isomers of, 3
CD49b antigen, 124
cDNA cloning, 8
Cecal ligation and puncture (CLP)-induced septic mice, 134
Cellular prion protein (PrP^C) expression, 216
cGMP phosphodiesterase, 21
C-1291G polymorphisms, 168
CREB. *See* cAMP-responsive element binding protein
Creutzfeldt-Jakob disease, 216
Cys-reactive fluorescent probes, 9
Cytokine/chemokine antibody array, map of, 120

D

DCs. *See* Dendritic cells
Delta-opioid receptor, 212

Dendritic cells
immune system, 122
mRNA, 116, 122
DNA synthesis, 38
Dopamine β-hydroxylase promoter, 145
δ-Opioid receptors, 30
DOR. *See* Delta-opioid receptor; δ-Opioid receptors
DraI polymorphisms, 168
DX5 antibody, 124

E

EAE. *See* Experimental autoimmune encephalomyelitis
ECLs. *See* Extracellular loops
E/DRY motif, 26
EGFR signaling, 65
EGFR transactivation, 64
Endogenous catecholamines, mechanism of, 2
Endoplasmic reticulum, 32, 79, 80
chaperone proteins, 81
export motifs, 82
Endothelial NO synthase (eNOS), 67
Epinephrine (Epi), 1, 2, 70, 113
ER. *See* Endoplasmic reticulum
ER-Golgi intermediate compartment (ERGIC), 80
Experimental autoimmune encephalomyelitis, 133
Extracellular loops, 25

F

Fluorescence resonance energy transfer, 57, 101, 103
based Gi biosensor, 107
based GPCR biosensors, 105
receptor activation/deactivation, kinetics of, 106
FRET. *See* Fluorescence resonance energy transfer
Functional MRI (fMRI), 171

G

GDP–GTP exchange, 52
Genome-wide association study, 178
GFP. *See* Green fluorescent protein
Gi-protein signaling, 110
Go-containing heterotrimers, 30
Golgi apparatus, 80, 87, 92
GPCRs. *See* G protein-coupled receptors
G-protein
activation kinetics, 109
G protein α-subunit i2 ($G\alpha_{i2}$), α_2-adrenoceptors involved in, 147
regulator (phosducin), genetic variation of, 153
rhodopsin, 24
signaling, 62
switching of, 54
G protein-coupled receptor kinases, 6, 52, 53, 166
agonist-dependent recruitment of, 23
G protein-coupled receptor kinase 2 (Grk2), 143
mammalian cells, 53
G protein-coupled receptors, 20, 39, 79, 80, 101, 139, 140
biosensors, 101
cell surface targeting of, 94
in central nervous system, 207
de novo complexes of, 37
dimerization, gold standard for, 28
ER-to-cell surface movement, 80, 84
experiment, principle of, 103–105
G protein-coupled receptor 3 (GPR3), 212
heterodimers, roles of, 28
ligand, 198
presynaptic/postsynaptic receptors, 140
signaling complexity, 21
signal transduction systems, 20, 102
G protein-coupled receptor signal, 36
G protein-coupled signal transduction systems, 20–21
G protein-independent signaling, 7, 55
Green fluorescent protein, 103
GRK-catalyzed receptor phosphorylation, 6
GRK5/6 enzymes, 69
GRK-phosphorylated receptors, 6
GRKs. *See* G protein-coupled receptor kinases
Gs heterotrimer, 21, 30
GTPases, 81
mutants, 84
in Rab and Sar1/Arf subfamilies, 89
GTP-binding proteins, 4
GTP-restricted mutant Sar1H79G, 89

GWAS. *See* Genome-wide association study

H

hCaR. *See* Human calcium receptor
Heart failure
  β_1AR blockade, 199
  βAR system and, 192, 194, 195, 198
  β-blockers, 23
  clinical features of, 132
  RAS inhibition, 199
  RGS2-deficient mice premature death, 197
  in vivo studies, 199
Heart ischemia-reperfusion injury, 123
Hepatic NK cells, 125
Heterozygous, 173
HF. *See* Heart failure
High mobility group box 2 (Hmgb2) protein, 132
Human calcium receptor, 90
Human diseases, pathogenesis of, 80
Human genome sequencing, 163
Hyperpolarization-activated and cyclic nucleotide-gated channel (HCN), 162

I

Ig Thy-1 (CD90), 129
Inflammatory cytokines, antibody array membrane for, 121
Inositol hexakisphosphate (IP6), 55
Intercellular adhesion molecule-1 (ICAM-1), 118
Intracellular loop 2 (ICL2), 26
Intracellular loop 3 (ICL3), 81
Irritable bowel syndrome (IBS), 180

K

Kir3 ion channels, 30, 31
Kupffer cell, preparations, 121

L

Ligand-gated ion channels, 29
Lipopolysaccharide, 123
Locus coeruleus (LC) neurons, 144
Long-term potentiation (LTP), in hippocampal neurons, 214

M

Macrophage inflammatory protein, 119
Madin–Darby canine kidney II (MDCKII) cells, role of ICL3 in, 88
Major depressive disorder, 168–170
MAPK. *See* Mitogen-activated protein kinase
Mast cell peroxidase, 123
Matrix-metalloproteinase, 61
MCP-1. *See* Monocyte chemotactic protein-1
MDD. *See* Major depressive disorder
Metabolism and type 2 diabetes, α_2-Adrenergic receptor in, 177–179
  glucose handling effects, 177–179
  insulin secretion effects, 177–178
  pancreatic islets, 178
  treatment of, 179
MI. *See* Myocardial infarction
MIP. *See* Macrophage inflammatory protein
Mitogen-activated protein kinase, 133
  biochemical assays for, 57
  signaling, 54
MMP. *See* Matrix-metalloproteinase
Monocyte chemotactic protein-1, 118
Mouse embryonic stem cells, hit and run gene targeting, 144
MPO. *See* Mast cell peroxidase
Murine macrophage RAW264 cells, 122
Muscle contraction, eliciting, 2
Muscle relaxation, 2
Mutant V2 vasopressin receptors, 37
Myelin basic protein (MBP), 133
Myocardial infarction, 193, 199

N

Natural killer (NK) cells, 124
Natural killer Tcells (NKTcells), 127
NE. *See* Norepinephrine
Nerve growth factor, 209
Nervous system, 2
Neurokinin NK1 receptors, 6
Neuropsychiatric disorders
  major depressive disorder, 168–170

psychiatric disorders, 170–171
schizophrenia, 170
suicide, 168–170
Neurotransmitter norepinephrine, 216
Neurotransmitters, 11
NGF. *See* Nerve growth factor
Nonrapid eye movement (NREM) sleep-promoting pathway, 146, 147
α_2-adrenoceptors involved in, 147
Norepinephrine, 1, 2, 58, 113, 163, 211
deficient mice, 152
extracellular signal-regulated kinase activation, 126

O

Olanzapine, schizophrenia, 170
μ-Opioid receptor, 29

P

PAMPs. *See* Pathogen-associated molecular patterns
Pathogen-associated molecular patterns, 117
Pattern recognition receptors, 117
PBL. *See* Peripheral blood lymphocyte
Peripheral blood lymphocyte, 128
Pertussis toxin treatment, 70
PKA. *See* Protein kinase A
PKC. *See* Protein kinase C
POA. *See* Preoptic-anterior hypothalamus
Post-traumatic stress disorder, 171
Prazosin treatment, 132
Preoptic-anterior hypothalamus, 149
neurons, 149
thermoregulatory center, α_2-adrenoceptors in, 150
Pre/post-synaptic α_2-adrenoceptors
α_2-adrenoceptor-mediated antinociception, 148
α_2-adrenoceptor-mediated sedation and hypnosis, 145–147
α_2-agonists, cardiovascular effects of, 153–154
behavior and depression, 152–153
cognitive functions, 149–152
hypothermia, 148–149
Presenilin 1 (PS1), 206
Prostaglandin E2 (PGE2) receptors
cAMP formation, 209
inducer of inflammation, 212
POA neurons, 149
Protein kinase A, 54
dependent phosphorylation of GluR1, 214
enzymatic activity of, 54
Protein kinase C, 121
PRRs. *See* Pattern recognition receptors
Psychiatric disorders, 170–171
PTSD. *See* Post-traumatic stress disorder

R

Rab GTPases, function of, 93
Rab isoforms, 32
Radiolabeled antagonists, 3
RAGE. *See* Receptor for advanced glycation end products
RAMPs. *See* Receptor activity modifying proteins
Reactive oxygen species (ROS), 193
Receptor activator of nuclear factor kappa-B ligand (RANKL), 122
Receptor activity modifying proteins, 81
Receptor for advanced glycation end products, 216
Receptor/G protein/effector (R/G/E) interactions, 31
Receptor homo/hetero-oligomers, asymmetric organization of, 34
Receptor tyrosine kinases, 29
Regulator of G protein signaling, 31
B/R4 subfamily members, 196
in cardiac, 196
RGS2-deficient mice, premature death, 197
Resonance energy transfer (RET), 28
Restriction fragment length polymorphism (RFLP), 164
RGS. *See* Regulator of G protein signaling
RGS proteins, 196
RGS proteins (RGS3-5), 196
RTKs. *See* Receptor tyrosine kinases

S

Schizophrenia, 170
Schizophrenia, olanzapine, 170
Scorpion venom, 122

Seven-transmembrane receptors, 51, 52
molecular architecture of, 52
signal transduction, 53
Spontaneously hypertensive rats (SHR), PBL isolated from, 128
Suicide, 168–170
SV. *See* Scorpion venom

T

T cell receptor, 124
T cells, 124, 125, 126, 130
TCR. *See* T cell receptor
TCR$\alpha\beta^{low}$ thymocytes, 128
TGN. *See* Trans-Golgi network
Threonine 383, in rat β-arrestin2, 65
Thymus-derived lymphocytes. *See* T cells
TIMP-1. *See* Tissue inhibitor of metalloproteinases 1
Tissue inhibitor of metalloproteinases 1, 118
T lymphocytes. *See* T cells
T4 lysozyme, 25
7TMRs. *See* Seven-transmembrane receptors
TNF-α secretion with Toll-like receptor (TLR)4, 121
Trans-Golgi network, 80
APP proteolysis, 210
to cellular destinations, 86
clathrin-coated vesicles on, 90
endosome compartments, 210
Type 2 diabetes, α_2-Adrenergic receptor in, 177–179
glucose handling effects, 177–179
insulin secretion effects, 177–178
pancreatic islets, 178
treatment of, 179
Tyrosine (Y54) residue, 66

U

Ubiquitin specific protease(s) (USP(s)), 71
Ubiquitin specific protease 33 (USP33), 71–72
Ubiquitin (Ub), 66
Urapidil treatment, 128

V

VEGF receptors, 38
Vesicular stomatitis virus glycoprotein, 83
VSVG. *See* Vesicular stomatitis virus glycoprotein